The Assessment of Learning in Engineering Education

The Assessment of Learning in Engineering Education

Practice and Policy

John Heywood
Trinity College, The University of Dublin

IEEE PRESS

WILEY

Library of Congress Cataloging-in-Publication Data is available.

ISBN: 978-1-119-17551-3

Printed in the United States of America.

10 9 8 7 6 5 4 3 2 1

In Memory of
Georgine Loacker
Doyen of Assessment

Contents

Preface

The core idea explored in this book is that of judgment. Judgment is something that all professionals are called upon to do simply because of what it means to be a professional. This book explores specifically how we judge engineering education. A great deal of time and effort worldwide is being put into redesigning engineering courses and programs to meet the accreditation requirements of various agencies that judge whether degree programs are producing sufficiently qualified engineers. These agencies focus on assessment as the mechanism for evaluating courses and student learning, but because such assessment has far-reaching impact, it is appropriate to examine both the agencies and their methods. In this text, I try to, in a somewhat nonlinear fashion, explore what it means to claim to be "professional" in one's role as a teacher. Beyond content knowledge and understanding, theories of learning, professionalism includes a defensible theory and philosophy of assessment for one role of a teacher is to judge student progress. The aim of the text is, therefore, to provide sufficient data for an engineering educator to acquire a defensible theory or philosophy of assessment.

Assessment is not a stand-alone topic best left to experts. Examining assessment is one way of focusing on the curriculum for it cannot be divorced from learning, instruction, and content. It is integral to the curriculum process yet, more often than not, it is the afterthought of the educational process. Focusing on assessment forces us to consider in detail the aims and objectives of programs and for whom they serve. This book focuses internally on the problems of the curricula we have, and externally on the dictates of the sociotechnical system in which we live; these are inextricably linked. The book brings into sharp relief the relative responsibilities of academia and industry for the development of their charges as professionals.

This book is not about particular techniques of measurement or a "how to" guide for instructors. These can be found in most books on assessment in higher education (Angelo and Cross, 1994; Heywood, 2000). Rather this book is about the validity of teaching and judging learning and therefore includes illustrations of what our colleagues are doing framed against the backdrop of accepted educational policies.

The late Sister Georgine Loacker, Professor of English at Alverno College and Chairperson of the College's Assessment Council to whom this is book is dedicated and I disagreed about many things but we did not disagree about the principles of assessment. She and her colleagues defined *assessment* as "a multidimensional attempt to observe and, on the basis of criteria, to judge the individual learner in action" that is the position taken in this book. In the case of professional students, it means judging them in *both* academic and workplace situations.

"To undertake assessment means sampling, observing, and judging according to criteria. It means developing a whole array of techniques to take into account the fullest possible range of human talents. And it means an ongoing commitment to dealing with these kinds of questions" (Alverno College, 1994). This book is about how the engineering community has and is answering these questions and how it has and is determining the criteria on which such judgments are made. In sum, it is about practice and how research and theory should inform practice and policy.

Readers may come at this from different perspectives. Administrators and policy makers may be more concerned with assessment and its role in accountability, whereas instructors may be concerned with the impact that assessment has on learning. Similarly, the focus of the former will be, in all probability, on the curriculum, whereas the latter are likely to be more focused on techniques. Here it is argued that the two are not separable from each other. The more one tries to separate them, the more one becomes misled because they become detached from the system of which they are a part. This book is an attempt to demonstrate that this is the case and frame assessment more holistically.

John Heywood

References

Angelo, T., and Cross, K. P. (1994). *Classroom Assessment Techniques*. San Francisco: Jossey-Bass.

Alverno College. (1994). *Student-Assessment-as-Learning at Alverno College*. Milwaukee, WI: Alverno College Institute.

Heywood, J. (2000). *Assessment in Higher Education. Student Learning, Teaching Programmes and Institutions*. London: Jessica Kingsley.

Acknowledgments

This book resulted from a request by Marcia Mentkowski to write a short commentary on developments in performance-based assessment in engineering education. Bill Williams encouraged me to develop the drafts into a book and Rich Felder provided valuable criticism. During the subsequent development, Bill, together with Mani Mina and Alan Cheville, spent many hours critiquing the text. Russ Korte drew my attention to important papers that had escaped my search. I am most grateful to them for their kindness and their no-holds-barred support. Meriel Huggard has helped me get the text ship-shape for which I am also most grateful.

A book of this kind is the result of many influences. Sir Charles Carter triggered my interest in assessment when, 50 years ago, he offered me a Leverhulme Senior Research Fellowship to investigate examining in universities. The work undertaken for this fellowship led the late Harry Edels to invite me to assist him and his colleagues in the development of a new university entrance examination in engineering science for the Joint Matriculation Board. The development of this examination resulted in a 20-year collaboration with the late Deryk Kelly. We were supported by George Carter and Glyn Price. It should be noted that the Board, supported by its Secretary Mr. R. Christopher and later Mr. C. Vickerman, departed from its usual organizational practices and provided a structure that enabled this unique curriculum development. Later, in Ireland, in collaboration with Seamus McGuinness and the late Denstone Murphy, we attempted to apply the model developed for engineering science to school examinations in history and mathematics.

During this period, Michael Youngman, Bob Oxtoby, the late Denis Monk, and I responded to the criticism that university and school curricular in engineering did not respond to what engineers actually do by undertaking a job analysis of what engineers actually did using a statistical methodology developed by Youngman. I was also invited to undertake other collaborations and investigations in industry by the late Bill Humble and the late Barry Turner. Throughout the last 30 years, I have had valuable discussions with John Cowan about assessment.

After my first book on assessment in higher education was published, Georgine Loacker contacted me and the developments at Alverno College came into my thinking. I hope that to use the title of their book that this contribution will provide "Learning That Lasts."

Taken together, these studies provided the setting for the development of this book and those I have mentioned undoubtedly influenced the direction that my thoughts on assessment have taken as did Alec Martin and Lester Parker. To these should be added James Freeman, who introduced me to research on assessment in medical education and showed me its bearing on engineering education, and to Catherine Griffin, my doctoral student, who undertook a longitudinal study of portfolio assessment among nurses with whom I had many mind-stretching discussions. To all whom I have mentioned— Thank You.

Finally, no acknowledgment would be complete without a thank you to all the authors whose work I have read, and whose contributions gave me so much pleasure. I regret that the scale of the exercise prevented me from drawing attention to many other studies that merited attention.

1

Prologue

1.1 General Introduction: The Functions of Assessment

Recently I had a cause to enquire of a friend how he was recovering from an operation on his heart. He mailed a reply, which said, "They opened my chest, split my sternum, pried my rib cage apart, turned off my heart and lungs and let a machine do the work, replaced my aortic valve with a device fashioned from a cow's pericardium, cut out a piece of my ascending aorta and replaced it with a Dacron tube, restarted my refurbished heart and lungs, pulled my sternum back in place, and stapled my chest back together. Miracle 1—I'm still alive after all that. Miracle 2—three weeks later, and I'm almost fully functional unaided and what mild aches and pains I have are managed well with gabapentin (a nerve pain pill) and Tylenol."

"These docs are magicians […]."

I am sure I would have felt the same. I am not sure that I would have considered it magic but I would certainly have thought it incredible even though hospital "soaps" lead me to believe that such operations are normal every day activity, much more exciting than operations on the brain! Be that as it may, the decision I would have to make, that is, to have or not to have the operation as my friend had to make would have been made on the basis of trust in the surgeons. Such trust is acquired from the understanding that the surgeons have considerable experience at doing such operations and have a not inconsiderable training that enables them to understand that experience as

The Assessment of Learning in Engineering Education: Practice and Policy, First Edition. John Heywood.
© 2016 The Institute of Electrical and Electronics Engineers, Inc. Published 2016 by John Wiley & Sons, Inc.

enabling learning so as to better utilize that experience in the future. That understanding is reinforced by the knowledge that at all the stages in that training the surgeons have been examined or assessed (as some prefer) in formal situations to ensure they can do the job. Moreover, we expect those examinations to be psychometrically reliable and valid so that we can safely assume that the candidate will perform like that in the future. When we go to the surgeon's clinic, we expect to see his credentials, for that is what the accumulated certificates are, hanging on a wall. Should we not expect that from engineering educators?

Fortunately, we do not often have to trust surgeons but there are others in whom we have to place continuing trust, for example, the members of our family. Like them are the teachers to whom we trust our children. In the United Kingdom and the United States, that trust expects the teachers to act *in loco parentis* in activities that go well beyond the classroom although this is not the case in some European countries like France and Germany, where the teaching role is a teaching role without any social attachments. A great deal more is expected of teachers in boarding schools. Just like the surgeons, the trust extended to teachers is helped by the knowledge that they have had a similar training although not as long. They have acquired the knowledge that will enable them to teach a specialism; we expect a person who teaches mathematics to have a qualification in mathematics. But just as we expect surgeons to have gained a high level of craft skill, so many of us expect teachers to have developed the craft of teaching, or as it is more properly called pedagogy. I say many rather than all because there are individuals, politicians among them who think teaching is an intuitive activity that anyone can do. Their expectations do not stretch much beyond the experience of the teaching in their own school which they took to be easy. They find it difficult to believe that there is a serious activity of pedagogical reasoning that requires training on which experience can be built. As Shulman (1987) wrote, any explanation of pedagogical reasoning and action requires a substantial number of categories (i.e., Comprehension; Transformation [preparation, representation, selections, adaptation and tailoring to student characteristics]; Instruction; Evaluation [including testing]; Reflection, and New Comprehensions). Fortunately, the "many" do expect teachers to have credentials that document they have been trained in the theory and practice of teaching, and that to include assessed practice in real classrooms. There is a creeping realization that teachers exert very powerful influences over our children like no other they will experience, and these experiences can be for good or ill.

One of the primary functions of examinations is to aid the credentialing process. Thus, before a person can become a consultant, they have to perform junior roles and be mentored by senior doctors who all the time are monitoring their performance. There may even be performance tests to be taken. All of these tests are to judge their competency both of knowledge and performance. Knowing that they have had years of training is the first step in establishing trust. Much less is required of teachers although some countries require a period of probation and in some countries they are regularly evaluated by government inspectors in their classrooms.

Examinations and tests—assessments—perform many interrelated functions. For example, while an important function of assessment is to ensure that the goals of the program are being met, the certification of that achievement provides an individual with a credential.

Credentials are also summative: they bring together all that has been learnt in training and they are gained only if a person demonstrates mastery of both skill and knowledge in some way or another. Examinations and tests (assessments) also function as motivational agents: they make some students very competitive, but all students benefit from the role of examinations and tests as formative agents, that is from the feedback they get about their performance the intention of which is to highlight their strengths and weaknesses. Related to the concept of credentials is the idea of a profession and belonging to a profession. In Britain and the United States, not so much in Europe, value is attached to belonging to a profession. Professions give prestige, status, and esteem (Hoyle, 2001) and in these countries, credentials initiate a candidate into the "tribe" and in some circumstances, they enable the "tribe" to regulate entry into itself. In the United Kingdom, groups seek professional status by increasing the level of qualifications required; for example, nurses are now required to possess a university degree in nursing. To be a professional is a valued goal, notwithstanding the sociological view that the term *profession* has lost its meaning (Runté, 1995).

There has been a long-standing debate about whether or not teaching is a profession. Heywood and Cheville and Heywood (2015) have been bold enough to ask if "engineering educators are professional." One outcome of the debate about the teaching profession has been a distinction originally drawn by Hoyle (1975, Exhibit 1.1) between restricted and extended professionalism that irrespective of whether teaching is profession or not indicates what we might expect from good and poor teachers. Logically, it would extend to teaching in higher and engineering education in particular.

In the United Kingdom, the issues of status, esteem, and power have continued to bother the engineering profession since the end of the Second World War. They were upset by a finding of Hutchings (1963) that entrants to engineering schools had lower A level grades (see Section 3.1) than those in the sciences and they have bothered about such differences ever since. Similarly, they believe and continue to believe that there is a shortage of qualified engineers. Currently, in the United States it is supposed that there is a shortage of candidates for STEM (Science, Technology, Engineering and Maths) courses.

My friend who is very distinguished in his field of activity went on to say, "I'm ashamed to call myself doctor—I can't do anything that even comes close." While I do not happen to believe that is the case I was rather facetious in my reply for I said "on this side of the Atlantic you would not have that problem because we call surgeons 'Mr' or 'Miss' not doctor which is reserved for physicians!" This is said to point out that there are considerable differences between the educational systems of Europe, the United States, and United Kingdom and therefore with many countries of the old British Empire, where Britain established the systems that they developed. This is particularly true of Australia and New Zealand (Yeung, 2014) and countries in Asia. Canada, in contrast, mirror the system in the United States. Because education structures vary from country to country, establishing data that is transferable is exceptionally difficult. Although everyone is concerned with the basic parameters of examining and testing namely, achievement, validity, and reliability, exogenous variables that are unaccountable influences on the

Restricted Professionality in Engineering Education	Extended Professionality in Engineering Education
Instructional skills derived from experience	Instructional skills derived from mediation between experience and theory
Perspective limited to immediate time and place	Perspective embracing broader social context of education
Lecture room and laboratory events perceived in isolation	Lecture room and laboratory events perceived in relation to institution policies and goals
Introspective with regard to methods of instruction	Instructional methods compared with those of colleagues and with reports of practice
Value placed on autonomy in research and teaching	Value placed on professional collaboration in research and teaching
Limited involvement in nonteaching professional and collegial activities	High involvement in nonteaching professional and collegial activities
Infrequent reading of professional literature in educational theory and practice	Regular reading of professional literature in educational theory and practice
Involvement in continuing professional development limited and confined to practical courses mainly of a short duration	Involvement in continuing professional development work that includes substantial courses of a theoretical nature
Instruction (teaching) seen as an intuitive activity	Instruction (teaching) seen as a rational activity
Instruction (teaching) considered less important than research	Instruction (teaching) considered as important as research
Assessment is a routine matter. The responsibility for achievement lies with the student	Assessment is designed for learning. Achievement is the coresponsibility of the institution, instructor (teacher), and student

EXHIBIT 1.1. Eric Hoyle's characteristics of extended and restricted professionals among schoolteachers adapted for teachers in higher education (Hoyle, E. (1975). Professionality, professionalism and control in teaching, in V. Houghton et al. (eds.), *Management in Education: The Management of Organizations and Individuals.* London: Ward lock in Association with Open University Press. See also Hoyle, 2001).

data often make it difficult to ascertain what is actually happening within the system, its teachers, and its students (Berliner, 2014).

Apart from the basic functions discussed earlier in this section and the difficulties of making comparisons, there are, I think, two issues that are common to most assessment systems. The first is illustrated by the text in Exhibit 1.2. It is the opening paragraph of a book that I published on *Assessment in Higher Education* in 1977 (Heywood, 1977). I did not put the last sentence in bold as it is here. In spite of changing structures in the United Kingdom and Ireland, I find that colleagues have an affinity with the picture in that Exhibit. I had titled the chapter after a weekly political satire televized by the BBC and hosted by the late David Frost called "Not so Much a Programme, More a Way of

> *"Senate approved the syllabus for our new degree course at its last meeting. X will act as Director of Studies and co-ordinate the timetable for the new degree (Head of Department, January, 1975)"*
>
> *"Please let me have your questions for the Part I examination (of the new degree course) by the end of next week at the latest (Director of Studies February, 1975)"*
>
> *"Hell, I'd forgotten about the exam. Let's see what I set in this course in the other degree programme last year... (Teacher, February, 1975)."*
>
> The Director of Studies got his questions several weeks later and the students were sensible enough to look up the questions set by the lecturer in previous examinations in similar courses, and with a bit more luck, few will fail. **Examinations are the great afterthought of the educational process.**

EXHIBIT 1.2. Extract from Chapter 1 of Heywood, J. (1977). *Assessment in Higher Education*. Chichester, UK: John Wiley & Sons.

Life." Another distinguished colleague in the United States told me the other day that he rushes around at the end of the semester putting together his tests for the end of term. And several colleagues apologized for delay in replying to my e-mails before Christmas because they had to mark tests. Examinations (tests) are a nuisance, and they remain a nuisance!

When I published that book, there was no text on assessment in higher education available in the United Kingdom. However, in the United States, Paul Dressel had edited a useful text on *Evaluation in Higher Education* in 1960. The chapter on "learning" by J. Saupé occupies an important place in my memory system for the very reason that he defines learning. It is that definition that informs this text: "Learning is that process by which experience develops new and reorganizes old responses" (conceptual frameworks). His views of the principles by which learning is achieved are shown in Exhibit 1.3 and still resonate, although McKeachie (1974) pointed out that learning is an interaction between a variety of complex variables that Saupé's list does not seem to convey.

In the 1960s, *Universities Quarterly* published an important issue on the topic as well as another on the topic of wastage (then termed "dropout(s)" in the United States), that is the proportion of students withdrawing before they had completed their courses. I believed that there was a need for a theory of assessment in higher education that looked at all the factors that contributed to achievement. There are many. By 1987, it was apparent that the book needed to be substantially revised and a second edition was published in 1989 (Heywood, 1989): 10 years later the story repeated itself and a completely new book was published in 2000 (Heywood, 2000). Whereas I used to think I was providing a theoretical basis for discussions about assessment, I think in today's world it might have been better called a philosophy. I argued that assessment should not be the afterthought of the curriculum process but integral to it and it is that philosophy that underpins this study. Philosophy has a major role to play in determining the aims

1. *"Without appropriate readiness a learning experience will be inefficient. Learning will not occur"* (relates to the entering characteristics of students and the assessment of prior knowledge).
2. *"Learning will more effective when motivation is"* (as opposed to extrinsic: relates to formative assessment and self-assessment).
3. *"Learning proceeds much more rapidly and is retained much longer when that which has to be learned possesses meaning, organization and structure"* (relates in the first instance to the learning of concepts; also relates to the planning of tuition).
4. *"The learner learns only what he himself does"* (a reason for the use of inquiry/discovery-based learning)
5. *"Only those responses that are learned are confirmed."*
6. *"Transfer can only occur to the extent that students expect it to."*
7. *"Transfer will occur to the extent that students expect it to."* (5 and 6 relate to both the design of instruction and the design of assessment).
8. *"Knowledge of a model(s) of problem finding and solving or aspects of critical thinking can contribute to its improvement."*

EXHIBIT 1.3. Saupé's conditions that lead to successful learning. Adapted from Saupé, J. (1961). Learning. In: Dressel, P. L. (ed.), *Evaluation in Higher Education*. Boston, MA: Houghton Mifflin.

of a program since our values and beliefs determine those aims, and those values and beliefs are supported by our personal philosophy.

At the time of writing, engineering education is, in some degree, in turmoil. A number of leading engineering educators are looking for a new vision, a new curriculum that would emerge in different ways in different parts of the world. This book seeks to establish what we have learnt about assessment, its design, implementation, and validation so that it can contribute to the understanding and making of that vision. Necessarily, it surveys developments but it is not a textbook as such, yet may serve as a handbook. The other point about which I think there can be universal agreement is the total inability of people in the system to agree a common set of terms, the most critical of which is assessment.

1.2 Health Warning: Ambiguities in the Use of the Term "Assessment"

"Assessment" seems to have become a term that means what you want it to mean. For example, it has more or less replaced the term "evaluation" that was commonly used in the 1960s and 70s, not it would seem for any significant reason. As is seen in section 1.8, in the late 1960s the Committee of Vice-Chancellors and Principals (CVCP) in the United Kingdom began to use the term *assessment* as a cover for examinations and coursework (CVCP, 1969). By the end of the 1980s, Hartle (1986) had distinguished between six usages of the term "assessment" in the United States. Sadler (2007), an Australian, observed "that many of the terms we use in discussion on assessment and grading are used loosely. By this I mean we do not always clarify the several meanings a given term may take even in a given context, neither do we necessarily distinguish various terms from one another when they occur in different contexts. For example, the terms 'criteria' and 'standard' are often used interchangeably." Yokomoto and Bostwick (1999) chose "criteria" and "outcome" to illustrate problems posed by the statements in ABET EC 2000 (now called ABET Engineering Criteria). In this text, there is room for plenty of confusion between terms such as "ability," "appraisal," "assessment," "capability," "competency," "competences," "competencies," "criteria," "criterion-referenced," "evaluation," "objectives," "outcomes," "objectives," and "performance-based," all of which will be used in the text that follows. Semantic confusion abounds and there is disagreement about plurals, for example, "competences" or "competencies." While some just represent changes in usage, as for example "assessment" for "evaluation" others hide different theories, for example the "inside" and "outside" theories of competence development discussed section 1.3. Larkin and Uscinski (2013), writing of the evolution of curriculum assessment in a particular program, felt it necessary to define their terms in a separate section in their paper (see Appendix A).

Writing of performance assessment, the concept that was the driving force behind this text, McGuire (1993), a distinguished medical educator, argued that the use of the term *performance* is a naïve disregard of plain English. We are not concerned with performance *per se*, "rather we are concerned about conclusions we can draw and the predictions we can make on the basis of that performance. And this is a high inference activity." One of the reviewers of this text suggested that the original title that was *"Performance Based Assessment in Engineering: Retrospect and Prospect"* should be changed because he or she feared that engineering educators would think it had been written for industrialists and not for them. To be clear, this book is addressed to both.

Our emotions are expressed through language and the terms used can cause us to be very emotive, hence the care with which they have to be chosen. In the British Isles, if university lecturers were called *"instructors"* and they would be very offended. Instruction is what training officers do when they teach apprentices craft skills! Similarly in collegiate universities "tutor" has a different meaning to its use in the United States, so I have used "teacher" where I would sometimes have used "tutor."

Related to the concept of "performance" is the concept of "competency." But even this term has its difficulties. Griffin (2012) wrote that "whilst the competence vocabulary may be 'jargon' to some, it offers clarity to the notion of competency and competence as is evident in management literature (Young, 2002). In contrast, the medical and nursing literature offers the reverse, for example, some authors separate performance from competence, whilst others argue that there is direct relationship and others, still use 'competence' when they mean 'competency' and vice versa (Talbot, 2004). Defining the terms is more than semantics as it has implications for teaching, learning, assessing and the granting of licence to practice to medical and nursing practitioners" as we shall see. The same applies to engineering and engineers, as well as to other professions and their practitioners. According to Griffin, the prevailing trend in medical and nursing literature is that performance is what the practitioner does and competence is what the practitioner is capable of doing.

One of the functions of assessment is the measurement of achievement that *ipso facto* is a measure of a specific performance. That seems to be the way it is taken by the engineering community for a crude count of papers presented at the *Frontiers in Engineering Conferences* between 2006 and 2012 yielded some 80 papers that had some grammatical form of the term "assessment" in their title that might be relevant to this text although only one them included the term "performance." Eleven included the term "competence," which is clearly related to performance. But most of them were European in origin. Two years later, the 2014 proceedings of the American Society for Engineering Education contained a few papers from the United States that used competence synonymously with outcomes. It is evident small changes in language usage are taking place all the time.

Therefore, while this text clearly relates to the testing of academic achievement of student learning, that is only relevant in so far as it predicts the performance of individuals in the practice of engineering, and many attempts have been made to do this (e.g., Carter, 1992; Nghe, Janecek and Haddaway, 2007; Schalk et al., 2011). Much of the recent concern with the assessment of performance in the United Kingdom and elsewhere has been driven by politicians seeking accountability in order to try and demonstrate that value is added to the students and the economy by the education they have received (Carter and Heywood, 1992). This is nothing new; moreover, it is terribly difficult to do (Berliner, 2014).

While the term "assessment" is now commonly used to describe any method of testing or examining that leads to credentialing it embraces "formative" assessment that while not contributing directly to credentials helps students improve their "learning," and "self-assessment" that may or may not contribute to those credentials (George and Cowan 1999). It may be carried out in real time (Kowalski, Kowalski, and Gardner, 2009). An American-centered view on the techniques of assessment written for engineers is found in the *Cambridge Handbook of Engineering Education Research* (Pellegrino, DiBello, and Brophy, 2014). In Appendix A of this text, I have tried to provide an every person's guide to the changing terminology in the field of assessment with the exception of the psychometric terms of reliability and validity. The latter is considered below.

1.3   The Assessment of Persons for the Professions

Given that the purpose of professional study is preparation to undertake work in a specific occupation, a major question is the extent to which a curriculum prepares or does not prepare students to function satisfactorily in that occupation. Assessments that make this judgment may be made at both the level of the program and the level of student learning. In recent years, engineering education in the United States has had a significant focus on the program level for the purpose of accreditation. In the Health Care professions, both doctors and nurses receive clinical training and are assessed on their performance during this training. Assessments are made on the basis of judgments they make about "real" patients and rubrics for their assessment date back to the 1960s (Freeman and Byrne, 1973; see Section 3.5). The problems they faced continue to be relevant. Similarly, when teachers are trained, they have to conduct live classes of pupils in the age range that they (the students) want to teach (e.g., elementary [primary], High School [secondary] that have to be assessed) (e.g., Dwyer, 1994; Educational Testing Service., 1995). In both cases, "clinical" work is integrated with the curriculum in one way or another. Over the years, schedules for the assessment and self-assessment of the performance of teachers have been developed (Roberts, 1982: Reeves, 2009).

There is no equivalent "clinic" for engineering students except in cooperative programs ("sandwich" in the United Kingdom) where the students interweave, say 6 months of academic study with 6 months of industrial training throughout a period of 4 years. In that case, given that they are given jobs that are appropriate to the professional function, the equivalent of a "clinical" experience is provided and their professional performance can be assessed. This is what this writer has always taken performance-based assessment to mean, a view that has some similarity with that of Norman (1985). Miller (1990), who has had a great influence on medical education, put forward a pyramid model for the assessment of medical students. At its base is "KNOWS" (knowledge) that supports "KNOWS HOW" (competence) and that in turn supports "SHOWS HOW" (performance), and "DOES" (action) at the apex. He argued that competence was the measurement of an examinee's ability to use his or her knowledge, this to include such things as the acquisition of information, the analysis and interpretation of data, and the management of patient problems (Mast and Davis, 1994; see also Norman, 1985). Thus, while competence embraces the whole of the pyramid, university examinations at the time, only assessed competence from the "KNOWS HOW" perspective. Miller took the view of Freeman and Byrne (see Section 3.5) and others that no single assessment method could provide all the data required to judge if a person was able to deliver the professional services required of a physician. This same principle informed the designers of the engineering science examination described in Section 3.1. Rethans et al. (2002) took a different view. They thought that Miller's pyramid was no longer useful and suggested that the top two sections should be inverted thereby classifying performance and competence into the "DOES" section and the "SHOWS HOW" section. This approach, known as "The Cambridge Model," defines all assessments taken under examination like settings as competence-based assessments, and all assessments of actual practice as performance based. In this case, Freeman and Byrne's (1973) scale of overall competence in the practice-based situation is performance based (see Section 3.5). Griffin (2012) considered that the examples given by Rethans et al. "bear no resemblance to a cohesive model for assessment of competence or performance. The validity and reliability of some of the methods would also need to be established to avoid instances where doctors are wrongly classified." Unfortunately, Griffin makes no reference to Freeman and Byrne's assessment practices (see Section 3.5) that would seem to provide a cohesive model for postgraduate education that could be used as an exemplar. But she does take up an issue that is avoided in discussions of this kind namely, incompetency.

It is not a function of this text to consider the social psychology of incompetence that is, with the way that incompetence is dealt or not dealt with by professions. It is, however, the task of competency-based measures to state what incompetency is. In my experience, there is in Ireland, at least there was in my day, a desire not to want to fail intending teachers except in very severe circumstances. Griffin (2012) reports some evidence that some student nurses passed their clinical assessment without demonstrating competence, and she notes that it seems that clinical assessments may not always be recognized by third-level institutions as being important within the nursing program. A key purpose of self-assessment is that students learn what they are incompetent in (see Section 8.9). This has implications for the design of questionnaires that require respondents to make judgments about their capabilities.

More generally, it seems that the "inside" view of competency has been taken by the engineering profession. Competency in this view is something within the person, the quantity of which can be measured. It depends on habits of mind, including attentiveness, critical curiosity, self-awareness, presence, and a willingness to recognize and correct errors that is, to (self)-evaluate (Griffin citing Epstein and Hundert, 2002). The corollary is that if it can be measured it can be taught. Thus, many academics and industrialists believe that the competences industry requires can be taught without any assistance from industry. This position suggests that the acquisition of competence is not a developmental process, but this is rather contrary to observation as is rather well explained by Sternberg's (2005) model of the development of abilities into competencies, and competencies into expertise (see Sections 3.2 and 5.8). Students are experts at many levels. They may have a hobby such as photography in which they have the highest level of expertise. Young children often have more expertise with computers than their parents. It is evident that even in academic studies, students develop expertise as many papers on concept learning in engineering show (see Fordyce, 1991, in particular, and Cowan, 2006). There should be an increase in expertise with each additional year in the learning institution.

Griffin (2012) argues that the idea that competence is located in the individual is deeply entrenched in Western conceptions of intelligence and the mind (see Section 10.1). It is this theory that governs the thinking of engineering educators, more especially organizations like ABET (see Chapter 9 and also Section 6.8, which are based mainly on this premise). Even within this framework, there are differences of opinion about what a competence is (see Section 5.1). But there is also an "outside" view that suggests that the self has to be considered within the context that a competence is acquired. Sandberg (2000) has demonstrated that the acquisition of specific competences within an industrial organization is very much dependent on the context in which they are acquired (see Section 10.2). Taken together, these have a bearing on the structure and design of courses in engineering as well as their assessment (see Chapter 12).

It is not surprising that among the professions especially medicine that there should have been several studies on what it means to be professionally competent. From the literature, we are led to the view that competence is a complex, if not elusive, concept that is necessary for understanding the purposes of the education of professionals. Its assessment is, therefore, likely to be equally difficult. Given the closeness if not equivalence of "outcomes" with competence, the same finding applies.

1.4 The Engineering Profession

To become a professional engineer in the United States or a Chartered Engineer in the United Kingdom, the graduate is required to have completed a number of years of approved experience. In many other countries, there are similar requirements. Many of those in the old British Empire follow the pattern established in the United Kingdom, for example, Australia and Ireland. But there is no requirement for engineering educators to have professional status. The changing patterns in the workforce described in Chapter 12 suggest that new approaches to credentialing will be required as well as assessment of the work undertaken throughout a person's career.

Industrialists in the United Kingdom hoped that the Colleges of Advanced Technology (CATs) that were created in 1956 would produce engineers more suited to industry than the products of the engineering science courses offered by the universities that they thought prepared students for research. The electrical industry, in particular, had made a very heavy investment in them but it was disappointed with the results (Heywood, 2011). G. S. Bosworth, one of their leaders, published a paper in *Universities Quarterly* (Bosworth, 1963), in which he called for engineering courses to be more creative (see Section 2.3). Subsequently, he gave up the idea of changing the undergraduate curriculum, and in a report of an official committee that he chaired, rested his hopes that the needs of industry would be met by postgraduate programs (Bosworth, 1966; Heywood, 2013). Since the majority of engineering students do not pursue such programs, it is pertinent to ask how, if at all, engineering educators can predict the capability that a person will bring to an engineering task. Or, for that matter the occupation they have entered. To be fair, in colleges, students do a lot of laboratory practice and project work that is relevant. However, such predictability presupposes that something is known about where students go and what they do when they graduate.

Until this decade, engineering educators have paid very little attention to professional practice although this is now being rectified (Williams and Figuieredo, 2013; Trevelyan, 2014). Throughout the second half of the twentieth century, there has been a strong presumption among academics that university examinations predict subsequent behavior. Equally, at regular intervals, industrialists and the organizations that represent industry have complained that graduates, not just engineering graduates but all graduates, are unprepared for work (performance) in industry. To put it in another way, the examinations being offered did not have predictive validity. What is remarkable, as will be demonstrated in the sections that follow, is the persistence of these claims over a very long period of time (50 years), in spite of a lack of knowledge by all the partners of what it is that engineers actually "do." Nevertheless, from time to time during the last 60 years, attempts to find out what they do either directly or indirectly have been made. Each has contributed to the picture we have, which is not inconsequential (see Chapters 4 and 11).

If education is associated with what is learnt in college and training is associated with what is learnt in industry in order to become an engineer who can practice, it is clear that education cannot be divorced from training. It is also self-evident that the "complete" preparation of an engineer for practice cannot be accomplished only in a course of university study. Industry has a vital role to play in that training. One aim of this text is to demonstrate that that is the case.

Assessment has become a preoccupation of many engineering educators. At the time of writing the 2013 ASEE Conference proceedings, search engine yielded 273 entries for the term *assessment*, and most had the term in the title of their paper (http://www.asee.org/). There were 272 in 2014. There were also sessions on assessment at the Frontiers in Education Conferences and there are contributions in the growing number of journals in engineering education. The material is overwhelming.

Challenging the view that more rigorous research is the key to achieving educational reform, Felder and Hadgraft (2013) write: "we believe that if engineering education research were stopped completely right now (which we are in no way advocating) and engineering faculties could be induced to put into practice everything we know about teaching and learning from past research, cognitive science and experience, then we would achieve innovation with impact to an extent beyond the wildest dreams of the most idealistic reformers. The question then becomes, how can we do that?"

I believe that this is as true of assessment as it is of teaching and learning. My contribution to the answer to their question is to show the truth of their proposition through this survey of the development of assessment during the last 60 years with a view to suggesting directions it might take in the future as an integrated part of the curriculum process. Can we learn from the past ideas, philosophies if you will, that can enable us to judge where we are in the present? It is the contention of this study that we can.

1.5   The Development of Higher and Engineering Education as Areas of Academic Study in the 1960s

The early 1960s saw the beginnings of research in higher education in the United Kingdom, and in particular in technical and technological education (Heywood and Ann Abel, 1965). In the United States, Nevitt Sanford (1962) had edited *The American College* and Alexander Astin had begun his mighty studies of the impact of college on students (Astin, 1968). In the United Kingdom, there were rumblings that university teaching was not all that it should be and by 1964 the Universities of Essex, Lancaster, and Manchester had research units in various aspects of higher education. At Lancaster, a university that opened its doors to students in 1964, the Vice Chancellor (President) created a two-person department of higher education, the first in the country. This writer was one of those two persons and was appointed to do research in the area of university examinations (the term *assessment* was only just beginning to be used). The other person was appointed to research university teaching methods. A researcher also in the area of examinations was appointed at the University Essex. He (Roy Cox) became widely known for a much-cited review of research in examinations that was published among other papers on the same topic in *Universities Quarterly* that had been brought together by a working party of the newly founded (1964) Society for Research into Higher Education (SRHE) (Cox, 1967). When Bradford College of Advanced Technology was given the status of a technological university, it created an education department that undertook numerous studies of the university at work, including teaching and learning in technology (Cohen, 1968; Smithers, 1965).

1.6   Assumptions About Examinations: Reliability

It should be appreciated that at the time there were no textbooks published in the United Kingdom on teaching and learning in higher education let alone assessment and examinations which, to a limited extent, was made good by a monograph on group teaching (Abercrombie, 1964). It showed that there was much more to teaching than lectures, and in this respect by a much-quoted Penguin publication on the use and abuse of lectures (Bligh, 1971). However, two books on assessment related to higher education had been published in the 1930s. They reported investigations into the reliability of examinations. They were written by Sir Philip Hartog (Principal of the University of London) and E. C. Rhodes (a distinguished statistician) and published in 1935 and 1936, respectively (Hartog and Rhodes, 1935, 1936). Their results suggested that university examinations were not very reliable in the sense used by psychometricians. Reliability in this case, or "consistency" as Steven Wiseman, a distinguished psychologist preferred to call it, is the measure of the extent to which a test or examination gives consistent results with repeated applications. This is the reliability of a test in time. In the world of psychometric testing, much attention had focused on "consistency," but a test that is "consistent" is not necessarily valid. So when I was interviewed for the job at Lancaster, I suggested that we knew a lot about reliability (or unreliability!) and that we should now begin to look at questions of validity, and it was the issue of validity that guided my subsequent interest in examinations research.

The degree of consistency is determined by the method of correlation and the particular inconsistency that Hartog and Rhodes considered related to the fact that when two examiners are given the same essays to mark, not only do they assign different marks for the same performance but they are also likely to produce different distributions for the same group of candidates. I found that there could be different distributions as between the subjects of the curriculum (Heywood, 1977, p. 31). It could be that some subjects were more difficult than others or that the marking was more objective in some subjects than others. Either way, in some subjects the probability of getting a high grade was better than that in other subjects, a fact that was of particular concern to the National Union of Students (1967, 1968). Most recently, Lorimer and Davis (2014) have reported on a 10-year longitudinal study of mathematics, and (over 6 years) engineering assessments administered to preengineering students who show a remarkable degree of consistency when responding to the Force Concept Inventory.

The Hartog and Rhodes study led some to condemn essay examinations, although Philip Vernon (1965), another distinguished psychologist, rose to their defense. He noted, as had other reviewers, weaknesses in some of the experiments that were conducted, but in contrast to those reviewers he was impressed by the smallness of the discrepancies rather than their largeness. He found that in the best conducted examinations the median disagreement between any two examiners was not more than 3%. However, Vernon argued that Hartog and Rhodes had to be taken seriously because less thorough examinations are shown to be deplorably unreliable: "that in the absence of a scheme of instructions drawn up and applied by experienced examiners much worse discrepancies may arise; that when the average and dispersion marks are not standardised, gross differences may appear in the proportions of credits, passes and fails etc. which are awarded and that even a percentage error may make all the difference between a pass and a fail, or a first and a second class."

These concepts were not part of the language of examining used by teachers. In the many examiners meetings I attended, I found that across the spectrum of subjects there

Number of Items in Objective Test	Standard Error
<24	2
24–47	3
48–89	4
90–109	5
110–129	6
130–150	7

EXHIBIT 1.4. The standard error for items (questions) that can be scored 1 or 0 that are intended to measure performance within a reasonably generous time limit due to P. B. Diedrich (1960). *Short Cut Statistics for Teacher Made Tests.* Princeton, NJ: Educational Testing Service. McVey (1976) attempted to calculate the standard error of a problem-solving examinations in electronics and arrived at a standard error of 8 for a typical 3-hour paper.

was little if any understanding of the concept of "standard error." Suppose a candidate scores a mark at a borderline grade level such as 59 instead of 60, or 69 instead of 70, assessors would stand by their mark, irrespective of the fact that in any mark there is an inbuilt error, and this even applies to objective tests (see Exhibit 1.4). Moreover, in any group of examiners, there are those prepared to mark "up" and others who will always mark "down."

In defense of reliability, it was also pointed out that the final year classes in many subjects might be rather small and the teacher may know the students very well. Double marking that some universities now insist on is fraught with difficulty. McVey (1975), who examined scripts in electronic engineering, showed that marks left on a script by the first examiner could influence those of the second examiner. In any case who is right? McVey's investigations also showed that examinations in electronic engineering that might be thought to demonstrate a high level of reliability were not as reliable as expected. McVey also showed that the standard error was an important factor in marking (McVey, 1976a, 1976b). Engineering educators continue to be interested in reliability (Allen et al., 2008).

1.7 Myths Surrounding Examinations

University examinations were shrouded in myth. Oppenheim, Jahoda, and James (1967) listed 20 assumptions that were made about university examinations that for most part could be tested empirically for their validity. These are shown in Exhibit 1.5. They were, of course, specific to the system of examining in Britain, although some are universal and some have changed or are being pursued. The fourth and fifth assumptions are one of the reasons for this text and the pursuit of the engineering education to relate assessment to performance in the real world beyond academia. Nowadays in the United Kingdom, much more use is made of coursework assessment (seventh assumption). The 10th assumption was never true of engineering because the professional institutions have Royal Charters that enable them to examine at this level. Attempts have been made

1. The assumption that university examinations can include some so-called imponderables such as "quality of mind," "independent critical thinking," breadth, etc., in their assessment.
2. The assumption that "quality" of academic performance is rateable on a single continuum for first-class honors to failure.
3. Whereas many courses include a good deal of practical work, and a few approach some kind of apprenticeship training scheme or sandwich (cooperative) course, we usually pay less attention to those aspects of the course when examining.
4. To some extent, it is assumed that examination performance is a mock–real-life performance.
5. The assumption that each examinee should have individual responsibility for his own performance; we do not expect collaboration or teamwork, no matter how common this may be in real-life performance.
6. The assumption that a student who fails has only himself to blame for not working hard enough, or for being stupid, or in some other way.
7. The assumption that the proper place for examinations is at the end of certain courses—not later or sooner.
8. The assumption that university teachers should also be university examiners and university selectors.
9. The assumptions about the impartiality of examination.
10. The assumption that the university should have the sole right to examine at this level.
11. The assumption that the use of external examiners prevents bias.
12. The assumption that forced regurgitation of knowledge under stress is predictive of future performance.
13. Assumptions concerning mental growth and development and the acquisition of an "educated mind."
14. The assumption that pressure is required.
15. The assumption that anxiety is necessary.
16. The assumption that examination results should be distributed in a certain way.
17. We are forced to make, and then retract, all kinds of assumptions about the comparability of degree from university to university and from country to country.
18. The assumption of the need for uniformity in undergraduate examinations: all students in a given year group must pass the same examination paper, and we do not allow examinations to be tailored to individual needs.
19. The assumption that "learning" is to be valued for its own sake and not merely as a preparation for career and financial gain.
20. The assumption that the outside world wants the results of university examinations or takes much notice of them.

EXHIBIT 1.5. Twenty assumptions underlying the use of university examinations in the United Kingdom listed and discussed by Oppenheim, A. N., Jahoda, M., and James, R. L. (1967). Assumptions underlying the use of university of examinations. *Universities Quarterly*, 21(3), 341 (also cited in Heywood, J. (1977). *Assessment in Higher Education.* Chichester, UK: John Wiley & Sons).

to improve external examination but questions remain (Warren-Piper, 1994), and there have been worldwide attempts to ensure the comparability of degrees, for example, the ABET Engineering Criteria, the Bologna Agreement, and the Washington Accord (17th assumption) (Bucciarelli, Coyle, and McGrath, 2009). What industry, of for that matter society, wants from examinations remains an "open" question. At the time of writing, the UK division of Ernst & Young (EY) the professional services organization announced a transformation of their recruiting policy because their own research had shown that screening graduates on academic performance alone was too blunt an approach to recruitment. They had found that success in higher education was not correlated with performance in subsequent professional qualifications. Therefore, they proposed to evaluate candidates for their potential through the use of online tests. Only at the final stage would the academic performance be made known to the recruiters. While they would still value academic achievement and maintain high intellectual standards, they hoped that this would enable the organization to become more socially inclusive. It will be of no small interest to see if EY achieves this goal. PwC is also selecting graduate trainees by aptitude and behavioral tests (The Times, 2015).

The reader should be reminded that one of the difficulties when talking or writing about the assessment of student learning is that the assessment systems to be found in countries such as the United Kingdom and the United States are so different that they are difficult to understand if you do not work in those systems. Moreover, they are subject to change. In the early 1960s, the stereotype of assessment in the United States was of objective tests (known to the public as multiple-choice questions) set during and at the end of courses and its equivalent in the United Kingdom of 2- to 3-hour written examinations set at the end of the year in each of the subjects studied, grades being distributed in the former at the top end of the mark spectrum while the latter were neo-norm referenced.[1] Objective tests can meet the criteria specified by Vernon. The two tests commonly used for selection in the United States (Scholastic Aptitude Test [SAT] and the American College Testing Program) are standardized tests and have very high reliability. An objective test almost inherently has a higher reliability than an essay test but teacher-designed tests that have not been designed and piloted properly may not be as good as those that are. Ager and Weltman (1967), who were participants in the SRHE group, concluded that "no single examination technique is completely satisfactory in terms of both reliability and validity," and for this reason they thought a variety of techniques should be used. For the quite different reason that anxiety could be reduced, university medical officers also recommended the use of a variety of techniques (Malleson, 1964; Ryle and Lungi, 1968; Ryle, 1969). It was not appreciated that if a distributed model of assessment was used that the redistribution of stress in the system might have an equally negative effect on those who preferred terminal examinations. The practice of examining is very much the art of compromise. But these studies did not relate their findings to validity and the purposes of university examinations in any detailed or considered way. However, support for multitechnique approach is also to be found in a detailed analysis of recent literature together with changes in the techniques he uses for teaching and assessment by Parsons (2008). Apart from encouraging a variety of styles of assessment, he encouraged the use of open-book examinations because of the memory support they gave students, opportunities for students to comment on the assessment at or around the time of assessment, and the relaxation of time constraints on assessment activities.

Currently in the United States, a mixed techniques approach is often cited that makes use of direct and indirect methods of assessment. By indirect methods are meant such things as exit and other interviews, archival data, focus groups, and written surveys and questionnaires. In addition to all kinds of examinations, Sundarajan (2014) includes simulations, behavioral observations, and performance appraisal as direct measures. The problem with mixed approaches, as it is with continuous assessment, is that they can easily overload students to the extent they are unable to cope with the work they are given. This seems to be an aspect of testing that has received little consideration (Myllymaki, 2013).

1.8 The Introduction of Coursework Assessment

However, concerns about university examinations in the United Kingdom not least the view that students could have an off-day on the day they took their examination led to changes in the 1960s notably to the introduction of what was commonly called "continuous assessment" that in effect was the periodic assessment of some form of work done during the course (Coursework Assessment). There is a correspondence with what happens currently in American programs where assessments are made during each course that include tests, quizzes, home examinations, and homework. It may incorporate self-assessment.

In the United Kingdom, it was introduced in a haphazard way such that natural justice could be compromised. For example, a student could be required to undertake a lot of coursework for which he or she would receive no marks in the final collation of marks, whereas in other courses the students might be told that 10% of their final mark would contribute to their final score (grade). The variations in practice in one university are shown in Exhibit 1.6. The study summarized in this exhibit did not take into account the procedures adopted by the Open University. Nowadays, courses are a mix of the traditional 2 or 3-hour written papers and continuous assessment. Homework is not set. The weighting between the two varies considerably across the British Isles. It could be as many as 80% for coursework or equally for the examinations. One program in Design Technology (for teachers) relies solely on coursework assessment for its grade. There have been some radical innovations in engineering courses and there is an affinity between some practices of continuous assessment and mastery learning (e.g., Cole and Spence, 2012; see Section 6.4).

Today continuous assessment is used in a number of institutions across the world and is seen as a means of improving performance. At the University of Oulu in Finland, its purpose "is to help the students themselves to become more effective self-assessing, self-directed learners and it is based on cognitive, constructivist, and socio-cultural theories" (Myllymaki, 2013).

Continuous assessment is also used by universities seeking accreditation from ABET that are located outside of the United States, where a range of instruments is used to evaluate the level of achievement of the program's educational objectives

---

Cumulative assessment
A proportion of the mark in a 3 (or 4)-year course is arrived at from scores achieved during the first and second years. The proportion of marks allowed for the first year is usually smaller than that for the second.
For example, 1st year = 12% of the years marks; 2nd year = 24% of the years marks; 3rd year = 64% of the years marks.
The marks may be achieved by combinations of coursework assessment and examinations, or examinations alone. For purposes of certification, they could be equated to 100%.

Diagnostic coursework
An early assessment of student coursework to determine those in difficulty and the nature of the difficulties.

Informal coursework assessment
Most tutors make a judgment about the qualities of their students while a course is in progress. Such judgments are often used to moderate the marks given to candidates at the final meeting of examiners with the external assessor. Students are not informed about such procedures.

Formal Coursework Assessments
The characteristics of such systems are that the students know how coursework will contribute to their final mark. There are several systems (the terminology is the writers):
 (i) *Fixed percentage schemes*: In such schemes, coursework is formally assessed and contributes a fixed percentage to the total degree mark. In England, some universities operating such procedures do not impose restrictions on departments. The proportions awarded depend on the value ascribed by individual departments to coursework assessment.
 (ii) *Positive moderation schemes*: These are schemes in which coursework is formally assessed but the result is used only to raise a candidate's mark.
 (iii) *Formal requirement schemes*: are those that require satisfactory performance in coursework, before a person can either obtain a final degree or sit at a final examination.

> The scheme adopted depends very much on the department (or faculty) objectives in initiating coursework assessment procedures. Broadly speaking, there are two categories that can be described as *supplementary* and *complementary* in terms of the information sought by examiners about a candidate.
>
>> (1) In schemes that provide *supplementary* information, the examiners hold that the coursework assessment is measuring the same abilities (qualities) as the written examination. It is, therefore, a check on the written examination and, as such, is used as a means of moderating the final mark.
>> (2) Coursework work assessment that provides *complementary* information to the examination is thought to measure different qualities to those measured in written papers.

EXHIBIT 1.6. Types of coursework in operation in English Universities in the late 1960s. Heywood, J. (1969) in Committee of Vice Chancellors and Principals and Association of University Teachers report of the 1969 Spring Conference. *The Assessment of Academic Performance*. London, Committee of Vice Chancellors and Principals (also cited in Heywood, J. (1977). *Assessment in Higher Education*. Chichester, UK: John Wiley & Sons).

and outcomes. They embrace course assessment (Abu-Idayil and Al-Attar, 2010; Al-Nashash et al., 2009; Christofourou and Yigit, 2008). It seems to me that they are more like Continuous Quality Improvement than the continuous assessment of learning.

Because the basic parameters of reliability and validity apply in every system of assessment, the propensity for one system to learn from another and vice versa is considerable.

1.9 Rethinking Validity

The application of the coursework assessment mark to the final mark gave rise to an important debate about the value of the single mark (assumption 2 of Exhibit 1.5) and some academics suggested that a candidate's performance should be recorded in a profile form. For example, practical abilities in science and technology should be separated from the theoretical. There began a move to find ways of measuring these complementary abilities. The lack of clarity about what the purposes of coursework were was a factor that encouraged the objectives movement to promote its wares in the United Kingdom. At the same time it raised serious questions about the validity of what was being done. Much of what went on seemed to rely on face validity—"if it looks right, it is right" but as Exhibit 1.7 shows validity is a much more complex concept than that (Wigdor and Garner, 1982) and measures are made of content, predictive, criterion, and construct validity. Content validity is the extent to which a test measures the content or skill that it is supposed to measure. Criterion (or concurrent) validity is closely related to predictive

Face validity: The extent to which an assessment appears to be measuring the variable it is intended to test, for example, visual inspection of the items or questions in comparison with declared objectives; do they, for instance, measure analysis?)
Content validity: The extent to which a test measures the content (or skill) which it is supposed to measure.
Predictive validity: The extent to which an assessment predicts future performance (e.g., degree grades as a predictor of work performance). Criterion validity is similar.
Criterion validity: Comparison of an assessment designed to evaluate performance in a task with an alternative evaluation (e.g., a test designed to predict driver performance compared with actual observations by a skilled judge). It predicts what an individual or group with particular scores on a test will perform on the criterion measure that will have been chosen to be close to the issue of interest. Cronbach (1971) calls this concurrent validity (see also Kline, 1993).
Construct validity: The extent to which an assessment measures the content (aptitude, attitude, and skill) it is intended to assess, and predicts results on other measures of content, aptitude, attitude, and skill as hypothesized. Wigdor and Garner (1982) write that it is a "scientific dialogue about the degree to which an inference that a test measures an underlying trait or hypothesized construct is supported by logical and scientific analysis" (see also Kline, 1993). "Intelligence" is one such construct.

EXHIBIT 1.7. Traditional measures of validity.

Ideological validity (Ridgeway and Passey, 1992). Refers to the educational moral, philosophical, and political values that are implied by the use of any particular assessment scheme. Related to this is *Stick and Carrot validity*, which assesses the extent to which an assessment system can be used to control the education system.

Generative validity. Refers to changes in behavior that occur because of a particular set of measures. The way in which particular measures influence the direction of the curriculum and teaching (Ridgeway and Passey, 1992).

Tentative generative validity. Identifies likely directions of change and their inherent value (Ridgeway and Passey, 1992).

The Corruption coefficient. Measures the extent to which scores can be manipulated without changing what is actually being measured (e.g., raising coursework scores without benefiting a student's understanding) (Ridgeway and Passey, 1992).

Experiential validity of the curriculum (Mentkowski, 2000). The student's observation that he or she has grown in some valued ability (or abilities) through his or her learning experiences.

EXHIBIT 1.8. Other concepts of validity resulting from the rethinking of validity in the 1990s.

validity, since it is the comparison of an assessment designed to evaluate performance in a task with an alternative evaluation (see Section 3.2; Cronbach, 1971; Kline, 1993). Construct validity is the extent to which an assessment measures the content (aptitude, aptitude, and skill) it is intended to assess and predicts results on other measures of content, aptitude, and skill as hypothesized. Wigdor and Garner (1982) write that it "is scientific dialogue about the degree to which an inference that a test measures an underlying trait or hypothesised construct is supported by logical and scientific analysis."

Gipps (1994) has pointed out that the definition of validity has been taken beyond that of measuring what a test measures to become an assessment of the evidence available to support test interpretation, and the potential consequences of test use. This indicates that validity is not a value neutral concept as is demonstrated by Ridgeway and Passey (1992; see Exhibit 1.8) which takes us into the realm of personal, institutional, and societal goals, and in terms of the quality assessment of higher education, the process itself.

The rethinking of the concept of validity in the 1990s challenges some of the assumptions listed in Exhibit 1.6 not least the view that student abilities can be assessed and reported unidimensionally (Rogers, 1994). It has an important bearing on the development of competency-based assessment.

1.10 Wastage (Dropout): The Predictive Value of School Examinations for Satisfactory Performance in Higher Education

As previously indicated, in the 1960s in both the United Kingdom and the United States, there was much concern with the predictive reliability and validity of the entrance tests and examinations used for the selection of students to university programs. This continues to be the case (e.g., Howell, Sorensen, and Jones, 2014). Interest in selection led the CVCP to develop a British equivalent of the SAT called the Test of Academic Aptitude with the intention of its result being combined with the results of General Certificate of Education (A Level) subjects to give a better prediction of university performance (see Chapter 3, Note 1). In spite of it being relatively successful, little was heard of it in subsequent years. One reason for this was that irrespective of the correlations between the grades of some A level subjects and the final grade awarded by the university being quite small some subjects, particularly the science subjects depended on school subjects to provide the content base for further study at university. This is why degree programs could be completed in 3 years because sixth-form studies, as the last 2 years of schooling were called, were highly specialized and the equivalent of the first year of study in many other university systems, including the United States.[2] The Engineering Science subject that is one of the subjects of discussion in Chapter 3 was of this kind. At the same time the GCE Advanced level examinations could be regarded as relatively successful when measured against an annual wastage rate from the universities that was constant at between 13 and 14% (University Grants Committee [UGC], 1968). Looked at in this way 85% of the intake completed and passed their examinations, most of them within a 3-year time limit. Nevertheless, looked at either from the perspective of the system or that of the "failed" individual, the costs could be considered to be high, but as Vaizey (1971) showed that there is no easy way of calculating such costs.

It is not surprising, therefore, that much interest should focus on wastage (dropout) (now more commonly considered from the perspective of retention). Investigators sought to understand the reasons for withdrawal from courses and examination failure. They found a complexity of reasons that ranged from getting married to psychological diffi- culties related to examinations. Numerous studies related to wastage were completed in the 1960s in the United Kingdom and a number of them related to technological studies and by 1971 there was a substantial literature on the topic (UGC, 1968; Miller, 1970; Heywood, 1971).

Considerable variations were found between the subjects. The failure rate in engi- neering and technology was approximately twice that of any other subject grouping except that of architecture. Between the different institutions, the highest failure rates for academic reasons in engineering and technology were in the ex-Colleges of Advanced Technology that had become universities in 1966.

These ranged from 24 to 40%. The lowest failure rates for academic reasons were at Oxford (2.6%), Cambridge (4.1%), and Birmingham (7%). Of those who persisted to their final examinations, only 1.7% of the candidates failed. Simple inspection of the data suggests that those institutions with the lowest rates were able to select candidates with higher "A" level grades than those institutions with the highest rates. Moreover, many students in those institutions were in 4-year programs.

Important methodological issues were discussed by American (Lavin, 1967) and British investigators (Entwistle, 1970; Kelsall, 1963). Lavin (1967) issued the reminder that a "significant relationship between predictor and criterion does not necessarily establish that the predictor is a causal determinant of the criterion." Kelsall (cited in Heywood, 1971) pointed out that for "any population of applicants in one year, we can never, in the nature of things, compare the academic performance of those admitted by the selectors somewhere with that of those not admitted by the selectors anywhere. For all we know, therefore, existing methods may be highly efficient in weeding out those with the poorest chance of success in university studies. We can only judge the overall efficiency of selection at most within the population of those admitted to some university somewhere in Britain." This is why, at the time, there was interest in cross-institutional studies initiated by Noel Entwistle that focused on persons with similar if not identical entry qualifications to courses designed to be of similar standard in Colleges of Education. Entwistle also noted elsewhere that correlation analyses can be misleading because they produce descriptions of the average successful student. They do not provide a method for comparing every real student with every other real student for the purpose of establishing the differences in test score profiles as between successful and unsuccessful students. At the time the new technique of cluster analysis held out some hope that this might be achievable; Brennan using this technique on Entwistle's data (Entwistle, 1970) identified different types of science and arts students and different types of successful student. Numerous UK and US studies looked at factors that might contribute to success or failure such as personality, mental health, motivation, and study habits.

Today in the United States, much interest in engineering education has focused on the reasons for transfer to other programs of study. In Seymour and Hewitt's (1997) seminal work, the focus is on why students do or do not persist. They found a sizeable problem. Among higher than average ability students, there was a loss of between 40 and 60% of science, maths, and engineering students. Moreover, they were not surprised to find that faculty wanted "to marginalize the issue of wastage" given the size of the problem. They argue that to improve "the retention among women and students of colour, and to build their numbers over the longer term is to improve the quality of the learning experience for all students" [...] Though faculty sometimes like to begin a program of reform with discussions of curriculum content and structure, this is unlikely to improve retention unless it is part of a parallel discussion of how to secure maximum student comprehension, application and knowledge transfer, and give students meaningful feedback on their academic performance" (Seymour and Hewitt, p. 394). The design of assessment, formative and summative, is central to such a reappraisal as are the aims and objectives they are supposed to obtain. Investigations into retention continue (e.g., Bernold, Spurlin, and Anson, 2007; Walsden and Foor, 2008).

1.11 Factors Influencing Performance in College Courses

It seems that many factors influence performance in college. In so far as engineering was concerned, Furneaux (1962) described the results of a personality test given to first-year mechanical engineering students at Imperial College. The tests were designed to differentiate between extraversion (unstable and stable) and introversion (unstable and stable). He found that there were significant differences in academic performance between the groups and that those with tendencies toward neurotic introversion tended to do better in examinations than those tending to extraversion (see Section 2.5). By the end of the 1960s, other studies had been completed that arrived at a similar conclusion. Child (1969) investigated the 1966 intake at Bradford University (503 men and 103 women) to compare personality intelligence and social class. While introversion was a characteristic of these students, they tended to extraversion when compared with norms for university students obtained from the Eysenck Personality Inventory. In what must be one of the earliest references to women scientists in the UK literature, Child found that they were the most extravert of all the groups. He suggested that this might be because extraverts are less susceptible to social conditioning and in consequence were less concerned about what others might think of irregular or unfeminine career choices. His findings about social class did not differ from other studies of technological courses. A large number of students in these courses, particularly in the ex-Colleges of Advanced Technology that had become universities, came from working-class homes. Relating this to Furneaux's study that suggested that neuroticism is a measure of drive level, Child suggested that "if this is the case, students from the homes of semi-skilled and unskilled workers tend to have higher drive levels than other students specialising in the sciences."

Entwistle and Wilson (1970), in a survey of the literature, concluded that while unstable students in mechanical engineering examinations were more successful they were atypical of university students as a whole. As a result of a large-scale study among students in universities, colleges of education and polytechnics they were led to conclude "the possibility of their being distinct differences in the relationship between neuroticism and academic performance that may explain some of the contradictions in previous findings." Some years later, Kline and Lapham (1992) found no differences between engineering students and other students in five British universities in respect of extraversion and emotional stability.

In the United States, Elton and Rose (1974) studied engineering students, using the *Omnibus Personality Inventory*. They found a significant difference between engineering students on the dimension of intellectual disposition. Strangely, an absence of high intellectual interests was found among those who persisted. They suggested that the faculty might consider a second experiment with the objective of developing new avenues of professional competence for the 25% of engineering students who withdrew. In another study of students in residence, they were led to conclude that personality differences among students sharing accommodation could enhance or impede achievement. Notwithstanding criticisms of its psychometric qualities, the Myers Briggs Type Indicator (MBTI) was used by Smith, Irey, and McCauley (1973) to investigate the personal qualities that might influence performance. In a later article, McCauley (1990) argued that people skills were undervalued by engineering educators as measured by the Feeling dimension of the MBTI. The MBTI became a test that was favored by personnel selectors in industry, and there are still references to its use by engineering educators.

Studies continue to report that Feeling is important, for example, among Chinese students (Zhang et al., 2014; see also Heywood, 2005). Recently, a study of 103 Dutch engineers (mean age = 48.4 years) using the Five Factor Personality Inventory[3] (Hendricks et al., 1999) found that this group was somewhat more extraverted than the population as a whole, yet, almost paradoxically they were more autonomous and less friendly than ordinary people, which might be problematic in interpersonal relations (van der Molen, Schmidt, and Kruisman, 2007). They would need to learn to be more "agreeable" (agreeable/quarrelsome being one of the five factors). The authors pointed out that their findings had some similarity with Chinese engineers who were found to be more emotionally stable and conscientious than a comparison group (Dai, 2003).

On both sides of the Atlantic, there was interest in study habit inventories. While there was no reported study of their use with engineering students, marked differences were found between preclinical and clinical students in medicine. Preclinical students tried to memorize work recently covered but clinical students did not. While there were other differences, the investigators did not find any clear relation between these differences and subsequent academic performance (Malleson, Penfold, and Sawiris, 1967). Entwistle and Wilson developed a study habit inventory and administered it together with the Eysenck Personality Inventory and found that among graduates of a Diploma in Education course that the study methods scale distinguished clearly between the worst students and the remainder. The extraversion scale sorted out the successful students from the rest. In another study, Entwistle and Entwistle (1970) concluded that good study habits made a partial contribution to an introvert's success.

The more important work on assessment was to be conducted later in Sweden and Britain, the vocabulary of which has entered into higher education discourse. Marton and Säljö (1984) distinguished between deep and surface approaches to understanding. They concluded that the strategies found, apart from anything else, were indicative of different perceptions of what is wanted from learning. From the perspective of assessment design, there is no guarantee that a student will perceive the demands of assessment in the same way as the tutor. Moreover, the design of assessment and the method of instruction can cause either deep or surface learning.

Entwistle and Ramsden (1983) developed an *Approaches to Study Inventory* that yielded four factors. *Meaning orientation* had high loadings on the deep approach associated with comprehension learning and intrinsic motivation: whereas the *reproducing orientation* was highly loaded on the surface approach, operation learning and improvidence were associated with fear of failure and extrinsic motivation. A nonacademic factor *nonacademic orientation* related to disorganized approaches to study, and *achieving orientation* to strategic approaches to study. Previously Entwistle, Hanley, and Ratcliffe (1979) had identified a *strategic approach* to learning in which students try to "manipulate the assessment procedures to their own advantage by careful marrying of their efforts to the reward system as they perceive it."

Twenty years earlier, Nevitt Sanford (1962), in the United States, brought together a number of authorities to give psychological and social interpretations of higher learning. There were sufficient authors to create a substantial volume. Among them was Christian Bey (1962), who, in a chapter titled "A Social Theory of Intellectual Development," suggested three types of student relationships with assessment. First, there were those who were academically oriented and who worked hard to obtain academic rewards. They would seem to have some affinity with the "strategic learners" identified by Entwistle. Presumably they would vary their learning from deep to surface as a function of their perception of what was required. Second, there were those who were intellectually oriented: it would seem that they would be likely to be deep learners: and third, there were those who were socially oriented. Twenty or more years later, Taylor distinguished between four orientations that she called, academic, vocational, social, and personal that is somewhat different. She distinguished between intrinsic and extrinsic persons in each category. Thus, an intrinsic person with a personal orientation is concerned with broadening or self-improvement, whereas an extrinsically motivated person in this category would look for proof of capability (Gibbs, Morgan, and Taylor, 1984). The point here is not the orientations of students but the fact that these orientations pointed toward the need for a social theory of intellectual development. In 1967, Sanford published a study with the title *Where College fails. A Study of the Student as Person.* These studies pointed to the need for assessment to take into account the "person."

These limited remarks are intended to show that consideration of assessment is very much more than setting examinations and tests. Satisfactory performance is dependent on a whole mix of factors that, if adjusted, may not necessarily solve problems of performance. The factors influencing student performance extend to the organization of the institution and its values as expressed through its faculty, as well as myriad aspects of a student's life [style] (Berliner, 2014). Any discussion of assessment has to take them into account, as is done, for example, in Chapter 7.

1.12 Assessment: Results and Accountability

One of the few parameters that can be used to measure the performance of an academic institution is examination (test) results. Politicians love league tables that are based on results and they often grievously misinterpret the data. In both the United States and the United Kingdom, there have been concerns about grade inflation. That is, substantial increases in the number of top and near top grades awarded. Very often politicians take this to mean that standards have fallen. But, in all probability the same politicians have said that standards need to be raised 3 or 4 years earlier. Universities are dammed if they do increase the number of better grades and dammed if they do not. This contradiction highlights the need for data that are collected, such as examination scripts, to be retained for a number of years so that past and present standards can be evaluated.

The point here is not to debate the accountability issue but rather to point out that in the 1960s a positivist philosophy came into play and higher education came to be viewed as an instrument for producing people to do particular jobs—engineers to do engineering. In Britain, there was, for example, a National Board for Prices and Incomes that took a particular interest in the productivity of universities or at least how to measure it. The Board wrote "We have received very little evidence of experimentation with new methods of teaching which would economise on the number of staff required. It is possible that the nature of control exercised from the outside has made it difficult for a university to experiment with different combinations of capital or labour, or with different combinations of staff which might lead to an improvement in teaching methods, helping it to determine the nature of the teaching staff required" (cited by Heywood, 1971).

It will be noticed that for the Board improvement in teaching is equated to reducing the number of teachers not with possible improvements in teaching or student learning *per se.* It will also be noticed that there is an expectation that universities should prepare students for their careers and select teaching staff to that end. Although the Board had a short life, subsequent governments listened to industry and influenced the agencies responsible for higher education to cause them to create corporate cultures within universities. At the same time it reflects a utilitarian view of what the aims of education should be. Not surprisingly, no official report on higher education that followed contained an adequate discussion of the aims of higher education. In such a climate, performance (competency)-based assessment would have an appeal. It is not surprising, therefore, that during this period, all across the Western world, without any specific announcement the universities began to reverse their aim to one of the primacy of research over the primacy of teaching. The performance of faculty came to be measured by the number of papers published in elite journals, and a conflict between research and teaching emerged to the detriment of teaching. This change also marked a change in the complex relationships between academia and industry.

1.13 Assessing the Learner

Assessment was and remains for many the afterthought of the educational process—a necessary evil. In the formative years of the study of higher education, the power of assessment to influence student learning and teaching was shown. Assessment was shown to be a major component of the curriculum process. Studies of assessment could not be confined simply to the measurement of performance in examinations since many factors, including several in the affective domain, contributed to that performance. Present-day thinking about assessment has to be judged by the extent it has become part of the curriculum process and taken into account all these other factors especially those from the affective domain that contribute to performance. The way in which engineering schools take into account the "person" is as important to the understanding of the impact of assessment on learning as anything else.

The curriculum process should examine the aims of education on which that curriculum is based in detail. The evidence collected suggests that very little has changed.

So far I have failed to define assessment. Georgine Loacker to whom this book is dedicated and I disagreed about many things but we did not disagree about the principles of assessment. It is appropriate, therefore, to take the definition of assessment from a book written under her direction in 1985 for the students of Alverno College. It reads "assessment: a multidimensional attempt to observe and, on the basis of criteria, to judge the individual learner in action" (Alverno College, 1985). In the case of students studying for a profession (e.g., engineering) that means, judging them in both academic and workplace (industrial training) situations. "To undertake assessment means sampling, observing, and judging according to criteria. It means developing a whole array of techniques to take into account the fullest possible range of human talents. And it means an ongoing commitment to dealing with these kinds of questions" (Alverno College, 1985). This book is about how the engineering community has and is answering these questions and how it has and is determining the criteria on which such judgments are made.

E. J. Furst (1951), one of the authors of *The Taxonomy of Educational Objectives* (see Section 2.5) and an expert in evaluation, wrote that every teacher should have a defensible theory of learning that in this writer's view would derive from a defensible philosophy of education. The intention of this book is to provide its readers with information about theory and practice that will enable them to develop their own theory or philosophy of assessment. This chapter has begun that process by describing the different meanings that have been given to the terms associated with assessment and in particular those of competence, outcomes, and performance. By definition, a professional education implies that much of it is about preparing the student to be a professional, which in turn implies that the curriculum is to some extent a reflection of what engineers do. Surprisingly, engineering education has been regularly criticized during the last 70 years for not knowing what it is that engineers do. The chapter that follows opens a discussion on the relationship between academia and industry.

Notes

1. Forms of norm-referenced grading are in use in American engineering courses and Seymour and Hewitt's (1997) seminal study on retention in STEM courses thought that norm-referenced grading fostered a competitive atmosphere that alienated otherwise capable students. Wolf and Powell (2014) reported that *left-of-center* grading was more likely to discourage females than peers. The students reported that it caused stress and frustration and focused their attention on the system than on their own learning. In the United Kingdom, it has been the system and students were brought up to it from elementary school onward. But the introduction of competency-based systems of assessment has meant that school examining authorities give marks up to 100%. At school level, politicians to the right have taken against continuous assessment and modular courses and there is a move back to more traditional examining
2. In England, there is no such thing as a liberal arts degree. Students will take mostly a single subject like history or engineering for 3 years; this would include any requisite studies. For example, engineering would require mathematics. Today engineering has been extended to a fourth year at the end of which a master's degree is awarded. Professional recognition as a chartered engineer now requires applicants to have a master's degree.
3. The Five-Factor Personality Inventory is also known as the Big Five and is proving attractive to engineering educators. The five domains are extraversion, agreeableness, conscientiousness, emotional stability, and openness. It is considered to be a better test than the MBTI. See Rhee, Parent, and Oyamont (1992) for its use in engineering. They cite a meta-analysis by Ozer and Bet-Martinez (2008) in support of its use.

References

Abercrombie, M. L. J. (1964). *Aims and Techniques of Group Teaching*. London: Society for Research into Higher Education.

Abu-Idayil, B., and Al-Attar, H. (2010). Curriculum assessment as a direct tool in ABET outcomes assessment in chemical engineering. *European Journal of Engineering Education*, 35(5), 489–505.

Ager, M., and Weltman, J. (1967). The present structure of examinations. *Universities Quarterly*, 21(3), 272.

Allen, T., Reed-Rhoads, T., Murphy, T. J., and Stone, A. D. (2008). Coefficient Alpha: an engineers interpretation of reliability. *Journal of Engineering Education*, 97(1), 87–94.

Al-Nashash, H., Khaliq, A., Qaddoumi, N., Al-Assaf, Y., Assalah, K., Dhaouadi, R., and El-Tarhuni, M. (2009). Improving electrical engineering education at the American University of Sharjah through continuous assessment. *European Journal of Engineering Education*, 34(1), 15–28.

Alverno College (1985). *Assessment at Alverno College.* Milwaukee: Alverno Productions.

Astin, A. (1968). *The College Environment* Washington, DC: American Council on Education.

Berliner, D. C. (2014). Exogenous variables and value-added assessment: a fatal flaw. *Teachers' College Record*, 116(10). http:/www.tereord.og. Accessed March 24, 2014.

Bernold, L. E., Spurlin, J. E., and Anson, C. M. (2007). Understanding our students. A longitudinal study of success and failure in engineering with implications for increased retention. *Journal of Engineering Education*, 96(3), 263–279.

Bey, C. (1962). A social theory of intellectual development. In: N. Sanford (ed.), *The American College*. New York: John Wiley & Sons.

Bligh, D. G. (1971). *What's the Use of Lectures*. London: Penguin.

Bosworth, G. S. (1963). Toward creative activity in engineering. *Universities Quarterly*, 17, 286.

Bosworth, G. S. (Chairman of a Committee) (1966). *The Education and Training Requirements for the Electrical and Mechanical Manufacturing Industries*. Committee on Manpower Resources for Science and Technology. London: HMSO.

Bucciarelli, L. L., Coyle, E., and McGrath, D. (2009). Engineering education in the US and UK. In: S. H. Christensen, et al. (eds.), *Engineering in Context*. Aarhus: Academica.

Carter, G. (1992). The fall and rise of university engineering education in England in the 1980s. *International Journal of Technology and Design Education*, 2(3), 2–21.

Carter, G., and Heywood, J. (1992). The value-added performance of electrical engineering students in a British university. *International Journal of Technology and Design Education*, 2(1), 4–15.

Child, D. (1969). A comparative study of personality, intelligence and social class in a technological university. *British Journal of Educational Psychology*, 39, 40

Christofourou, A. F., and Yigit, A. S. (2008). Improving teaching and learning in engineering education through a continuous assessment process. *European Journal of Engineering Education*, 33(1), 105–116.

Cohen, L. (1968). Attitude deterioration and the experience of industry. *The Vocational Aspect of Education*, 20, 181.

Cole, J. S., and Spence, S. W. T. (2012). Using continuous assessment to promote student engagement in a large class. *European Journal of Engineering Education*, 37(5), 508–525.

Cox, R. (1967). Examinations and higher education. A review of the literature. *Universities Quarterly*, 21(3), 292.

Cowan, J. (2006). *On Becoming an Innovative University Teacher*, 2nd edition. Buckingham: Open University Press.

Cronbach, L. J. (1971) Test validation In: R. L. Thorndike (ed.), *Educational Measurement*, 2nd edition. Washington, DC: American Council on Education.

Committee of Vice-Chancellors and Principals. (1969). *The Assessment of Academic Performance*. London: Committee of Vice-Chancellors and Principals.

Dai, X. (2003). A study of the occupational characters of different occupational groups. *Chinese Journal of Clinical Psychology*, 10, 252–255.

Dressel, P. L. (ed.) (1961). *Evaluation in Higher Education*. Boston, MA: Houghton Mifflin.

Dwyer, C. A (1994). *Development of the Knowledge Base for the PRAXIS III: Classroom Performance Assessments. Assessment Criteria*. Princeton, NJ: Educational Testing Service.

Educational Testing Service. (1995). *Teacher Performance Assessments. Assessment Criteria*. Princeton, NJ: Educational Testing Service.

Elton, C. F., and Rose, H. A. (1966). Within university transfer. Its relation to personality characteristics. *Journal of Applied Psychology*, 50(6), 539.

Entwistle, N. J. (1970). Students and their academic performance in different types of institution. Part of a report to the Joseph Rowntree memorial trust cited in Heywood (1971).

Entwistle, N. J., and Entwistle, D. (1970). The relationship between personality, study methods and academic performance. *British Journal of Educational Psychology*, 40, 132–143.

Entwistle, N. J., Hanley, M., and Ratcliffe, G. (1979). Approaches to learning and levels of understanding. *British Educational Research Journal*, 5, 99–114.

Entwistle, N. J., and Wilson, J. D. (1970). Personality, study methods, and academic performance. *Universities Quarterly*, 24(2), 147–156.

Epstein, R. M., and Hundert, E. M. (2002). Defining and assessing professional competence. *Journal of the American Medical Association*, 287(2), 226–235.

Felder, R. M., and Hadgraft, R. G. (2013). Educational practice and educational research in engineering: partners, antagonists, or ships passing in the Night. *Journal of Engineering Education*, 102(3), 339–345.

Fordyce, D. (1992). The nature of student learning in engineering education. *International Journal of Technology and Design Education*, 2(3), 23–40.

Freeman, J., and Byrne, P. S. (1973). *The Assessment of Post-Graduate Training in General Practice.* Guildford, UK. Society for Research into Higher Education.

Furneaux, W. D. (1962). The psychologist and the university. *Universities Quarterly*, 17, 33–47.

Furst, E. J. (1958). *The Construction of Evaluation Instruments.* New York: David MacKay.

George, J., and Cowan, J. (1999). *A Handbook of Techniques for Formative Evaluation: Mapping the Students Learning Experience.* London: Kogan Page.

Gibbs, G., Morgan, A., and Taylor, E. (1984). The world of the learner. In: F. Marton, D. Hounsell, and D. Entwistle (eds.), *The Experience of Learning.* Edinburgh: Scottish Academic Press.

Gipps, C. V. (1994). *Beyond Testing. Toward a Theory of Educational Assessment.* London: Falmer.

Griffin, C. (2012). A longitudinal study of portfolio assessment to assess competence of undergraduate student nurses. Doctoral Dissertation, University of Dublin, Dublin.

Hartle, T. W. (1986). The growing interest in measuring the education achievement of college students. In: C. Adelman (ed.), *Assessment in Higher Education.* Washington, DC: US Department of Education.

Hartog, P., and Rhodes, E. C. (1935). *An Examination of Examinations.* London: Macmillan.

Hartog, P., and Rhodes, E. C. (1936). *The Marks of Examiners.* London: Macmillan.

Hendricks, A. A. J., Hofstee, W. K. B., and de Raad, B. (1999). The Five-Factor Personality Inventory (FFPI). *Personality and Individual Differences*, 27, 307–325.

Heywood, J. (1971). A report on wastage. *Universities Quarterly*, 25(2), 189–237.

Heywood, J. (1977). *Assessment in Higher Education.* Chichester, UK: John Wiley & Sons.

Heywood, J. (1989). *Assessment in Higher Education*, 2nd edition. Chichester, UK: John Wiley & Sons.

Heywood, J. (2000). *Assessment in Higher Education. Student Learning, Teaching, Programmes and Institutions.* London: Jessica Kingsley.

Heywood, J. (2005). *Engineering Education. Research and Development in Curriculum and Instruction.* Hoboken, NJ: IEEE/John Wiley & Sons.

Heywood, J. (2011). Higher technological education in England and Wales 1955–1966. Compulsory liberal studies. In: Proceedings of Annual Conference of the American Society for Engineering Education. Paper 635.

Heywood, J. (2013). Higher technological education and British policy making: a lost opportunity for curriculum change in engineering education. In: Proceedings of Annual Conference of the American Society for Engineering, June 2013. Paper 6331.

Heywood, J., and Cheville, A. (2015). Is engineering education a professional activity? In: Proceedings of Annual Conference of the American Society for Engineering Education, June 2015. Paper 12907.

Heywood, J., and Ann Abel, R. (1965). *Technical Education and Training. Research in Progress. 1962–1964.* NFER Occasional Publication, No 8. Slough, National Foundation for Educational Research.

Howell, L. L., Sorensen, C. D., and Jones, M. R. (2014). Are undergraduate GPA and General GRE percentiles valid predictors of student performance in an engineering graduate program? *International Journal of Engineering Education*, 30(5), 1145–1165.

Hoyle, E. (2001). Teaching, prestige, status and esteem. *Educational Management and Administration* 29(2), 139–152.

Hutchings, D. G. (1963). *Technology and the Sixth Form Boy.* Oxford, UK: Oxford University Department of Education.

Kelsall, R. K. (1963). University student selection in relation to subsequent academic performance. *The Sociological Review*. Monograph No 7.

Kline, P. (1993). *The Handbook of Psychological Testing.* London: Routledge.

Kline, P., and Lapham, S. (1992). Personaility and faculty in British universities. *Personality and Individual Differences*, 13, 855–857.

Kowalski, S. E., Kowalski, F. V., and Gardner, T. Q. (2009). Lessons learned when gathering real-time formative assessment in the university classroom using Tablet PC's. In: ASEE/IEEE Proceedings of Frontiers in Education Conference, T3F-1 to 5.

Larkin, T. L., and Usinski, J. (2013). The evolution of curriculum assessment within the physics program at American University. In: Proceedings of Annual Conference of the American Society for Engineering, June 2013. Education Paper 6739.

Lavin, D. E. (1967). *The Prediction of Academic Performance* New York: John Wiley & Sons.

Lorimer, S., and Davis, J. A. (2014). Consistency in assessment of pre-engineering skills. In: Proceedings of Annual Conference of the American Society for Engineering, June 2014. Paper 9596.

Marton, F., and Saljö, R. (1984). Approaches to learning. In: F. Marton, D. Hounsell, and N. J. Entwistle (eds.), *The Experience of Learning.* Edinburgh: Scottish Academic Press.

Malleson, N. (1964). *British Student Health Services.* London: Pitman.

Malleson, N., Penfold, D. M., and Sawiris, Y. (1967) Medical students' study. Time and place. *British Journal of Medical Education*, 1, 169.

Mast, T., and Davis, D. A. (1994). Concepts of competence. In: D. A. Davis and R. D. Fox (eds.), *The Physician as Learner. Linking Research to Practice.* Chicago, IL: American Medical Association.

McCauley, M. M. (1990). The MBTI and individual pathways in engineering design. *Engineering Education*, 66(7), 729–716.

McGuire, C. (1993). Perspectives in assessment. In: J. S. Gonnella et al. (eds.), *Assessment Measures in Medical School Residency and Practice.* New York: Springer.

McKeachie, W. J. (1974). Instructional psychology. *Annual Review of Psychology*, 25, 163–193.

McVey, J. (1975). The errors in marking examination scripts in electronic engineering. *International Journal of Electrical Engineering Education*, 12(3), 203.

McVey (1976a). Standard error of a mark for an examination paper in electronic engineering. *Proceedings Institution of Electrical Engineers*, 123(8) 843–844.

McVey, P. (1976b). The "paper error" of two examinations in electronic engineering. *Physics Education*, 11, 58–60.

Miller, G. E. (1990). The assessment of clinical skills/competence/performance. *Academic Medicine*, 65, 563–567.

Miller, G. W. (1970). *Success, Failure and Wastage in Higher Education.* London: Harrap.

Myllymaki, S. (2013). Incorporation of continuous student assessment into lectures in engineering education. *European Journal of Engineering Education,* 38(4), 385–393.

National Union of Students. (1967). *Briefing on Wastage.* London: National Union of Students.

National Union of Students. (1968). *Executive Report on Examinations.* London: National Union of Students.

Nghe, N. T., Janacek, P., and Haddaway, P. (2007). A comparative analysis of techniques predicting academic performance. In: ASEE/IEEE Proceedings of Frontiers in Education Conference T2G-7 to 12.

Norman, G. R. (1985). Defining competence. A methodological review. In: V. R. Neufeld and G. R. Norman (eds.), *Assessing Clinical Competence.* New York: Springer.

Oppenheim, A. M., Jahoda, M., and James, R. L. (1967). Assumptions underlying the use of university examinations. *Universities Quarterly*, 1(2), 89.

Ozer, D. J., and Benet-Martinez, V. (2006). Personality and the prediction of consequential outcomes. *Annual Review of Psychology*, 57, 401–402.

Parsons, D. (2008). Is there an alternative to exams? Examination stress in engineering. *International Journal of Engineering Education*, 24(6), 1111–1118.

Pellegrino, J. W., DiBello, L. V., and Brophy, S. P. (2014). The science and design of assessment in engineering education In: A. Johri and B. M. Olds (eds.), *Cambridge Handbook of Engineering Education Research.* New York, NY: Cambridge University Press.

Reeves, D. B. (2009). *Assessing Educational Leaders. Evaluating Performance for Improved Individual and Organizational Results* Thousand Oaks, CA: Corwin.

Rethans, J. J., Norcini, J. J., Barón-Maldonado, M., Blackmore, D., Jolly, B. C., LaDuca, T., Lew, S., Page, G. G., and Southgate, L. H. (2002). The relationship between competence and performance: implications for assessing practice performance. *Medical Education*, 36(10), 901–909.

Rhee, J., Parent, D., and Oyamot, C. (2012). Influence of personality on senior project combining innovation and entrepreneurship. *International Journal of Engineering Education*, 28(2), 302–309.

Ridgeway, J., and Passey, D. (1992). An international view of mathematics assessment: through the glass darkly. In: M. Niss (ed.), *Investigations into Assessment in Mathematics Education.* Dordrecht: Kluwer.

Roberts, R. F. (1982). *Preparation for Teaching Practice.* Liverpool, City of Liverpool College of Higher Education cited in Heywood, J. (1982). *Pitfalls and Planning in Student Teaching.* London: Kogan Page.

Rogers, G. (1994). Measurement and judgment in curriculum assessment systems. *Assessment Update*, 6(1), 6–7.

Ryle, A. (1969). *Student Casualties* London: Allen Lane.

Ryle, A., and Lunghi, M. (1968). A psychometric study of academic difficulty and psychiatric stress in students. *British Journal of Psychiatry*, 114(506), 57.

Runté, R. (1995). Is teaching a profession. In: G. Taylor and R. Runté (eds.), *Thinking about Teaching. An Introduction* Toronto: Harcourt Brace.

Sandberg, J. (2000). Understanding human competence at work. An interpretive approach. *Academy of Management Journal*, 43(3), 9–25.

Sadler, D. R. (2007). Perils in the meticulous specification of goals and assessment criteria. *Assessment in Education: Principles, Policy and Practice*, 14(3), 387–392.

Sanford, N. (ed.) (1962). *The American College. A Psychological and Social Interpretation of the Higher Learning*. New York: John Wiley & Sons.

Saupé, J. (1961). Learning. In: P. Dressel (ed.), *Evaluation in Higher Education*. Boston, MA: Houghton Mifflin.

Schalk, P. D., Wick, D. P., Turner, P. R., and Ramsdell, M. W. (2011). Predictive assessment of student performance for early strategic guidance. In: ASEE/IEEE Proceedings of Frontiers in Education Conference, S2H-1 to 5.

Seymour, E., and Hewitt, N. M. (1997). *Talking about Leaving. Why Undergraduates Leave the Sciences.* Boulder, CO: Westview Press.

Shulman, L. S. (1987). Knowledge and teaching: foundations of the new reform. *Harvard Educational Review*, 57(1), 1–22.

Smith, A. B., Irey, R. K., and McCauley, M. H. (1973). Self-paced instruction and college students' personalities. *Engineering Education*, 63(6), 435–450.

Smithers, A. G. (1965). Occupational aspirations and expectations of engineering students. *British Journal of Industrial Relations*, 7, 415.

Sternberg, R. J. (2005). Intelligence, competence and expertise In: A. J. Elliot and C. S. Dweck (eds.), *Handbook of Competence and Motivation*. New York: Guilford Press.

Sundarajan, S. (2014). A strategy for sustainable student outcomes assessment for mechanical engineering program that maximizes faculty engagement. In: Proceedings of Annual Conference of the American Society for Engineering Education, June 2014. Paper 9251.

Talbot, N. (2004). "Monkey see, Monkey do." A critique of the competency model in graduate medical education. *Medical Education*, 36(6), 587–592.

The Times. (2015). Exam results just don't add up, says accountancy firm; August 3. Authored by Andrew Ellson.

Trevelyan, J. (2014). The Making of an Expert Engineer. Leiden, CRC Press.

University Grants Committee (1968). *An Enquiry into Student Progress*. London: University Grants Committee.

van der Molen, H. T., Schmidt, H. G., and Kruisman, G. (2007). Personality characteristics of engineers. *European Journal of Engineering Education*, 32(5), 495–501.

Vaizey, J. (1970). The costs of Wastage. *Universities Quarterly*, 25(2), 139–145.

Vernon, P. (1965). *The Measurement of Human Abilities*. London: University of London Press.

Walsden, S. E., and Foor, C. E. (2008). What's to keep you from dropping out? Student immigration into and within engineering. *Journal of Engineering Education*, 97(2), 191–206.

Warren-Piper, D. (1994). *Are Professors Professional. The Organization of University Examinations*. London: Jessica Kingsley.

Wigdor, A. K., and Garner, W. R. (eds.) (1982). *Ability Testing: Uses, Consequences and Controversies* Washington, DC: National Academies Press.

Williams, B., and Figueiredo, J. (2013). Finding workable solutions: Portuguese engineering experience. In: B. Williams, J. Figueiredo, and J. P. Trevelyan (eds.), *Engineering Practice in a Global Context. Understanding the Technical and the Social.* London: CRC Press/Taylor and Francis.

Yeung, A. T. (2014). Education and training requirement for a practicing civil engineer to qualify in Australia and the United Kingdom. In: Proceedings of Annual Conference of the American Society for Engineering Education, June 2914. Paper 8227.

Yokomoto, C. F., and Bostwick, W. D. (1999). Modelling the process of writing measurable outcomes for EC 2000. In: ASEE/IEEE Proceedings of Frontiers in Education Conference, 2, 11b-18-22.

Young, M. (2002). Clarifying Competency and Competence. Henley On Thames. Henley Management College.

Zhang, D., Yao, D., Cuthbert, N., L., and Ketteridge, S. (2014) A suggested strategy for teamwork teaching in undergraduate engineering programs particularly in China. In: ASEE/IEEE Proceedings of Frontiers in Education Conference, pp. 537–544.

2

Assessment and the Preparation of Engineers for Work

Studies of the impact of organizational structure on innovation and performance showed that often qualified engineers had poor communication skills. Engineers were required to speak several different "languages" to perform their work. Organizational structures were shown to influence attitudes and values, and by implication performance to the extent of modifying competence. The findings also have implications for the preparation of engineers for management and leadership roles.

In the 1950s, university education in the United Kingdom was considered suitable for training engineers and technologists for research and development but it was held that education and training for industry needed to be enhanced and sandwich (cooperative) programs were developed to provide an alternative degree program in the technical college sector to those offered by the universities. A brief description of the Diploma in Technology and Colleges of Advanced Technology that were set up to achieve this program is given. Assessment was mostly an afterthought although there were one or two serious attempts to devise assessment strategies for the period of industrial training. The evidence pointed to "curriculum drift," that is the creation of syllabuses as the mirror image of those offered by higher status institutions, in this case the universities. In the 1960s, criticisms were made of these programs by industrialists that led to the discussion of at least one alternative model for the education of engineers for manufacturing engineering that was based on recent developments in the educational sciences. It remains relevant.

The Assessment of Learning in Engineering Education: Practice and Policy, First Edition. John Heywood.
© 2016 The Institute of Electrical and Electronics Engineers, Inc. Published 2016 by John Wiley & Sons, Inc.

Throughout the sixties, the engineering profession was concerned with criticisms that scientists and technologists lacked creativity. There were problems about what it is, how it should be taught, and how it should be assessed. These questions continue to be asked. Felder listed the kind of questions that should elicit creative behavior in engineering. Recent work by engineering educators suggests some strategies for assessment including the Consensual Assessment Technique. The search for reliable, valid, and inexpensive (in time and money) techniques continues.

An analysis of examinations taken by mechanical engineering education students suggested that they primarily tested one factor. It was suggested that if examinations were to be improved, there was a need for their objectives to be clarified. The chapter ends with a discussion of some of the factors that influence change. Apart from the issue of status, lack of knowledge of the curriculum process and student learning contributed to the failure to take a new approach.

2.1 Engineers at Work

If it is intended that the curriculum should prepare engineering students for work, it is necessary to understand what qualified engineers do at work. When there was interest in developing a curriculum to produce engineers for industry in the 1950s and 1960s in the United Kingdom, no studies had been reported that specifically investigated the work that engineers undertook for the purpose of relating it to the curriculum. Yet curricula were proposed to prepare engineers specifically for work in industry, but these were based on unfounded assumptions and experience of day-release education (i.e., the release of apprentices and other employers for study on 1 day per week at a technical college on programs for craftsmen, technicians, and technologists). Nevertheless, two studies that focused on organizational structure and innovation provided limited information that could have been of help to curriculum designers, and more particularly the design of assessment. One study by Burns and Stalker (1961) was of innovation in the Scottish electrical industry. The other by Louis Barnes (1960) was a comparative study of two electronics departments in the United States whose organizational structures differed significantly from each other. Simply put, both studies showed that hierarchical organizations or divisions within them were not best placed to innovate when compared with less hierarchical and more open systems of organization. Organizational structure placed different kinds of stress on the management (leadership, if you wish) of the two systems and must, therefore, have influenced performance.

One of the interesting findings of the Burns and Stalker inquiry was the difficulty that research and development (R&D) personnel had in communicating with those responsible for manufacturing. They worked in two different cultures that arose in no small way from differences in the training and experience of the two groups. Engineers seemed to converse in several different "languages." Throughout the 50 years that have elapsed since these studies, industrialists have regularly complained about the poor communication skills of graduates (ASME, 2010; Talbot et al., 2013; Trevelyan, 2014). Education systems do not seem to have contributed to its resolution in spite of the attempts of many engineering educators to remedy this situation. A feature of the Barnes study is the way in which it shows how different organizational structures condition different attitudes among the employees to their work, a finding that has implications for the way that intending managers and executives are prepared (trained) for work.[1] Problems associated with the structure of organizations continue to be experienced. They have implications for the assessment of performance for the beliefs, values, and attitudes associated with a particular organizational structure and are likely to influence that performance (see also Chapter 10). One would presuppose that attention would have been paid to communication and management/leadership issues in the engineering curriculum, but this was not the case in the UK universities, although it was often the case in Diploma in Technology programs offered in the Colleges of Advanced Technology (CATs) between 1956 and 1966 where some form of liberal/general study was compulsory (Heywood, 2011; see Section 2.2). This diploma was developed specifically to prepare engineering students for industry by means of an alternative curriculum. Given that a major report on Higher Technological Education (Percy, 1945) was satisfied that university degree programs could produce the R&D personnel that were required, it is not surprising, therefore, that the examinations analyzed by Furneaux (see Section 2.5), and therefore the curriculum, all focused on the study of engineering science and its divisions.

2.2 An Alternative Approach to the Education and Training of Engineers for Industry

From time to time during the last 60 years, industry in the United Kingdom has complained that higher education institutions are not producing the type of graduate it needs. It is with this issue that the story begins when in 1945, immediately after the Second World War, a government committee on higher technological education (the Percy Committee), while expressing satisfaction with the products of university engineering programs especially as they prepared their graduates for R&D, expressed dissatisfaction with the preparation of engineers for industry by the technical colleges. The training they received by part-time release from industry on 1 day per week plus evening study would not provide the high-quality engineers that industry required. So in 1955, the government agreed to develop a system of higher technological education in the technical education sector. In England and Wales, it would develop eight (eventually nine) of the existing technical colleges into CATs which would award a diploma as a result of a curriculum oriented toward the needs of industry that was "sandwich" structured. The term used in the United States that has more or less the same meaning as sandwich is "cooperative." The programs would last for 4 years and the student would alternate between 6 monthly periods in industry and 6 monthly periods in college. The diploma was to be equivalent of a degree and accredited by an agency established for that purpose (National Council for Technological Awards). Unfortunately, the government did not restrict the award of the diploma to these colleges and allowed other colleges that were suitably resourced to offer programs for the diploma. They also allowed those CATs that offered programs for external degrees of the University of London (i.e., open to students other than those in the university's constituent colleges) to continue to offer those programs. Similarly, instead of insisting that all of the colleges offered the same sandwich structure, they were allowed to offer the structures they favored. One variant was to begin the 6-month rotation in college rather than industry. Another variation was to begin the program with 1 year in industry and to include only two periods of 6 months each during the course that followed. Yet another variation was to join these two 6 months together and send the student for a year in industry after his or her second year of academic study.[2]

The Diploma, which began in 1956–57, was characterized by the compulsory provision of Liberal/General Studies (Heywood, 2011), substantial project work by individual students particularly in the final year, and a sandwich (cooperative) structure, the most popular arrangement being the 6 months in industry followed by 6 months in college cycled over 4 years as indicated. It was offered primarily in eight CATs (later nine) and one or two other colleges that had experience of degree-level work.

Underpinning the sandwich formula is a philosophy of "integration," the idea that what is learnt in college should inform what the student does in practice, that is, in the industrial training period. In theory, this concept has implications for the structure of sandwich programs. Should a student begin with a period in industry or in college? The investigations that were made of the effectiveness of the industrial periods threw little light on this issue although some of the characteristics of integration, for example, the "self-illustrative" when a student becomes aware of a relation between past and present learning experiences, came to be understood. Fundamental to this experience was "awareness." Today we would infer the need for a reflective capability (see Chapter 7.7). The same applied to assessment that depended on reports on what students had achieved within training schemes approved by the colleges. Yet, it is in industrial training that it should be possible to judge the performance of a student in practice, in the same way as medical students are judged by their performance in a clinical situation. There was little attempt to develop procedures for the assessment of performance in the industrial period rather tutors and their industrial counterparts focused on providing the students with a good experience (Jahoda, 1964; Heywood, 1967). However, at Brunel College of Advanced Technology, a Department of Industrial Training was established that had as its aims the improvement of industrial training and the integration of the academic with the practical. By 1966, it had taken up the issue of the assessment of industrial training that continued after it had become a university in that year. "We should," wrote Rakowski (1990) of that institution, "be testing the student's ability to adjust to the industrial context, to test his technical judgment, to test his professional knowledge as opposed to his academic knowledge, to test his originality, how conscientious he is in his work." So in the mid-1980s, the Special Engineering Program at Brunel (Clark et al., 1985) introduced a 12-item Industrial Training Performance questionnaire alongside other methods of assessments to try and get a standardized procedure across all the companies where students were placed. It is shown in Exhibit 2.1: each item was rated on a 5-point scale. Its success clearly depended on the ability of the company to provide training that would enable a student to develop these qualities (Rakowski, 1990).

Controversially the marks obtained for industrial training were incorporated within the final marks for the award of a degree. "The introduction of assessment means that criteria for assessment are communicated to the students: that is; they now become aware of those aspects of professional conduct and performance that matter in the real life

---

1. General ability to carry out assigned tasks
2. Technical competence
3. Problem-solving ability
4. Willingness to accept responsibility
5. Initiative
6. Self-organization
7. Perseverance
8. Capacity to innovate
9. Interpersonal skills
10. Quality of oral reporting
11. Quality of written reporting
12. Quality of log book

EXHIBIT 2.1. Twelve-point industrial training performance questionnaire. Each item was rated on a 5-point scale. (From Rakowski, R. T. (1990). Assessment of student performance during industrial training. *International Journal of Technology and Design Education*, 1(2), 106–110).

work situation. This acts as a stimulus and motivation for continued high performance" (Rakowski, 1990).

During the last 10 years, there has been renewed interest in cooperative courses (see Section 10.2) and internships that has yielded several reports (e.g., Nasr, Pennington, and Andres, 2004; Pierrakos, Borrego, and Lo, 2008). Nagel et al. (2012) evaluated 107 students, mostly engineers, who participated in a summer internship in industry in which they completed what they called an "industry experience project." They identified seven major stages in the pursuit of the project's goals and these clearly correspond to those of a design project. These investigators were concerned like Jahoda (1963) in the 1960s to "understand what students learn during industry experiences as a means of also understanding how students learn and how they use and transfer complex problem solving skills in new settings." They wanted to find out what would be of benefit to college courses particularly in respect of the solving of complex problems. They were able to establish the approximate time spent by the students on the various stages of the project and they found, as might be anticipated, that the least time was spent on project management. They also found that students spent less time on tasks where they were likely to need help. The majority of the students were challenged by the experience with 10% of them finding it "very challenging." While the majority of the students had to find a lot of knowledge to complete the project, that knowledge was not considered too difficult.

Nagel et al. (2012) drew the conclusion that "The problem solving process in industry projects seems to be quite complex, but also somewhat structured. For engineering interns, or even entry level engineers, this model seems to be appropriate." This is consistent with educational research that suggests "problem-based learning experiences with a moderate degree of structuredness, authentic [...] and adapted to the students' cognitive

A	B
Engage students to continuously learn new knowledge and skills. The dynamically changing learning environment not only challenges students but also enables them to learn important problem-solving skills.	Increased technical skills
Moderately well structured with most of the project steps being relatively well defined, but most project stages are still considered to be challenging to very challenging.	Exposure to the world of work, thereby improving professionalism
Enable students to apply STEM knowledge, but also gain a substantial amount of new STEM knowledge. This is especially true for engineering knowledge.	Application of theoretical knowledge to real-world situations
Enable students to grow technically, cognitively, and affectively through experience of multicognitive operations in the context of continuously changing and dynamic team environment.	Improvement in time management skills
Involving a significant amount of teamwork and mentorship […]	Increase in self-worth, maturity, sense of responsibility, and feelings of accomplishments.

EXHIBIT 2.2. Effects of internships and part-time work/evening study on students as reported by Nagel et al. (2012) Understanding industry experiences: from problem solving to engineering students' learning gains. In: ASEE/IEEE Proceedings of Frontiers in Education Conference, pp. 927–932, and Golding, P., et al. (2008). Cooperative education: an exploratory study of its impact on computing students and participating employers. In: ASEE/IEEE Proceedings of Frontiers in Education Conference, F3A-1 to 6.

development and prior knowledge, and amenable to problem examination from multiple perspectives" yield effective learning (Schuurman, Pangborn, and McClintic, 2008; Sugrue, 1995). An Irish experience of teaching problem-solving models/decision making to young teenagers in postprimary education found that often middle- and low-ability students prefer the structure of problem-solving models like the Wales and Stager Guided Design Model (Heywood, 2008). It seems, therefore, that experience of industry coupled with appropriately designed problem-based learning could help cognitive development beyond the early stages of cognitive developmental models like Perry's and King and Kitchener's (see Sections 7.4 and 7.5). The findings of Nagel et al. in respect of the industry experience are shown in Exhibit 2.2 (column A).

The University of Technology in Jamaica experimented with a system that was more or less the norm in the United Kingdom in the 1950s. Students, in this case computing students, worked by day and studied by night (Golding et al., 2008). This structure is by no means new, neither are the problems they faced. In the United Kingdom for very many students, colleges and companies were separate entities and little attempt was made to integrate study. The problem that concerned the government at the time was the high levels of wastage (dropout) experienced in these courses and much of the research done at that time focused on this problem, and saw the production of reports that led to an increase in block release and sandwich education to replace part-time study and, ultimately to the full-time education of those who wished to be professional engineers. Many of the problems that Golding and his colleagues found would have been found in the United Kingdom in that early period. Some students had bad experiences, whereas others had good ones. Students would mostly be doing tasks for which they were capable unlike the students in Golding and colleagues' survey, where it was found that initially many of them were given tasks that had technical requirements beyond their capability.

Golding et al. (2008) reported that the benefits to industry were "(1) provided extrà manpower for companies to assign to tackle specific tasks/jobs, (2) identified potential permanent employees, and (3) improved corporate image." They do not mention the value to industry more generally of the improvement in professionalism that the students considered to be a benefit of the program (see Exhibit 2.2, column B).

Project work that included the teaching of design was apparently very successful and university programs began to include project work as part of their curriculum (de Malherbe, 1961; Hayes and Tobias, 1964/65 Pullman, 1964/65). Deere (1967) drew attention to the individual and group projects thought to be needed by *students* in engineering at the University of Reading. Heywood (1961), on the basis of his observations of sixth formers (grades 10–12) completing projects, outlined a principle of learning that corresponded with the idea of discovery learning promoted at the same time by Jerome Bruner.

The development of reliable and valid measures of course work assessment was not a priority. However, at the University of Bath, the former Bristol College of Advanced Technology, a criterion-based scheme of assessment for mechanical engineering had been put in place by the early 1970s (Black, 1975). Since then as project work has become more important in the teaching of design and the development of team skills so has its assessment in both the United Kingdom and the United States (Abraham, Greene, and Marches, 2007; Dartt, McGrann, and Stark, 2009; Northrup and Northrup, 2006; see Chapter 8).

Clearly, the purpose of the Diplomas in Technology was to provide a credential that indicated that diplomate (graduate) could perform the tasks of a "technologist" as opposed to a "technician." An organizational system was set up to achieve this goal. The credential was obtained by instruments that measured (assessed) student learning obtained within certain specified institutional structures and through certain pedagogical techniques that were both validated by a National Council for Technological Awards and not by individual colleges. It was found that that these arrangements were in conflict with the system of higher education *per se* and had a negative impact on the achievement of desired goals. For example, the award of a diploma rather than a degree was a consequence of the British social system and its undervaluing of the technical and vocational that continues to this day. Able pupils would be steered in the direction of university degree programs because the diploma was perceived to have lower status (Heywood, 1969). But it also had an effect on the design of the curriculum (Heywood, 2014).

One consequence of some importance was that many teachers in the colleges sought to design curricula and examinations that mirrored those of the universities, a process called "curriculum drift" a form of which was apparently evident at the time in the

United States (McConnell, 1962; Riesman and Jencks, 1962). Many teachers in the CATs argued that if the diploma was equivalent to a degree, then it should be called a degree. It followed that if they were offering degree equivalent qualifications their institutions should become universities, and so they did in 1966. The point is that even though it might have been possible to design and implement an alternative qualification of merit, the pressures in the system impeded that from happening. But there was little or no understanding of the concept of the curriculum other than that it consisted of discrete chunks of knowledge. It remains a feature of institutional life: for example, the belief of some faculty that a combined arts and engineering degree should be a feature of engineering programs is more likely to become a reality if models like Bucciarelli's recently circulated proposals show signs of success because of his MIT pedigree.

At that time, however, industrialists complained about the curriculum being offered by these colleges, a notable critic being G. S. Bosworth (see Section 2.3).

2.3 Toward an Alternative Curriculum for Engineering

As indicated, by 1960 industrialists were expressing dissatisfaction with the Diploma. Many wanted the 6-month structure to be double-banked so that the firm was not overloaded with students for one half of the year only. They wanted to be able to distribute their training places throughout the year. But there were also a number of industrialists who felt that the curriculum was an imitation of that offered in the universities and not what industry required. One of them, G. S. Bosworth, published a paper in *Universities Quarterly* on creative activity in engineering (Bosworth, 1963). Bosworth gave up his endeavors to reform the undergraduate curriculum, and as chairman of a subcommittee of the Committee on Manpower Resources for Science and Technology, he focused on postgraduate training as a means of resolving the issue. "Industry and education should collaborate in devising 'matching sections' of education and training designed to ensure the efficient conversion of the university graduate into an effective industrial technologist, and which will attract a higher proportion of the ablest graduates into industry" (Bosworth, 1966). The relationship between industry and education continues to be problematic to this day. At the time, in response to a request from the Vice Chancellor of the University of Lancaster, Bosworth's model of the curriculum was developed by B. T. Turner and a group of educators and industrialists coordinated by Heywood as the basis for the development of a Department of Engineering at the University. The original model that focused on the education of professional engineers for design and manufacture combined project (problem)-based learning with traditional studies was probably the first to have a base firmly grounded in the educational theories that were available at the time including Furneaux's research and *The Taxonomy of Educational Objectives*[3] (Heywood et al., 1966). Its aims and objectives were derived in an activity that Furst, one of the authors of *The Taxonomy*, called "screening" (Furst, 1958; Heywood, 1981). It attempted to incorporate the synthesizing rationale and deal with the properties of systems that Bosworth and subsequently his committee thought essential.

2.4   Creativity in Engineering and Design

A striking feature of Bosworth's paper was the title that called for "creativity" in engineering education. Similarly, in the United States, an equally striking feature of de Simone's list of objectives for the preparation of engineers (Exhibit 2.3) is the inclusion of creativity with the need for a program to keep it up. "Creativity" was one of the buzz words of the 1960s and 1970s, it has not gone away. In the United States, there were several American studies of creativity among scientists (Getzels and Csikszentmihalyi, 1967; Taylor and Barron, 1963). There were two important reviews (Cattell and Butcher, 1968; Freeman, Butcher, and Christie, 1970). A few papers considered how to develop creativity among engineering students (Mackinnon, 1961; Porcupile, 1969), but much of the talk about creativity was linked to design, and design and project work (Deere, 1967; Hayes and Tobias, 1964/65) although at least one or two papers linked it to research (Gregory, 1963; Kreith, 1961). Whether or not design could be taught was the subject of controversy and teaching design to postgraduates was something that was quite new (Cooley, 1967; de Malherbe and Wistreich, 1964; Keith-Lucas, 1963). Much attention was paid to the method of systematic design (Archer, 1965). One paper at least studied the criteria that engineers had for creativity (Sprecher, 1959).

Recognition and formulation of the problem	Sense of urgency, sensing what is important; multidisciplinary problems; making realistic assumptions; relating to other similar problems; asking pertinent questions; thinking in terms of analogies.
Ability to use full range of engineering methods	For example, reliability and other probability- based procedures. Optimizing techniques; simulation and modelling; trial-and-error approaches; gathering information from books, people, nature, experiment, decision-making techniques; metaphorical analogizing; brainstorming.
Consciousness of values and costs	Particularly, social and economic values; estimating costs; trade-off between capital and operating costs.
Appreciation of process of innovation	The process itself; its importance in enterprise growth; role in economy.
Cognizance of human factors of engineering	Awareness of and a start to competence in dealing with them; development of ability to communicate; sell ideas and understand motivation of others.
Critical point of view	The challenging of presuppositions; keen powers of observation; continually seeking improvements, insight.
Capacity for self-development	Ability to relate self to world; planning for continuing education; awareness of the psychology of creativity; program for keeping it up.

EXHIBIT 2.3. Educational objectives in the preparation of engineers. Data from Teare described by de Simone (1967, October 4) cited by Gregory, S. (ed.) (1972). *Creativity and Innovation in Engineering*. London, Butterworths, p. 9.

The topic seems to have acquired popularity among the public from Guilford's (1959) theory of intelligence in which are to be found the concepts of convergent and divergent thinking together with the association of divergent thinking with creativity, and also from a study of creativity among schoolchildren in the United States by Getzels and Jackson (1962). In the United Kingdom, the engineering institutions became concerned with the issue when a psychologist Liam Hudson (1966) published a report that said that students studying the arts and the humanities in high school sixth forms (grades 10–12) were likely to be more creative than those studying science subjects. Prior to that engineering in the United Kingdom had been criticized for the poor state of design and education for designers in the report of an official committee chaired by G. B. R. Feilden (1963). While Hudson received some criticism, notably that his findings were based on pencil-and-paper tests, his results worried the engineering establishment and the Council of Engineering Institutions (CEI) set up a working party on Creativity and Engineering jointly with the Design Research Society. They subsequently held a symposium and the book of the proceedings continues to be of interest (Gregory, 1972).

Gregory and Monk (1972) reviewed models and definitions of creativity several of which related to architecture. They classified theories of creativity into *intellectual* (and listed 14 since Plato), *motivational* (with 14 in the list including McClelland's need for achievement and Cattell's multiple personality factors), and a *combination of the two* of which they found 9, the last of which was leadership style. They also noted three that belonged to a category of *chance*. From Jung, the idea that the creative act will forever elude the human understanding brings us to the heart of the debate—is it or is it not a competency? Or, can it be taught? The debate resolves around one's definition of creativity and/or creative behavior.

McDonald (1968) drew attention to an important distinction between *originality* and *creativity* that has an important bearing on assessment, "Creative is the label we apply to the products of another person's originality." Assessment is concerned with evaluating improvements in originality defined as a behavior "which occurs relatively infrequently, is uncommon under given conditions, and is relevant to those conditions." It has to take into account context as well as aptitude. McDonald's distinction between originality and creativity had a bearing on the approach that the developers had to the assessment of coursework projects in the engineering science examination (described in Section 3.1) where they understood that what they were seeking was originality in design.

In this approach to the subject of creativity, it would seem to be a general domain and originality something that is more specific. Whether or not creativity is a general domain became the topic of much debate. In 1998, the *Creativity Research Journal* debated the evidence for creativity as a general domain that embraced "a set of skills, aptitudes traits and propensities, and motivations that can be productively employed in any domain—or conversely, whether the skills, aptitudes, traits, propensities and motivations that lead to creative performance vary from domain to domain" (Baer, 2010). My perception is that many of the participants at the CEI symposium would have thought that creativity was something that was domain general. Baer, who took the domain specific view in the 1998 publication, suggested in a recent review that the most likely development would be a hierarchical model that included domain specific and domain general elements. In his model (Baer and Kaufman, 2005) the first level, the *initial requirements* are domain general factors that have influence in all domains. He gives "intelligence" as an example. His second level is composed of *general thematic areas*. He identified seven such areas: math/science, problem solving, artistic/verbal, interpersonal, artistic/visual, entrepreneur, and performance. The third level is called *domains*. They are the subdomains that contribute to the general thematic areas, and at the final level are the *microdomains* or specific tasks within the domains.

A question for engineering educators is whether or not engineering design is a specific domain or whether it is embraced by problem solving. In this respect, is problem finding a separate domain to problem solving or a subdomain? McDonald (1968) considered that it was as important as problem solving. Is it the creative and/or original dimension of problem solving? A "yes" answer has considerable implications for assessment and in particular the design of examination/test questions.

McDonald pointed out that research suggested that individuals specifically trained to be original in problem-solving situations generally manifest more creative and original behavior than do untrained individuals. There is some evidence that stimulating idea production yields more ideas, some of which may be original. He was referring to activities like brainstorming (Parnes and Meadow 1959), nominal group techniques (George and Cowan, 1999), and learning that depended on undirected discussion of subject matter not obviously linked to the learning task (Abercrombie, 1960). Hesseling (1966) suggested providing some form of *chocs des opinions* to avoid what others have called *set mechanization* in our thinking (Luchins, 1942; cited by McDonald, 1968): that is the application of a formula we have learnt in every problem-solving situation even though it may be inappropriate. There was also interest in T-Groups (Whitaker, 1965) and business games for the purpose of teaching design. Another method was that of an intensive activity in which several groups of individuals with widely diverging backgrounds are asked to solve the same design problem. The groups meet from time to time to review where they are at with a view to changing attitudes or increasing the perception they had of their own work. In the author's experience of this kind of activity, different designs were produced.

At the time it was possible to summarize the results of research on group work (Heywood, 1972). A group may be more effective:

1. When the problem is more complicated but subdivided.
2. When group members share their knowledge and skills in the service of solving the problem.
3. When the group has an organization for coordinating its efforts.
4. When its leader stimulates cooperative endeavors.
5. When the group is able to minimize the individual's frustration in problem-solving situations.

They seem to be as relevant today as they were then as work on cooperative learning during the last 25 years has shown. Often it can improve achievement as a recent South African study has shown (Swart and Sutherland, 2014). The importance of the environment and context in which a person is expected to be original was highlighted at the CEI symposium as it is in both the Barnes and the Burns and Stalker studies of organization. It was reinforced by engineer and psychologist Whitfield (1975), who in a later publication showed how divergent and convergent thinking processes are brought into play in the problem-solving (design) process. Similarly, the environment and context are likely to impact on the acquisition of the competences required by the organization (see Chapter 10).

The 1970 symposium coincided with the publication of Edward de Bono's popular study of *Lateral Thinking*, which was taken up in at least one engineering course for civil engineers (Al-Jayyousi, 1999). At the symposium, Monk (1972) described a scheme for teaching problem solving and creativity. Since then there have been several reports of courses and pedagogical interventions such as one in creative engineering at Prairie View A & M University (Warsame, Biney, and Morgan, 1995), the University of Sydney (Baillie and Walker, 1998), the University of Virginia (Richards, 1998), and in Canada at McMaster, where Woods and his colleagues pioneered problem-based learning in engineering (Woods, 1994).

Apart from the problem of defining creativity, the central problem of this text is assessment and how creativity might be assessed. Baillie and Walker (1998) pointed out that not only did they have to change the technique of assessment away from a traditional examination to a "power test" (untimed examination; Baillie and Tookey, 1997) in order to assess deep learning, but they had also to change the techniques of teaching. Teachers, argued Christiano and Ramirez, (1993; Ramirez, 1993) have to become facilitators. Similarly in Sweden, Berglund et al. (1998) reported that examinations had to be changed to increase student creativity. Like Baillie and Walker, he followed in the footsteps of Marton and wanted to encourage deep learning. He and his colleagues replaced the traditional examination in computer architecture with public seminars on programming projects. If creativity is to be developed through a design project, a totally different approach to assessment will be required especially if it is expected to be fostered through teamwork.

Felder (1985, 1987), in much-quoted papers on creativity in engineering education, described how he had tried to encourage problem solving and creativity through different approaches to assessment. His ideas for questioning apply as much to the classroom as they do to tests. He suggested:

1. Questions that call for ideational fluency (where what counts is the quantity of possible solutions, not of quality), flexibility (variety of solutions), and originality.
2. Questions that are poorly defined and convergent.
3. Questions that require a synthesis of material that transcends subject or disciplinary boundaries.
4. Questions that require evaluation, in which technical decisions must be tempered with social ethical considerations.
5. Questions that call for problem finding and definition in addition to, or instead of, problem solving.

Tests of creative thinking can be used to determine fluency and flexibility and originality in the evaluation of courses. Waks and Menrler (2003) used such tests with a SAM (Something about Myself) questionnaire together with a general questionnaire when they sought to identify the pattern of engagement in creative thinking during a final project that took more than a year to complete. The research clearly evaluated the course at the same time.

Baer (2010) points out that the most widely used tests of creativity, that is, those that assess divergent-thinking, assume domain generality. They are not free from theoretical bias and do not measure "actual creativity." It is for this, apart from theoretical reasons, that there is much interest in the "Consensual Assessment Technique" developed by Amabile (1996) since it is based on actual creative endeavors and is held to measure of creativity. It is independent of theoretical models of creativity. Experts in a particular domain are asked to rate the products in a domain by comparing them one with another. Baer likens the procedure to committees that make the Academy Awards or the Nobel Prizes. "Experts rate the creativity of a set of artifacts by comparing them one to another. They are given no instruction—it is important that they use their own expert sense of what is creative in a domain and work independently." It would be essential that the conditions are standardized so that the raters are independent. It has been found that expert judges are likely to produce different ratings (Plucker and Makel, 2010).

The *Consensual Assessment Technique* has been used by engineering educators (e.g., Buelin-Biesecker and Wiebe, 2013). Buelin-Biesecker and Wiebe evaluated two different approaches to teaching creativity to middle school students the work of which was evaluated by seven expert raters. They found a high level of consistency among the raters. Their findings add to the growing body of knowledge in the field that suggests that creativity can be reliably assessed. They argue that engineering teachers should include creativity as a specific objective in their classrooms.

This method of assessment was also used at the Universidad de las Americas Puebla (Mexico) as part of a multidimensional approach to the assessment of creativity in a course on Chemical, Food and Environmental Engineering Design. The theoretical framework was provided by Sternberg and Lubart's (1992) "Investment Theory of Creativity" (ITC) (Husted et al., 2013). The resources that contribute to creativity in ITC are intellectual processes, knowledge of domain, intellectual style, personality, motivation, and environmental context. They used a *Creative Thinking Value Rubric* published by the Association of American Colleges and Universities (Rhodes, 2010) for their measurements.

At the present time, there is much exhortation about the need for engineers to be "innovative, inspired, and original in learning how to do more with less" (Farnsworth et al., 2013). Among others promoting this idea is the American Society of Civil Engineers (American Society for Civil Engineering, 2007). There is a small but steady flow of papers on creativity at the major engineering conferences (see Section 8.6).

Farnsworth and his colleagues report on laboratory courses at the University of Texas–Tyler designed to foster critical thinking and creativity through inquiry-based learning. Other reports have based their development on a considerable understanding of the various theories of creativity that have been developed in the last 50 years (Kozbelt, Beghetto, and Runco, 2010). One example is the derivation of a measure of creativity as a construct and a process in an introductory course to engineering design as part of an attempt to improve learning through self-evaluation (Aboukinane et al., 2013). The search for valid, reliable, and inexpensive in time and money techniques continues. A range of techniques exists that may be of value in engineering classrooms as reported by Plucker and Makel (2010).

If the widespread use of problem- and project-based learning especially in team situations is a measure of how far engineering education has come since Bosworth's criticism, it has come a long way. But the recent literature, while indicating a range of possibilities, is still full of exhortation. This view suggests that a close look should be taken at a study by Furneaux (1962) that supports Bosworth's proposition that university curricula had limited objectives. What did they achieve and what should they have achieved? (The reader may like to look at Section 9.7 at this stage.)

2.5 Furneaux's Study of a University's Examinations in First-Year Mechanical Engineering: The Argument for "Objectives"

As indicated in Section 1.11, Furneaux's study highlighted the role of personality in the academic achievement of engineering students. Students who tended toward introversion did better than those who tended toward extraversion. Could this be the reason why engineers like working on their own as several recent studies report (Trevelyan, 2014)? Could this be why some observers believed that engineering students lacked creativity? Or, is it due to the fact that students were not exposed to activities that would enable them to be creative? Be that as it may, it is the second part of Furneaux's study that is of interest here.

The report of this part of his investigation seems to have been totally neglected. He found that when the end of the first-year written examination results (in this case year I) were factor-analyzed only one major factor was to be found across the written-subject examinations excluding engineering drawing that evidently measured something different. Furneaux thought this factor represented an "examination passing ability." Whatever it was, it raised a whole host of questions about the validity and value of engineering examinations. They could not be said to be very valid if validity is defined by a range of skills used by engineers. For example, what was the point of setting six written examinations to test the same "thing"? It was suggested that what might be being measured was the ability to apply mathematical techniques to a problem in applied science for the purpose of obtaining a single solution (Heywood, 1968). In the same year Black (1962) surveyed university physics papers and concluded that most were of a common pattern and tested the same kind of ability (see also Lister, 1969). This issue, although better understood, continues to provide faculty with a challenge particularly in the design of what are called "fuzzy" (ill-structured or wicked) questions (Nagel et al., 2012). A major question for engineering educators is: "are we sure that we know what our examinations and tests actually test?" Therefore, Furneaux made the point that those responsible for the design and assessment of engineering courses should declare their objectives and design the curriculum system so that those objectives could be obtained. In that way, we should have some idea about what is being tested.

"The division which causes the greatest trouble in the design office is that between on-the-board design and off-the-board design. Although the conflict takes many forms, perhaps the classic situation occurs in an electronic drawing office staffed by experienced draughtsmen who have served an apprenticeship and taken HNC (Mechanical). Work comes into the office in the form of crudely drawn circuit diagrams sponsored by recent graduates who have often never heard of a specification. There is almost a complete lack of communication between the two sections."

EXHIBIT 2.4. From Turner, B. T. (1969). *Management Training for Engineers*. London: Business Books, p. 245. The drawing office was considered to be low status by young graduates so they went into engineering management without much experience of design (see also Monk, 1972; Youngman et al., 1978). It will be appreciated that much has changed in the organization of engineering firms since then. Turner gives detailed diagrams of how firms were organized at the time (pp. 46–47).

In order to obtain such objectives, much more information would have to be obtained about what engineers actually do in practice, a point that is increasingly understood (Trevelyan, 2014). As indicated previously, in terms of studies like those of Barnes and Burns and Stalker, it is clear that the engineering courses of the time did not concern themselves with either communication or management. Clearly laboratory (coursework) provided technical skills for subsequent practical work and its management, and more especially for those concerned with quality control. Equally, even though engineers by and large did not go into drawing offices, they needed to be able to interpret drawings (see Exhibit 2.4).

In any case, MacFarlane-Smith (1964) pointed out the value of engineering drawing to the development of spatial ability that more generally had intrinsic value for the study of mathematics and other engineering subjects, a fact that continues to be recognized (Manning and Hampshire, 2011; Onyancha and Kinsey, 2011). Conversely skill in sketching is important in graphics and design (Hammond, 2007). Macfarlane Smith pointed out that the education system, at least for the bright students, emphasized academic subjects that would exercise the left-hand side of brain but not the right-hand side where spatial ability was located. This, he believed, accounted for the shortage of students for engineering courses. In recent years, as Snyder and Spenko (2014) have reported, improvements in spatial visualization have been shown to improve or help learning in mathematics (Maloney et al., 2012; Sorby et al., 2012), and in science and engineering curricula in Europe (Potter et al., 2006). It would seem that the potential of the computer-assisted development of spatial ability is considerable but as yet this has not been found to be the case, and the mixed approach of traditional technical drawing and three-dimensional software modeling did not achieve the improvement desired. But this is unlikely to be the end of the story.

Turner (1969, p. 31) relying on the available knowledge of the brain published a picture of a face divided down the middle. The expectation of an engineer was that he or she would have to use both sides of the brain. Underneath the left hemisphere, Turner has the major heading "things" and underneath the right-hand hemisphere he has the title "People." His heading to the illustration is "The Schizophrenic Engineer." He gives four reasons why engineers so often fail as managers. The first is because they fail to understand people; the second, because they desire perfection, the price of which is often prohibitive; third, they cannot cope with the stress and strain of management; and fourth, the belief that their engineering ability will enable them to fill their new role; in consequence, they reject adequate pretraining. Some authorities might ask (rhetorically), "what has changed?"

It is not surprising that engineering educators in the United Kingdom had little knowledge of the objectives approach that had been championed by Ralph Tyler because the book that was the culmination of this idea *The Taxonomy of Educational Objectives* (commonly known as the *Bloom Taxonomy*) was not published in England until 1964 (Bloom et al., 1956, 1964-Exhibit 2.5).[3] Tyler (1949) had asserted that the four basic tasks of the educator were:

1. The determination of the objectives that the course should seek to obtain.
2. The selection of learning experiences that will help bring about the attainment of these objectives.

A. Principal Cognitive domains of the Taxonomy	B. Subcategories of the Domain of Comprehension	C. Domain of Comprehension Subcategory of "Translation"	D. Examples of Abilities and Skills Related to the Category of Translation
Knowledge Comprehension Application Analysis Synthesis Evaluation	Translation Interpretation Extrapolation	Comprehension as evidenced by the care and accuracy with which the Communication is paraphrased Or rendered from one language to the Form of communication to another. Translation is judged on the basis of faithfulness and accuracy, that is on the extent to which the material in the original communication is preserved, although the form of the communication has been altered.	The ability to understand nonliteral statements (metaphor, symbolism, irony, exaggeration). Skill in translating mathematical verbal material into symbolic statements and vice versa.

EXHIBIT 2.5. The domains of the 1956 taxonomy of educational objectives (A) showing the domain of comprehension with its subcategories, the introduction to the subcategory of translation and two elements of that subcategory. Currently, subabilities of the type shown in D are often called learning outcomes. They are preceded by the phrase "The ability to" and followed by a verb requiring action, e.g., "to identify"; to communicate: to analyze: to make." (Bloom, B., et al. (1964). *The Taxonomy of Educational Objectives. I. The Cognitive Domain*. London: Longmans Green).

3. The organization of those learning experiences so as to provide continuity and sequence for the student and help him integrate what might otherwise appear as isolated experiences.
4. The determination of the extent to which objectives are being attained.

At the time engineering educators would have been surprised to find that this was what their role was considered to be!

By 1968, the list of educational objectives shown in Exhibit 2.3 had been given to a subcommittee of the US Senate (de De Simone, 1967, 1968, cited by Gregory, 1972; Shoop and Ressler, 2011). The first attempt to design a curriculum in the United Kingdom using an objectives approach was, it seems, an examination in engineering science used for entry (selection) to university, which will be described in Section 3.1.

2.6 Discussion

The paths that change takes can be understood only in the light of its sociocultural history. In Britain, this cultural history extends to the industrial revolution and before. During the industrial revolution, many of the innovations were made by individuals who had no scientific training but had skills in craft. During the nineteenth century, a need for a more formal system of training emerged which by the end of the First World War was producing through a combination of evening study and sometimes day release study from the company (i.e., 1 day per week) qualified (recognized) engineers. Experience was highly valued. A national system of technical colleges developed to support this system and some offered study at degree level. Some employers, in particular the large employers, developed highly valued apprenticeship schemes not only for craft and skilled workers but for the engineers they believed they required (e.g., Rolls Royce). In parallel from the 1830s onward they developed a system of engineering education in the universities that was primarily science based.

As indicated previously, by 1944, as the Second World War was drawing to a close, the Percy Committee took the view that the products of the universities were adequately prepared to establish a national R&D base. However, it thought that engineers trained via the part-time route were inadequately prepared to meet the needs of industry, and that new approaches to the curriculum were required. These included full-time study so that these engineers would acquire the science and mathematics required for work in industry, and a more sophisticated approach to training. In effect, the colleges developed for this purpose had the opportunity to develop new curricular at the highest level through a new diploma that was taken to be the equivalent of a degree and awarded by an independent organization. Many of the academics in these colleges wanted them to be given degree awarding status, that is, university status. One consequence was that the curriculum in the diploma courses drifted to mirror that offered by the universities. The search for status governed the direction that curricular took and not the wishes of many employers. Some of the large employers, particularly in the electrical industry, made their views known and made detailed suggestions about the curriculum that was needed. Eventually they came to believe that their ideas would be best implemented through postgraduate study.

Could much more have been expected? It seems doubtful for the very simple reason that prior knowledge of the curriculum process, and such matters as learning, was simply not existent in higher education at the time. Integration of the college and industrial periods was *ad hoc*, and in terms of the experienced gained largely successful but no attempt was made to establish a theory of integration. This problem remains to be tackled but now much more data is available and a theory possible.

The opportunity for this particular group of employers to work with a "new" university came in 1964, and in what might be regarded as a semi-Delphi study produced a nontraditional curriculum that was characterized by problem (project)-based learning. The group had available to them knowledge of recent developments in educational theory and practice in relation to the curriculum and substantial use of them was made in determining the curriculum. The authors were clear that they were enunciating a particular philosophy. However, the summary of these proposals undertaken by the university's vice chancellor was rejected by the senate because within the traditions in science nationally established it did not seem to them that the model was plausible. The vice chancellor subsequently revised the proposal and the university's senate agreed to the establishment of an engineering department and the appointment of a professor whose focus was engineering design. These were departures from tradition in the university system. Had the original scheme been accepted, there would have had to have been radical changes in assessment but it is doubtful that the authors of the report really understood the power of assessment to influence the curriculum and learning.

One of the reasons suggested for curriculum drift and the failure to develop an alternative curriculum is that the teaching staff had little or no knowledge of recent developments in education particularly as they related to the design of curriculum. This applied equally to faculty in the United States and in this respect Stice's paper (1976) on the topic in *Engineering Education* is seminal for both sides of the Atlantic (Froyd and Lohmann, 2013). He argued that determining one's objectives was the first step toward improving one's teaching.

While there were one or two attempts to assess industrial training and the beginnings of competency-based schemes were discernible, the role that assessment plays as a determinant of learning was not understood.

It will be seen that those who want to make changes in the American approach to engineering in the twenty-first century are faced with the same problem. They argue that the engineering science approach to the engineering curriculum that was developed immediately after the Second World War needs to change. It is quite clear that if they are to achieve change they have to provide knowledge from which new approaches can be developed. This means that an in-depth discussion about the aims of engineering within higher education has to be had if a philosophical base is to be established that is plausible. At the same time they have to understand the factors that influence the identity of engineering educators to be able to persuade engineering educators to change that identity, for threats to identity are as much a cause of resistance to change as any other. A major question is whether or not changes in assessment can assist the drive toward curriculum change.

Notes

1. The participants in this study undertook the Allport-Vernon Study of Values. This inventory seeks to show an individual's relative ranking of six human values: theoretical, economic, aesthetic, social, political, and religious.
2. It was into this complexity that I came as a research worker with the primary goal of evaluating the merits of the different sandwich structures. I was required to take into account the views of industry, faculty and students, and factors affecting supply and demand, which meant finding out what students and teachers in schools perceived these courses to offer. It also meant that I had to try and establish if the diplomas were really the equivalent of degrees, which I did within the constraint that I could not do a psychometric comparative study. I could, however, interview the external examiners. The hidden agenda in the remit like an insurance policy was in the small print. It was a hot potato. What the sponsors of investigation hoped was that I would establish that an "end-on" arrangement was the preferred structure. This meant that while students were in the 6-month period in industry, the colleges would take in another intake. This would enable industry to fill their training places throughout the year. It would also mean that the colleges would be open throughout the year. While faculty were overjoyed, management in the colleges and policy makers were far from pleased when I found that industry would not double the number of students on these courses but would redistribute the existing students across the year. Indeed, to make matters worse, I could find no shortage of qualified manpower, which made me even more suspect (Heywood, 1974). During the 50 years since these enquiries we have been told at regular intervals that there are shortages of qualified manpower; so, today the concern in the United States is with the output of STEM graduates. I learnt to be very suspicious of these forecasts a long time ago.
3. The *Taxonomy* for the Cognitive domain is probably the most referenced book in education in the second half of the twentieth century. It had a profound effect on the UK University Matriculation Boards who required their syllabuses to be accompanied by statements of aims and objectives. The purpose of *The Taxonomy* was to provide a common framework that could be used by all university examiners and so promote the exchange of test materials and ideas for testing. This was to be achieved by a hierarchy of objectives expressed in behavioral form since it was believed "that all educational objectives when stated in behavioral form have their counter parts in student behaviour" (Bloom, 1994). The base objective was knowledge in all its manifestations that were listed (e.g., knowledge of principles, concepts) then in ascending order—Comprehension, Application, Analysis, Synthesis, and Evaluation. There were many criticisms of *The Taxonomy* not least among them that it was not a true Taxonomy and certainly not hierarchical (Anderson and Sosniak, 2012). Notwithstanding these criticisms, some engineering educators continue to find it useful (Castles et al., 2009; Cheville et al., 2008, Highley and Edlin, 2009; Jones et al., 2009). *The Taxonomy* was revised in 2001 to take account of some of these criticisms (Anderson et al., 2001). Recent usages in engineering are discussed in Section 9.2.

   The group that developed *The Taxonomy* for the cognitive domain also produced a second volume for the affective domain under the leadership of David Krathwohl (Krathwohl, Bloom, and Masia, 1964). Notwithstanding the publication of the volume on the affective domain, Dressel (1971) an American Professor of Higher Education, was vociferous in his criticism of *The Taxonomy* because he believed that it underestimated the role of values in human behavior. While Bloom's group envisaged a volume for the psychomotor domain, this was left to others to suggest (e.g., Harrow, 1972). Recently Timothy Ferris (2010) of the University of Southern Australia has described a taxonomy for the psychomotor domain that should provide a clear pedagogical rationale for "expensive" laboratory work. His primary sevenfold classification is (1) recognition of tools and materials; (2) handling tools and materials; (3) basic operation of tools; (4) competent operation of tools; (5) expert recognition of tools; (6) planning of work operations; and (7) evaluation of outputs and planning means for improvement. It should be compared with an intensive course in manufacturing technology processes and materials described by Owen and Heywood (1990) which they gave to 15- to 16-year-olds. They may be criticized for not recognizing the psychomotor ability required in handling instrumentation. Ferris provides a short but useful introduction to the taxonomies. A complementary paper that is equally useful that describes a taxonomy for engineering practice is due to McCahan and Romkey (2014).

References

Abercrombie, M. L. J. (1960). *The Anatomy of Judgement*. Harmondsworth Penguin.

Aboukinane, C. (2013). Fostering creativity in engineering education through experimental and team based learning. In: Proceedings of Annual Conference of American Society for Engineering Education, June 2013. Paper 8221.

Abraham, J., Greene, C., and Marches, A. (2007). Work in progress—External assessment through peer-to-peer evaluation of capstone projects. In: ASEE/IEEE Proceedings of Frontiers in Education Conference, S1G-10/11.

Al-Jayyousi, O. (1999). Introduction to lateral thinking to civil and environmental education. *International Journal of Engineering Education*, 15(3), 199–205.

Amabile, T. M. (1996). *Creativity in Context. Update to the Social Psychology of Creativity.* Boulder, CO: Westview Press.

American Society for Civil Engineering. (2007). *Civil Engineering Body of Knowledge for the 21st Century. Preparing the Civil Engineer for the Future*, 2nd edition. Reston, VA: American Society for Civil Engineering.

American Society of Mechanical Engineers. (2012). *Vision 2030. Creating the Future of Mechanical Engineering Education.* New York, NY: American Society of Mechanical Engineers.

Anderson, L. W., Krathwohl, D. R., Airasian, P. W., Cruikshank, K. A., Mayer, R. E., Pintrich, P. R., Raths, J., and Wittrock, M. C. (eds.) (2001). *A Taxonomy for Learning Teaching and Assessing. A Revision of Bloom's Taxonomy of Educational Objectives.* Addison Wesley/Longman.

Anderson, L. W., and Sosniak, L. A. (eds.) (1994). *Bloom's Taxonomy a Forty-Year Retrospective.* Chicago, IL: National Society for the Study of Education, University Press of Chicago.

Archer, L. B. (1965). *Systematic Method for Designers*. London: HMSO.

Baer, J. (2010). Is creativity domain specific? In: J. C. Kaufman and R. J. Sternberg (eds.), *The Cambridge Handbook of Creativity*. New York, NY: Cambridge University Press.

Baer, J., and Kaufman, J. C. (2005). Bridging generality and specificity. The amusement park theoretical (APT) model of creativity. *Roeper Review*, 27, 158–163.

Baillie, C., and Walker, P. (1998). Fostering creative thinking in student engineers. *European Journal of Engineering Education*, 23(1), 33–38.

Baillie, C., and Tookey, S. (1997). The "power test": its impact on student learning in a materials science course for engineering students. *Assessment and Evaluation in Higher Education*, 22(1), 33–38.

Barnes, L. B. (1960). *Organizational Systems and Engineering Groups. A Comparative Study of Two Technical Groups in Industry.* Boston, MA: Harvard University, Graduate School of Business Administration.

Berglund, A., Daniels, M., Hedenborg, A., and Tengstrand, A. (1998). Fostering creativity through projects. *European Journal of Engineering Education*, 23(1), 45–54.

Black, J. (1975). Allocation and assessment of project work in the final year of the engineering degree course at the University of Bath. *Assessment in Higher Education*, 1(1), 35–53.

Black, P. J. (1962). University examinations. *Physics Education*, 3(2), 93.

Bloom, B. S., Engelhart, M. D., Furst, E. J., Hill, W. H., and Krathwohl, D. R. (eds.) (1956). *The Taxonomy of Educational Objectives. Handbook 1. Cognitive Domain*. New York: David Mackay. (1964) London: Longmans Green.

Bosworth, G. S. (1963). Toward creative activity in engineering. *Universities Quarterly*, 17, 286.

Bosworth, G. S. (Chairman of a Committee). (1966). *The Education and Training Requirements for the Electrical and Mechanical Manufacturing Industries*. Committee on Manpower Resources for Science and Technology. London: HMSO.

Buelin-Biesecker, J., and Wiebe, E. N. (2013). Can pedagogical strategies affect students' creativity? Testing a choice based approach to design and problem-solving in technology, design and engineering education. In: Proceedings of Annual Conference of American Society for Engineering Education, June 2013. Paper 6045.

Burns, T., and Stalker, G. (1961). *The Management of Innovation*. London: Tavistock.

Castles, R., Lohani, V. K., and Kachroo, P. (2009). Utilizing hands-on learning to facilitate progression through Bloom's Taxonomy within the first semester. In: ASEE/IEEE Proceedings of Frontiers in Education Conference, W1F-1 to 5.

Cattell, R. B., and Butcher, H. J. (1968). *The Prediction of Achievement and Creativity*. Indianapolis, IN: Bobs-Merrill.

Cheville, A., Yadav, A., Subedi, and Lundeberg, M. (2008). Work in progress—Assessing the engineering curriculum through Bloom's taxonomy. In: ASEE/IEEE Proceedings of Frontiers in Education Conference, S3E-15/16.

Christiano, S. J. E., and Ramirez, M. (1993). Creativity in the classroom. Special concerns and insights. In: ASEE/IEEE Proceedings of Frontiers in Education Conference, pp. 209–212.

Clark, C., Medland, A., Rakowski, R., and Wild, R. (1985). A new enhanced engineering programme for manufacturing industries. *International Journal Applied Engineering Education*, 1, 21.

Cooley, P. (1967). Creative design—can it be taught? *Chartered Mechanical Engineer*, 14, 228.

Dartt, K., McGrann, T. R., and Stark, J. T. (2009). ABET assessment of student initiated interdisciplinary senior capstone project. ASEE/IEEE Proceedings of Frontiers in Education Conference, M3E- 1 to 6.

Deere, M. (1967). Creative projects in university engineering. *New University*, 1, 30.

De Malherbe, M. C. (1961). Teaching design at the university. *Technology*, 5, 42.

De Malherbe, M. C., and Wistreich, J. G. (1964). An experiment in education for design. *Chartered Mechanical Engineer*, 11, 149.

De Simone, D. V. (1967, October 4). Statement by the Director, Office of Information and Innovation, National Bureau of Standards in "New Technologies and Concentration," Part 6 of Economic Concentration. Hearings before the subcommittee on Anti-Trust and Monopoly. US Senate Special report 26.

De Simone, D. V. (ed.) (1968). Education for Innovation. Oxford, Pergamon.

Dressel, P. L. (1971). Values, cognitive and affective. *Journal of Higher Education*, 42(5), 400–402.

Farnsworth, C. R., Welch, R. W., McGinnis, N. J., and Wright, G. (2013). Bringing creativity into the lab environment. In: Proceedings of Annual Conference of American Society for Engineering Education, June 2013. Paper 6691.

Feilden, G. B. R. (1963). *Engineering Design.* Report of a Committee. Department of Scientific and Industrial Research. London: HMSO.

Felder, R. M. (1985). The generic quiz. A device to stimulate creativity and higher level thinking skills. *Chemical Education*, Fall, 176.

Felder, R. M. (1987). On creating creative engineers. *Engineering Education*, 74(4), 222–227.

Ferris, T. L. J. (2010). Bloom's taxonomy of educational objectives: a psychomotor skills extension for engineering and science education. *International Journal of Engineering Education*, 26(3), 699–707.

Freeman, J., Butcher, H. J., and Christie, T. (1970). *Creativity. A Selective Review of Research.* London, UK: Society for Research into Higher Education.

Froyd, J. E., and Lohmann, J. R. (2013). Chronological and ontological development of engineering education as field of scientific in inquiry. In: A. Johri and B. M. Olds (eds.), *Cambridge Handbook of Engineering Education Research.* New York: Cambridge University Press.

Furneaux, W. D. (1962). The psychologist and the university. *Universities Quarterly*, 17, 33–47

Furst, E. J. (1958). *The Construction of Evaluation Instruments.* New York: David MacKay.

George, J., and Cowan, J. (1999). *A Handbook of Techniques for Formative Evaluation: Mapping the Students Learning Experience.* London: Kogan Page.

Getzels, J. W., and Csikszentmihalyi, M. (1967). Scientific creativity. *Science Journal*, 3, 80.

Getzels, J. W., and Jackson, P. W. (1962). *Creativity and Intelligence.* New York: John Wiley & Sons.

Golding, P., McNamarah, S., White, H., and Graham, S. (2008). Cooperative education: an exploratory study of its impact on computing students and participating employers. In: ASEE/IEEE Proceedings of Frontiers in Education Conference, F3A-1 to 6.

Gregory, S. A. (1963). Creativity in chemical engineering research. *Transactions of Institution of Chemical Engineers*, Paper 650, p. 70.

Gregory, S. A. (ed.) (1972). *Creativity and Innovation in Engineering.* London: Butterworths.

Gregory, S. A., and Monk, J. D. (1972). Creativity: definitions and models. In: Gregory, S. A. (ed.), *Creativity and Innovation in Engineering.* London: Butterworths.

Guilford, J. (1959). Three faces of intellect. *American Psychologist*, 14, 469.

Hammond, T. (2007). Enabling instructors to develop sketch recognition applications in the classroom. In: ASEE/IEEE Proceedings of Frontiers in Education and Conference, S3J–11 to 16.

Harrow, A. J. (1972). The Psychomotor Domain: A Guide for Developing Behavioral Objectives. New York: David Mackay.

Hayes, S. V., and Tobias, S. A. (1964/65). The project method of teaching creative mechanical engineering. In: Proceedings of the Institution of Mechanical Engineers, pp. 179.

Hesseling, P. (1966). *A Strategy for Evaluation Research.* Assen, the Netherlands: Van Gorcum.

Heywood, J. (1961). Research by sixth form boys. *Nature*, 191, 860–861.

Heywood, J. (1967). The effectiveness of undergraduate (dip.tech) industrial training. *International Journal of Electrical Engineering Education*, 5, 281.

Heywood, J. (1968). Technical education. In: H. J. Butcher (ed.), *Educational Research in Britain.* Vol. I. London: University of London Press.

Heywood, J. (1969). *An Evaluation of Certain Post War Developments in Higher Technological Education.* 2 Vols. Thesis, University of Lancaster, Lancaster.

Heywood, J. (1972). Short courses in the development of originality In: S. A. Gregory (ed.), *Creativity and Innovation in Engineering*. London: Butterworths.

Heywood, J. (1974). Trends in the demand for qualified manpower in the sixties and seventies. *The Vocational Aspect of Training*, 26(64), 65–72.

Heywood, J. (1981). The academic versus practical debate. A case study in screening. Institution of Electrical Engineers (IEE) Proceedings Part A 12897), 511–519.

Heywood, J. (2008). Instructional and Curriculum Leadership. Towards Inquiry Oriented Schools. Dublin. Original Writing for National Association of Principals and Deputies.

Heywood, J. (2011). Higher technological education in England and Wales 1955–1966. Compulsory liberal studies. In: Proceedings of Annual Conference of the American Society for Engineering Education. Paper 635.

Heywood, J. (2014). Higher technological education and British policy making: a lost opportunity for curriculum change in engineering education? In: Proceedings of Annual Conference of the American Society for Engineering Education. Paper 8689.

Heywood, J., Lee, L. S., Monk, J. D., Moon, J., Rowley, B. G. H., Turner, B. T., and Vogler, J. (1966). The education of professional engineers for design and manufacture (a model curriculum). *Lancaster Studies in Higher Education*, 1, 4–153.

Highley, T., and Edlin, A. E. (2009). Discrete mathematics assessment using learning objectives based on Bloom's taxonomy. In: Proceedings of ASEE/IEEE on Frontiers in Education Conference, M2J—1 to 6.

Hudson, L. (1966). *Contrary Imaginations*. London: Methuen.

Husted, S., Gutierrez, J. V., Ramirez-Corona, N., Lopez-Malo, A., and Palou, E. (2013). Multidimensional assessment of creativity in an introduction to engineering design course. In: Proceedings of Annual Conference of American Society for Engineering Education, June 2013. Paper 8789.

Jahoda, M. (1963). *The Education of Technologists*. London: Tavistock.

Jones, K. O., Harland, J., Reid, J. M. V., and Bartlett, R. (2009). Relationship between examination questions and Bloom's taxonomy. In: ASEE/IEEE Proceedings of Frontiers in Education Conference WIG-1 to 6.

Keith-Lucas, J. (1963). Teaching design to graduates. *Chartered Mechanical Engineer* 10, 103.

Krathwohl, D., Bloom, B., and Masia, B. B. (1964). *Taxonomy of Educational Objectives. The Classification of Goals. Vol. 2 Affective Domain*. London: Longmans Green.

Kreith, F. (1961). An experience in teaching engineering research and creativity to gifted undergraduates. *Journal of Engineering Education*, 51, 810.

Kozbelt, A., Beghetto, R. A., and Runco, M. A. (2010). Theories of creativity. In: J. C. Kaufman and R. J. Sternberg (eds.), *The Cambridge Handbook of Creativity*. New York: Cambridge University Press.

Lister, R. E. (1969). The aims of questions in A level biology examinations. *The School Science Review*, 50(172), 514.

Luchins, A. S. (1942). Mechanization in problem solving. The effect of "einstellung." *Psychological. Monographs* 248 (cited by Mcdonald, 1968).

MacFarlane-Smith, I. (1964). *Spatial Ability*. London: University of London Press.

McCahan, S., and Romkey, L. (2014). Beyond Bloom's. A taxonomy for teaching engineering practice. *International Journal of Engineering Education*, 30(5), 1176–1189.

McConnell, T. R. (1962). A General Pattern of American Higher Education. New York: McGraw Hill.

McDonald, F. J. (1968). *Educational Psychology*. Belmont, CA: Wadsworth.

Mackinnon, D. W. (1961). Fostering creativity in engineering students. *Journal of Engineering Education*, 52, 129.

Maloney, E. A., Waechter, S., Risko, E. F., and Fugesland, J. A. (2012). Reducing sex difference in math anxiety. The role of spatial processing ability. *Learning and Individual Differences*, 22, 380–384.

Manning, K. S., and Hampshire, J. (2011). Work in progress—Technical freehand sketching. In: ASEE/IEEE Proceedings of Frontiers in Education Conference, F1C-1/2.

Monk, J. D. (1972). Creativity in higher eduycation. In: Gregory, S. A. (ed.), *Creativity and Innovation in Engineering*. London. Butterworth.

Nagel, J. K. S., Pierrakos, O., Zilberberg, A., and McVay, S. (2012). Understanding industry experiences: From problem solving to engineering students' learning gains. In: ASEE/IEEE Proceedings of Frontiers in Education Conference, pp. 927–932.

Nasr, K. J., Pennington, J., and Andres, C. (2004). A study of students' assessment of cooperative education. *Journal of Cooperative Education*, 38(1), 13–12.

Northrup, S. G., and Northrup, D. A. (2006). Multidisciplinary teamwork assessment: individual contributions and interdisciplinary interaction. In: ASEEE/IEEE Proceedings of Frontiers in Education Conference, S2E-15 to 20.

Onyancha, R., and Kinsey, B. (2011). The effect of engineering major on spatial ability improvements over the course of undergraduate studies. In: ASEE/IEEE Proceedings of Frontiers in Education Conference, T1H-20 to 24.

Owen, S., and Heywood, J. (1990). Transition technology in Ireland. An experimental course. *International Journal of Technology and Design Education*, 1(1), 21–32.

Parnes, S. J., and Meadow, A. (1959). Effects of "brainstorming" instructions on creative problem-solving by trained and untrained subjects. *Journal of Educational Psychology*, 30, 171.

Percy, L. (Chairman of a Committee) (1945). *Higher Technological Education*. London: HMSO.

Pierrakos, O., Borrego, M., and Lo, J. (2008). Preliminary findings froma quantitative study. What are students learning during cooperative education experience. Proceedings Annual Conference American Society for Engineering Education

Plucker, J. C., and Makel, M. C. (2010). Assessment of creativity. In J. C. Kaufman and R. J. Sternberg (eds.), *The Cambridge Handbook of Creativity*. New York, NY: Cambridge University Press.

Porcupile, J. C. (1969). An inductive approach to teaching creativity. *Bulletin of Mechanical Engineering Education*, 8, 327.

Potter, C., van der Merwe, E., Kaufman, W., and Delacour, J. (2006). A longitudinal evaluation study of student difficulties with engineering graphics. *European Journal of Engineering Education.*, 31(2), 201–214.

Pullman, W. A. (1964/65). Teaching design to sandwich course students. *Proceedings of Institution of Mechanical Engineers*, 179, 100.

Rakowski, R. T. (1990) Assessment of student performance during industrial training placements. *International Journal of Technology and Design Education*, 1(3), 106–110.

Ramirez, M. R. (1993). The influence of learning styles on creativity. In: Proceedings of Annual Conference of American Society for Engineering Education. Paper 2225.

Rhodes, T. (ed.) (2010) *Assessing Outcomes and Improving Achievement. Tips and Tools for Using Rubrics*. Washington, DC: Association of American Colleges and Universities.

Richards, L. G. (1998). Stimulating creativity. Teaching engineers to be innovators. In: ASEE/IEEE Proceedings Frontiers in Education Conference, pp. 1034–1039.

Riesman, D., and Jencks, C. (1962). The viability of the American College. In: N. Sanford (ed.), *The American College*. New York: John Wiley & Sons.

Schuurman, M. K., Pangborn, R. N., and McClinties, R. D. (2008). Assessing the impact of engineering undergraduate work experience. Factoring in pre-work academic performance. *Journal of Engineering Education*, 97(2), 207–213.

Shoop, B. L., and Resler, E. K. (2011). Developing the creativity and innovation of undergraduate engineering students. *International journal of Engineering Education*, 26(4), 820–850.

Snyder, M. E., and Spenko, M. (2014). Assessment of students' changed spatial ability using two different curriculum approaches: technical drawing compared to innovative product design. In: Proceedings of Annual Conference of American Society for Engineering Education, June 2014. Paper 9841.

Sorby, P., Casey, P., Veurink, N., and Delaney, A. (2012). The role of spatial training in improving spatial and calculus performance in engineering students. *Learning and Individual Differences*, 26, 20–29.

Sprecher, T. B. (1959). A study of engineers' criteria for creativity. *Journal of Applied Psychology*, 43, 141–146.

Sternberg, R. J., and Lubart, T. I. (1992). Creative giftedness. A multivariate approach investment. *Gifted Child Quarterly*. 37(1), 7–15.

Stice, J. E. (1976). A first step toward teaching. *Engineering Education*, 67, 394–398.

Sugrue, B. (1995). A theory based framework for assessing domain specific problem solving ability. *Educational Measurement: Issues and Practice*, 14(3), 32–35.

Swart, A. J., and Sutherland, T. (2014). Cooperative learning versus self-directed learning in engineering: student preferences and implementation. In: ASEE/IEEE Proceedings of Frontiers in Education Conference, pp. 1466–1470.

Talbot, C., Alley, M., Marshall, M., Haas, C., Zappe, S. E., and Garner, J. K. (2013). Engineering ambassador network: professional development of the engineering ambassadors. In: Proceedings of Annual Conference of American Society for Engineering Education, June 2013. Paper 7435.

Taylor, C. W., and Barron, F. (eds.) (1963). *Scientific Creativity: Its Recognition and Development*. New York: John Wiley & Sons.

Trevelyan, J. (2014). *The Making of an Expert Engineer*. London: Taylor and Francis Group.

Turner, B. T. (1969). *Management Training for Engineers*. London: Pitman.

Tyler, R. W. (1949). Achievement testing and curriculum construction. In: E. G. Williamson (ed.), *Trends in Student Personnel Work*. Minneapolis: University of Minnesota.

Waks, S., and Merdler, M. (2003). Creative thinking of practical engineering students during a design project. *Research in Science and Technological Education*, 21(1), 101–120.

Warsame, A., Biney, P. O., and Morgan, J. O. (1995). Innovations in teaching creative engineering at freshman level. In: ASEE/IEEE Proceedings of Frontiers in Education Conference, 2C4–21 to 24.

Whitaker, G. (1965). *T Group Training ATM Occasional Paper No 2*. Oxford: Blackwell.

Whitfield, P. R. (1975). *Creativity in Industry*. Harmondsworth: Penguin.

Woods, D. R. (1994). *Problem Based Learning. How to Gain the Most from PBL*. Hamilton, Ontario: McMaster University Bookshop.

Youngman, M. B., Oxtoby, R., Monk, J. D., and Heywood, J. (1978). *Analysing Jobs*. Aldershot. Gower.

3

The Development of a Multiple-Objective (Strategy) Examination and Multidimensional Assessment and Evaluation

Partly in response to Furneaux's criticisms of examinations in engineering, and partly in response to the ideas contained in "The Taxonomy of Educational Objectives," a public (high school) examination in Engineering Science was designed to achieve multiple objectives, each objective thought to represent a key activity in the work of an engineer. Its underlying philosophy was that the ways of thinking of engineers were different from the ways of thinking of scientists even though engineers required a training that was substantially science based. The designers were also persuaded of the power of assessment to influence that philosophy and the changes they wished to bring about. The development of this examination is described together with the development of criterion-referenced assessments for practical and project work. Problems associated with the examination of design by written techniques and its assessment in coursework are identified and an interpretation is given. The examination and coursework assessment procedure exemplify a "balanced" system of assessment.

It was found that criterion-referenced assessments require careful design and need to be piloted if they are to be valid. It is also clear that determining what examinations and competency-based rubrics are actually measuring is a complex activity. It is all too easy to overload the curriculum and overburden teachers with assessment.

The Assessment of Learning in Engineering Education: Practice and Policy, First Edition. John Heywood.
© 2016 The Institute of Electrical and Electronics Engineers, Inc. Published 2016 by John Wiley & Sons, Inc.

The subjects' development was impeded by the same factors in the British culture that were outlined in Chapter 2 and although it was offered for 20 years it did not acquire the status necessary for it to be attractive to large numbers of students and teachers. However, arguments about its philosophy have resurfaced during the last decade and the procedures and techniques of assessment it developed offer a bench mark against which other studies can be evaluated. It was probably the first course in the United Kingdom to state specifically what attitudes and interests a course in engineering science should promote, although no systematic attempt was made to evaluate the success or otherwise of the course in developing these attitudes.

At the same time, but at the other end of the educational spectrum, in the vocational training of graduates for general practice in medicine, there was considerable emphasis on the assessment of the attitudes that future general practitioners should have and the "soft" ("professional") skills they should possess. The chapter concludes with a description of a multidimensional study of a 3-year program of vocational training of newly qualified medical practitioners about to embark on a career in general practice. Based on pre- and postcourse measurements, it is a study that can be used as a model for the evaluation of program goals. It is designed to try and understand the factors that influence student achievement (e.g., course organization, motivation, personality). Course organization is shown to be an important factor influencing the performance of weaker students.

In contrast to engineering science where the objectives were derived from a model of what it was thought the work of professional engineers was, the objectives of the general practitioner training program and its assessment were based on a substantial analysis of the work undertaken by general practitioners.

3.1 The Development of an Advanced Level Examination in Engineering Science (For 17/18-Year-Old High School Students): The Assessment of Achievement and Competency

One of those in the university departments of education in the United Kingdom who knew about *The Taxonomy* in the early 1960s was R. A. C. Oliver, who designed an "A" level for the General Certificate of Education in "General Studies" for the Joint Matriculation Board (JMB) that attempted to assess the domains of *The Taxonomy*.[1] Taken together with Furneaux's paper (1962), these and other studies influenced the designers of a new curriculum at "A" level in engineering science that had the intention of being an alternative to physics for university entry (Carter, Heywood, and Kelly, 1986). The philosophy behind the course was clearly stated by Professor Harry Edels, a member of the Board who was the venture's product champion (Edels, 1968). He argued that engineers thought differently to physicists because the problems they solved were different from those of physics. These differences continue to be discussed by engineering educators and philosophers (e.g., Goldman, 2004; van de Pol and Goldberg, 2011). Edels argued that if high school students are to be attracted to study engineering they need to be exposed to the ways of thinking of engineers. As a consequence of this and other arguments, the Board had agreed a syllabus in engineering science that was a statement of content to be taught. It was accompanied by example examination papers of a traditional kind (two, each of 3 hours' duration), and a traditional approach would be taken to the assessment of laboratory work.

It was immediately apparent that this approach would produce an examination that would, as with Furneaux's examinations, yield only one factor, probably similar to his. Moreover, it was unlikely that it would cause students to appreciate the way of thinking of engineers. The Board created a coordinating committee to develop the subject and this led to the design of what Heywood (1977) called a multiple-objective examination. The chief constraint on the design was the obligation on the examiners to show that it was equivalent to physics. Nevertheless, the new "A" level in General Studies showed a way forward and following the approach of the *Bloom Taxonomy* the examiners made a sharp distinction between knowledge and comprehension. The examiners considered that the traditional papers because of the type of question set tested analysis and certain problem-solving skills but the questions had to be designed within an engineering context. They found as Bloom et al. (1956) had predicted that his cognitive dimensions did not meet the needs of the subject. In the general statement of knowledge, understanding and abilities to be tested a domain of "communication" was included, the domain of "synthesis" was modified to incorporate "design" (i.e., Synthesis and Design), and the domain of "evaluation" was modified to include "judgment" (i.e., Evaluation and Judgment) (JMB, 1972). *The Taxonomy* made no provision for originality/creativity but a domain for this dimension was incorporated in the original notes for guidance on laboratory and coursework (see Appendix B). In comparison with current procedures for ABET (Accreditation Board for Engineering and Technology) accreditation in the United States while the Engineering Science examiners called the domains "abilities" their overall statement corresponds to a statement of program outcomes that also incorporates learning outcomes.

In the event, the two papers were divided into a series of subtests as shown in Exhibit 3.1. The first 3-hour session was divided up into a test of knowledge using multiple-choice questions, a comprehension exercise in which the candidate had to read an article from a journal and respond to questions about the article in short answers, and a project planning exercise. The rubric specifically states that the article may be on subject matter not specifically in the syllabus. This approach opened up the possibility of Cross Domain Transfer, an ability that is currently sought from engineering students, although it was not sought then. Each of these subtests was of 1 hour's duration. The second paper focused on analysis. Long- and short-answer questions were designed to correspond as much as possible with the kind of real-life constraints an engineer might encounter. Today it would be said that they were attempts to design "wicked" questions. This 3-hour paper was divided between long answers (40 minutes' duration) and short answers (15 minutes' duration).

It was believed that written examinations would not be able to assess all that was desired, for example, creativity, design, project planning and evaluation, and skill in investigation. To achieve this goal, a nontraditional approach to coursework and its assessment was introduced. During the 2 years of the course, the student would under-take a number of controlled assignments—that is, traditional laboratory experiments

Subtest	Objective and Technique of Assessment and Duration	% of Total Assessment
Written paper I (3 hours) Section 1	Knowledge and short chain problem solving. Forty objective items: 1 hour	13.5
Section 2	Comprehension exercise (candidates read article from a journal and answer questions on it that show an underlying understanding of the principles): 1 hour	13.5
Section 3	Project planning and design exercise: 1 hour	13.5
Written paper II (3 hours) Section A	Applications of engineering science (analysis and application): $1\frac{1}{2}$ hours, six from nine questions	20.0
Section B	Applications of engineering science (analysis and application): $1\frac{1}{2}$ hours three from six questions	20.0
Coursework (throughout the 2 years of the course) CW I	Two experimental investigations to be submitted by the candidate for assessment	
Coursework (50 hours of laboratory work) CW II	Project chosen and completed by the student	20.0

EXHIBIT 3.1. The multiple-objective approach to examining and assessment adopted for "A" level engineering science (circa 1972). Twenty percent of the available marks were available for all coursework. In addition, students were expected to keep a journal throughout the 2 years that showed they had completed a number of controlled assignments (traditional laboratory experiments), a variety of experimental investigations, and the work for the project (Carter, G., Heywood, J., and Kelly, D. T. (1986). *Case Study in Curriculum Assessment. GCE Engineering Science (Advanced)*. Manchester: Roundthorn.) See Appendix B.

conducted in a short period of time with the aid of very precise instructions that would cover each area of the syllabus. Reports of these would be kept in a journal. They would also complete a number of open-ended experimental discovery investigations (now often called "inquiry," e.g., Narayanan, 2013; Widmann, Self, and Prince, 2014). Each one was to be of no more than 12 hours' duration in the laboratory. Reports of four of them (later reduced to 2) were to be submitted for assessment. Finally, a report of a project of a maximum of 50 laboratory hours' duration was to be submitted. The rules did not prevent students from doing additional work at home. The rules allowed the moderators to inspect the work done by students *in situ*. (A similar approach to laboratory [course] work will be found in the University of Brighton's MEng program; Thomas and Izatt, 2003.) The objectives for the different types of engineering science coursework are given in Exhibit 3.2. The assessment would be criterion referenced.[2] A few of the items in

Controlled assignments

Controlled assignments are of short duration and normally accomplished within a 2-hour period; they are intimately connected with the subject matter. Pupils may work singly or in groups. Such assignments will:
(a) Reinforce and illuminate lesson material.
(b) Familiarize students with the use of scientific equipment.
(c) Develop a reliable habit of faithful observation, confirmation, and immediate record in a journal style.
(d) Introduce the techniques of critical review, analysis, deduction, and evaluation.
(e) Promote good style and presentation in the formal technical report.

Experimental investigations

An experimental investigation poses an engineering or scientific problem and involves the student in an analysis of the situation and an appropriate selection of procedures and techniques for solution. The end point of the particular investigation may or may not be known, but the means for its achievement are comparatively discretionary.

The time needed for an investigation of this type should normally lie in the range 6–12 hours.

A record of investigation should include:
(a) A clear account of the analysis of the problem.
(b) Brief report and comments on the work as the experiment proceeded.
(c) Comment upon the results.
(d) An appraisal of what has been achieved.

Projects

The project is a major undertaking for which it is suggested that 50 hours of laboratory time would be suitable. The pupil will be required to design a device or design and conduct an investigation to fulfil a specification and to evaluate the degree of fulfillment achieved.

Projects call for strong mental connective abilities rather than for craft skills and the time spent on construction or practical investigation should be kept at a minimum, the emphasis being on design and the formulation of problems, literature search in its widest sense, and evaluation.

EXHIBIT 3.2. Definitions of the components of course work in engineering science. Reproduced from *Notes for the Guidance of Schools on Engineering Science at the Advanced Level*, Joint Matriculation Board, Manchester (1972).

the rubric to which "yes" or "no" responses were required are shown in Exhibit 3.3. In addition to these exhibits, Appendix B includes the complete syllabus and extracts from the *Note for the Guidance of Schools*.

Overall this approach mirrored that of ABET, the chief difference being the terminology used. It differed in the sense that the subtests were designed to assess specific domain objectives for which reason it was termed a "multiple objective examination." It also resulted in the development of coursework assessment rubrics of a type commonly described by present-day practitioners the development of which remains of interest possibly as a benchmark.

Both the moderators and the teachers found the scales difficult to use. Inspection of the reports showed that answering "yes" or "no" did not allow for gradations of what

Two questions from the six questions that constituted part II of the rubric for assessment of the project
8 (a) Has the candidate discussed alternative solutions to problems arising during the course of the project? And (b) Has the candidate given acceptable reasons for the solutions adopted?
10. Has the candidate discussed the validity of the conclusions reached or evaluated the final product in terms of the original specification?

EXHIBIT 3.3. Extract from document ES/CWA (1970) *Engineering Science (Advanced) Coursework Assessment*. Manchester: Joint Matriculation Board. Separate rubrics were required for the experimental investigations. The document included detailed guidance and instructions for the assessors.

was reported. As some critics had envisaged dichotomous scales of this kind were not really suitable for deriving grades that could be incorporated into a norm-referenced system of scoring. They did not allow for shades of grey and could be used only for very simple operations. Neither were they necessarily valid. I recently met teachers of Higher National Certificate Engineering and Business who were experiencing the same difficulty. Students had to show that they had demonstrated each outcome to pass a component of the syllabus. They were not allowed to mark a good or exceptional answer as a pass if one of the outcomes was not covered by the student.

Because the numbers of students involved was small, it was possible to conduct an experiment in which the students were asked to determine the nature of the electrical component(s) between the two terminals of a sealed box using a range of equipment set out on a bench. Each student was given half an hour to plan his approach (there were no females) and half an hour to carry out the exercise.

The students understood that they were participating in an experiment and were not being assessed for the examination by their teacher and the two moderators who assessed them using the published rubrics. It was found that four of the criteria gave rise to considerable difficulty. In the following 2 years, the moderators asked the students to provide them with examples of what they thought would obtain an affirmative response to each of the questions from the assessors. One of the moderators and the teacher were asked whether or not they agreed with the student responses. As a result, the assessment schedule was divided into two parts. The first remained a hurdle and was criterion referenced; that is, the response could be only "yes" or "no" and some items were revised. A student had to be able to demonstrate the competencies that were demanded in part 1 for part 2 to be assessed. They were not made difficult but thought to be essential for a person to conduct practical work safely. Part 2 was based on scaled rubrics (see Exhibit 3.4). The immediate effect of moving to the new system in the year that followed was to elevate the distribution at the lower end of scale and so recognize some competence on the part of the weaker candidates. Over a 15-year period, the late D. T. Kelly found that the coursework component discriminated well between the candidates and was reasonably consistent from year to year (unpublished documents made available to this writer and Carter, Heywood, and Kelly, 1986).

Extract From the Scheme of Assessment in 1971	Grade
8. Execution	
In executing the plan, the candidate gave thorough consideration to realistic alternatives at every stage and made a reasoned selection of the optimum solution in each case.	3
gave consideration to realistic alternative solutions with inadequate reasons for selection.	2
gave some attention to the consideration of alternative solutions.	1
paid little attention to this aspect of the work.	0
9. Critical view	
In comparing the final product or outcome with the original specification, the candidate has produced a thorough objective discussion in which consideration has been given to all major aspects of the work including suggestions for further development and a critical appraisal of the conduct of the project with a clear indication of the lessons learnt.	3
a reasonable depth of discussion that, however, lacks either objectivity or coverage.	2
some significant comparison.	1
discussion of little significance.	0
Originality as defined in the notes for guidance; applies across the spectrum of assessment	
The development of the ability to: (a) Formulate hypotheses from given sets of observations. (b) Formulate experiments to test hypotheses. (c) Devise and improve upon experimental procedures. (d) Appreciate the relative importance of errors in differing situations.	
Originality as incorporated into the scaled domain for design	
9. Design activity	
In relation to the design for all or part of the project with respect to procedure or artefact, the candidate produced a markedly significant and original contribution.	3
An original contribution.	2
A new device by applying a standard design technique.	1
Little or no design activity during his work on the project	0

EXHIBIT 3.4. Extracts from the revised assessment procedure. Also shows how the domain objective of originality was incorporated in the domain of design activity. The procedure for the experimental investigations was also revised. *Engineering Science (Advanced) Course Work Assessment*. ES/CWA/1971. Manchester, Joint Matriculation Board (see Appendix B).

1. Recognize the existence of a problem.
2. Define the problem.
3. Select information pertinent to the problem.
4. Recognize assumptions bearing on the problem.
5. Make relevant hypotheses.
6. Draw conclusions validly from assumptions, hypotheses, and pertinent information.
7. Judge the validity of the processes leading to a conclusion.
8. Evaluate a conclusion in terms of its assessment.

EXHIBIT 3.5. Saupé's (1961) model of the steps in critical thinking in Dressel, P. (ed.), *Evaluation in Higher Education.* San Francisco, CA: Jossey-Bass.

The responsibility for the choice and conduct of the project was the student's. An early finding was that very often projects failed because they were not fully thought out. Often they were too ambitious at other times they were too vague. The result: incomplete and poorly done projects. Co-incidentally, it seemed that the shorter the title the more likely it was to be borderline. The title had to demonstrate a focus. These findings highlighted the importance of problem finding as a distinct process in the problem-solving activity (McDonald, 1968: see also Section 2.4). By 1972, it had become a requirement that each student should submit a project outline to the moderators. It would be accompanied by comments from the teacher(s) who would also certify that the work was that of the candidate. The form is shown in Exhibit 3.8. The external moderators advised either approval of the project without comment, or with suggestions for revision, or counseled rejection. Taken together, it is clear that coursework mirrored the model of critical thinking that Saupé had suggested in 1961, which is shown in Exhibit 3.5 As many subsequent illustrations in this study show, his categories continue to be relevant.

With minor modifications, the revised rubrics[3] were used until the subject was discontinued to become Physics B nearly 20 years later for lack of a large number of entrants, a consequence of the failure to gain status. When the rubrics were designed, the assessors quickly found that they had to make choices about which competencies to assess. Teachers could only cope with so many rubrics. They also understood that not everything they wished to measure could be measured so they had to decide what was important and what was not. Much time was spent in determining the nature of creativity and how it could be measured, if at all. As indicated in Section 2.4, it entered the syllabus as "originality."

Building on this experience, other examining authorities developed similar rubrics for the evaluation of projects in their examinations for engineering science. Now they are commonplace (e.g., McCormack, 2011).

George Carter, who was chairman of the coordinating committee, used the experiences gained to investigate laboratory assessment in undergraduate electrical engineering courses (Carter et al., 1980) and the table of aims and objectives shown in Exhibit 3.6 resulted. Inspection shows that many of the objectives not only meet the needs of industry but show that the laboratory process is concerned with that goal (items 11, 15, 16, and 17; Exhibit 3.6). It will also be noticed that the table focuses on the attitudes that laboratory work can foster; this follows from Engineering Science, where a hard-fought

Aim	Behavioral Objective Capable of Being Observed and Tested
1. To stimulate and maintain the students' interest in engineering	Student, likes working in the laboratory, is often to be seen there, arrives early, and leaves late.
2. To illustrate, supplement and emphasize material taught in lectures	Student uses lecture material, in laboratory problems and vice versa, has knowledge of methods learned.
3. To train the student to keep a continuous record of laboratory work	Student keeps well-laid-out notebook for this purpose rather than loose sheets of paper.
4. To train the student in formal writing of the experimental procedures adopted in laboratory practicals, and the writing of technical reports	Student hands in well-written reports on time, discusses them with tutors, and attempts to improve them.
5. To teach student how to plan an experiment so he derives useful meaningful data	Student comes to laboratory having read necessary references and with prepared plan of operation.
6. To give the student training in the processing and interpretation of experimental data	Student uses graphs and tables intelligently and draws fair conclusions from and deals sensibly with errors.
7. To train the student to use particular apparatus, test procedures, or standard techniques	Student shows competence in handling common laboratory equipment and learns how to use new equipment quickly.
8. To improve the learning/teaching process by improving the communication and rapport between staff and students	Student talks with staff, initiating discussion on the experiment and other matters.
9. To strengthen the student's understanding of engineering design, by showing him that practical work and design work must be integrated to achieve viable solutions to design problems	
10. To develop the student's skill in problem solving in both single- and multisolution problems	Student progresses from "dashing off in all directions" methods to be planned attacks on problems.
11. To provide each student with an opportunity to practice the role of a professional engineers so that he can learn to perform that role	Student exhibits responsible, truthful, and reliable attitude toward data, use of time, care of equipment etc.

EXHIBIT 3.6. (*Continued*)

Aim	Behavioral Objective Capable of Being Observed and Tested
12. To provide the student with a valuable stimulant to independent thinking	Student creates his own solution to problems and does not wait to be told what to do. In discussion, student puts his own points clearly.
13. To show the use of practical work as a process of discovery	
14. To demonstrate the use of experimental work as an alternative to analytical methods of solving engineering problems	
15. To help students understand that small models of plant or processes can aid greatly in the understanding and improvement in such a plant or processes	
16. To familiarize the student with the need to communicate technical concepts and situations—to inform and persuade management to certain courses of action—to disseminate technical expertise for the benefit of all	Student can explain clearly what he has done and why using proper technical concepts, using graphs, tables, sketches, etc., as seems most useful.
17. To help students bridge the gap between the university academic situation and the industrial scene, with its associated social, economic, and other restraints that engineers encounter	
18. To teach the student how accurate measurements made with laboratory equipment can be; to teach him how to devise methods that are precise when precision is required	Student can determine and report errors correctly, can devise more accurate methods of measurement, and demonstrate them. Student also guesses correctly and knows when to ignore errors and when to "round off" numbers.
19. To teach the student what "scientific method" is and how it is applied in an engineering laboratory	
20. To give the student confidence in his ability to imagine a concept or hypothesis, plan an experiment, t test it, carry out that experiment, and report its results to others	Student acts confidently yet sensibly and safely in the laboratory.

EXHIBIT 3.6. The aims of laboratory work and some behavioral objectives extracted from numerous sources by Jordan, T. A., and Carter, G. (1986). *Student Centered Learning in Engineering. Engineering Enhancement in Higher Education.* Monograph. Salford: University of Salford. See also Carter, G., et al. (1980). Assessment of undergraduate electrical engineering laboratory studies. *IEEE Proceedings*, 127, Part A, (7), 460.

Attitudes and interests that the course in Engineering Science should foster
1. The recognition of a need for a method which is organized, careful, and intellectually honest particularly in respect of experimental observation.
2. The acceptance of the need to consider the parallel social and economic bases of engineering.
3. An awareness of the advantages of deriving the more particular relationships from the basic concepts.
4. An awareness of the advantage of seeking parallels in other fields to relate one kind of phenomenon to another.
5. An awareness of the advantage of attempting to reduce a social, economic, or scientific situation to a simple system.
6. The recognition of the fact that it may be necessary to exercise judgment as well as reason when dealing with a problem.
7. The recognition of the fact that a perfect answer to a problem may not exist, and that the best available answer must be sought.
8. The recognition of the fact that not all the information necessary to tackle a problem may be available and that some which is available may not be relevant.
9. The acceptance of the fact that more than one way of thinking exists, and that different ways may be more appropriate to different problems or different stages of the problem.
10. The recognition of the fact that the required exactness of calculation may vary from case to case (e.g., from a preliminary quick "order of magnitude" estimate to a precise forecast of performance).

EXHIBIT 3.7. Statement of the attitudes and interests that the course in Engineering Science should foster. Joint Matriculation Board. (1973). *Notes for the Guidance of Schools on Engineering Science at the Advanced Level of the General Certificate of Education.* ES/N2. Manchester: Joint Matriculation Board.

battle in the Board led, most unusually, to a statement of attitudes that the examiners hoped would be developed from participation in the course (Exhibit 3.7). They proved to be somewhat controversial but a course designed to meet those objectives would clearly go some way to meet the criticisms of industry. Nevertheless, some industrialists criticized the model because it was based upon assumptions about what engineers do rather than "real" studies of what they actually did. This led to the investigation of what engineers were doing at work that is described in Chapter 4.

It should be noted that the school situation was different from that of the university in that, very often, a single or at most two teachers taught all the subjects in the course and assessed the coursework that was moderated by external assessors appointed by the examining authority to approve the marks of all the candidates from the different schools presenting for examination. The written examinations were set at the end of the 2-year course. At the same time, students took two or three additional subjects to the same standard that were also examined at the end of the 2-year period. Students of engineering science would normally include mathematics as one of these subjects. This structure enabled substantial degree programs to be offered by the universities, and the 3-year courses offered particularly in science subjects in England were dependent on those standards being obtained in high school. To achieve a multiple strategy approach to curriculum design implies substantial changes in the roles of teachers and their teaching.

3.2 Skills Involved in Writing Design Proposals and Practical Laboratory Work

As indicated in Section 3.1, it was found that students would be helped if they conducted a project planning exercise prior to approval being given for them to proceed. The guidelines are shown in Exhibit 3.8.

In order to test the skills involved in planning, the terminal written examination incorporated a subtest that described a problem (project) and the candidates (students) were required to present a plan for its solution. It was expected that this would cause them to repeat the skills they had learnt while planning their projects. There would, therefore, be a high correlation between the two marks.

Unfortunately, the Board's research unit did not find a satisfactory correlation between marks for project planning and evaluation in the project exercise and those for the examination that was supposed to model these skills. In other words, the criterion validity was low. However, interpretation is difficult since the assessors were given the correlations between the overall mark for coursework and the other components of the examination and the lowest correlation was found to be between these two activities. Nevertheless, each of three consecutive repetitions produced the same pattern of results, and the factorial analyses suggested that the subtests were measuring different things as was hoped (Heywood and Kelly, 1973). At first it was thought that this low correlation was because engineering design was not a requirement of the syllabus, and the students learnt project planning skills by osmosis. However, a decade later when Sternberg published his Triarchic theory of intelligence, another explanation became possible (Sternberg, 1985). Sternberg distinguished between three components of intelligence: *meta-components* that are processes used in planning, monitoring, and decision making in task performance; *performance components* that are processes used in the execution of a task; and *knowledge-acquisition components* used in learning new information. Each of these components is characterized by three properties that are duration, difficulty, and probability of execution. They are, in principle, independent.

It is evident that the project assessment schemes of this kind are concerned with the evaluation of meta-components. Elsewhere, Sternberg calls them appropriately "executive processes." We can see that a key difference between the project planning exercises and the written subtest is the time element. The two situations required the student to use different information-processing techniques. The written exercise is a different and new domain of learning for which training is required. In order for the skill to become an old domain, a high level of automatization is required so that the different processes in the meta-component are brought into play more quickly. That is to say they have at a certain level to become nonexecutive. The project and the written paper, while demanding the same meta-components, might be regarded as being at different levels in the decision making (experiential learning) continuum, as a function of the time available for their implementation. Some executive processing will always be required at the written paper level, and it is possible to argue that the task performance and stress that it creates are

Contents of the outline

Title

This should be a clear statement of the problem to be tackled. While the title should be brief, it must not be vague or so general that it does not convey the essence of the project.

Analysis of the problem

The problem to be dealt with in the project should be analyzed as fully possible. A general statement of the problem should be given, and where possible, quantities laid down together with the limitations under which you will be working, such as restraints of size, cost, use, and availability of workshop facilities and assistance. For example, if an engine test bed is to be constructed, the size and nature of the engine test bed and associated equipment should be stated, the use to which the engine is to be put should be given and the parameters to be measured should be listed. If the project is of a more investigatory nature, a similar analysis is required. For example, if it is concerned with an investigation into atmospheric pollution, the nature of the variables to be measured, the periods over which measurements are to be made, the factors likely to affect these variables, and the uses that might be made of the information gained should be stated.

Practical problems to be solved

Having considered the project in outline, you will be able to recognize the major practical problems that need to be overcome. These may be the design and manufacture of a piece of equipment or the design of experimental procedures, or both.

Possible solutions

It should be possible at this stage to see your way to solving these major practical problems in order that success can be achieved. It is therefore important that you should offer likely solutions to these problems. It may be that one solution is so obviously the best that a lengthy consideration of alternative approaches is unnecessary. In most cases, however, a number of alternative solutions will occur to you or will arise as a result of consultation with your teacher or other people. The final choice of a solution will, in most cases, depend on further work and consultation, and the use of appropriate references. Your outline should give the main direction of your ideas at the time of submission.

Resources

The choice of the best solution will also depend on the resources you have available. You should, therefore, list under appropriate headings, equipment, manufacturing facilities, and materials required, references, consultants, technical assistance available, and the approximate cost involved. Such headings will not be equally important for all projects.

Timetable

You will now be in a position to draw up an approximate timetable of operation. It does not help to make wild guesses about the number of hours you will need; it is better to work in weeks available and then split the period into component parts. Do not forget to list the time necessary for writing the final report. In planning your time, always assume that any task will take you much longer than you imagine on first consideration. It is also important to allow a certain degree of flexibility; if you draw a time sequence diagram, allow for a fair amount of variation.

EXHIBIT 3.8. Extract from the student guidance document and assessment schedule *Engineering Science Project Outline* (1972). Manchester: Joint Matriculation Board.

References
In submitting your project outline, list the books and articles you have read in connection
with the planning and also the individuals whose advice you have sought.

Future work
You are strongly advised to read the appropriate sections of the Notes for the Guidance of
Schools at all stages of the project, particularly during the planning period. When the
moderators have studied your outlines, they will forward their comments to your teacher. You
are strongly advised to follow any recommendations made by your teacher or moderator.

EXHIBIT 3.8. (*Continued*)

a more accurate reflection of the everyday activities of executives than the substantive
project.

Notwithstanding the validity of this interpretation, more generally these investigations also showed that when criterion and semi-criterion measures appear to have high
face validity, there is a need to ensure that there is congruence between student and
assessor perceptions of the items (see Section 7.3). The measurement of performance
is not as easy as it seems. Similarly, it cannot be assumed that the goals that assessors
have for multiple-strategy assessments are necessarily being met even when they appear
to have face validity. It is argued that these findings are axiomatic and apply generally to performance or competency-based assessments such as those required on the one
hand by the Accreditation Board for Engineering and Technology's Engineering Criteria
requirement and on the other hand for the assessment of project work in the humanities
(e.g., history, see Heywood, 1977, in which he set out in detail his theory of multiple
objective assessment).

3.3 A Balanced System of Assessment

Heywood was invited by a Government Committee on examinations in the Republic of
Ireland to apply the principles of this approach to examining intermediate examinations
in history and mathematics. The results of the action research that followed led the
Committee to propose a multiple objective approach to assessment "i.e. that as the
objectives of education are multiple, so the modes of their assessment should be diverse
and adapted to the objective they serve" (Intermediate Certificate Examination, 1975,
p. 88). Taken as a whole, the report of the committee shows that it was clearly seeking
what the US Committee on the Foundations of Assessment (National Research Council)
report in 2001 called a balanced system of assessment. That is a balance between
classroom (teacher) and large-scale external tests of the kind offered by ACT and SAT
(Pellegrino et al., 2001). (In UK terminology, between formal public examinations and
in this case, moderated teacher coursework assessment). Such systems are characterized
by comprehensiveness, coherence, and continuity.[4] The examination that the National
Research Council's committee found to approximate to a balanced system was Physics
B as Engineering Science had come to be known and developed in the late 1980s

(Heywood, Carter, and Kelly, 2007). There are many similarities between the standards-based system of grading reported by Carberry, Siniawaski, and Dionisio (2012) and the scheme reported here (see Section 10.5). The evaluations of engineering science support their findings that have been discussed elsewhere (Heywood, 2014).

Elliott (2014), who designed a flipped base course, argued for a balanced system of assessment by which he meant a balance between formative, interim, and summative assessments. The simplest form of flipped course is to ask the students to study the lecture material at home and do the homework in course time. However, e-learning enables the students to interact with recorded lectures prior to the class so that class time is freed up for interactive activities. Elliott found as others have before him that students are reluctant to depart from traditional instructional procedures so they had to be shown that the change was for their benefit. This included showing how the components assessment would respond to their efforts in preparation (formative assessment), classroom work (interim assessment), and homework assignments (summative assessments). The reduction in homework "would relieve some of the burden of normal assignments." The interim assessments took place during class time. "All summative assessments were completed individually, although students had to review collaborative work during the interim assessments and incorporate them into their deliverable where appropriate. This technique also gave the students the opportunity to revise and, if necessary, reject solutions that they felt were not correct, thus affording them the opportunity to analyze a variety of thought processes when composing their own solutions." Although the quantity of the assessments had increased Elliott's view that they would negatively impact student perceptions of the flipped course was not upheld as most of the students felt the quantity of all three types was just about right.

3.4 Pictures of the Curriculum Process

The 1960s oversaw a considerable change in the understanding of the curriculum. It came to be seen much more as process with accompanying products than simply a syllabus (i.e., a list of contents). Notice, at least among educationalists concerned with school education came to be taken of curriculum theory. However, higher education has remained largely untouched by these developments. Surprisingly, among engineering educators the curriculum remained largely "received" although there was some "restructuring" brought about by changing technology and advances in knowledge (Eggleston, 1977). Perhaps the biggest change has been the implementation of problem-based learning courses in some institutions.

Following Tyler (1949), those committed to an objectives approach became adept at illustrating (picturing) the curriculum process, and just as with "objectives" they each had to have their own model! This writer is no exception to that rule and it is with his model and with the considerations that led to its present form that this section is concerned. The current version of the model is shown in Exhibit 3.9. Earlier versions were criticized by Sister Georgine Loaker because they did not show the dynamic nature of the curriculum that had been revealed by studies of the developing curriculum at Alverno College, Milwaukee. Her ideas were subsequently expressed in *Learning That*

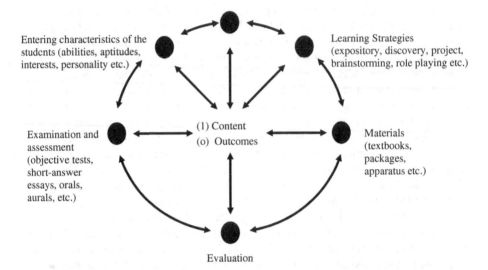

EXHIBIT 3.9. A model of the assessment–curriculum–learning–teaching process. Showing (1) the first phase in which the structure of the syllabus content is derived and (2) how the intended learning outcomes are a function of a complex interaction between all the parameters and allowing that there will also be unintended outcomes. The original model *in Enterprise Learning and Its Assessment in Higher Education* (Technical Report No. 20, Employment Department, Sheffield) referred only to the design of the syllabus while indicating that evaluation took care of the dynamic nature of the model. Professor Georgine Loacker of Alverno College suggested that this dynamism would be better expressed if the model also recorded the outcomes of the ongoing activity in the center. This model was reported in Heywood, J. (2005). *Engineering Education. Research and Development in Curriculum and Instruction.* New York: IEEE/John Wiley & Sons.

Lasts (Mentkowski and Associates, 2000). I have had to concede that it is very difficult, if not impossible, to illustrate the dynamic nature of the curriculum process without additional comment. A similar model that is directly related to course design to meet ABET requirements was described by Felder and Brent (2003; see Exhibit 3.10) in which the domain covers classroom assessment techniques (Angelo and Cross, 1993), tests, other measures, and surveys. Tests might include homework (Head, Owolabi, and James, 2013), concept inventories (Lorimer, 2013; Prince, Vigeant, and Nottis, 2013). Other measures might include assessments of intellectual development, learning styles (Miskioghi and Wood, 2014), self-efficacy inventories (Stickel, Harri, and Liu, 2014), and other personality measures like the MBTI (Brown et al., 2013). Davis et al. (2013)

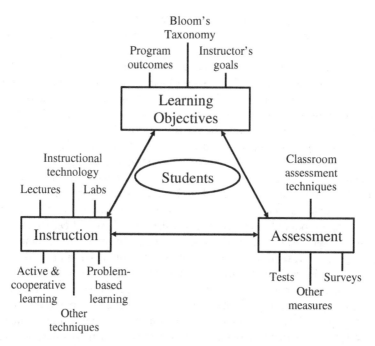

EXHIBIT 3.10. Felder and Brent's (2003) model of the process for designing courses to meet the ABET engineering criteria. Reproduced by kind permission of the authors and the *Journal of Engineering Education*.

have described a website ASSESS, which locates assessment and evaluation instruments relevant to engineering education.

My original model arose from the fact that in the engineering science development it became clear that contrary to the position held in England by the Examination Boards that their examinations should test their syllabus and not interfere in its teaching, that in fact, *examinations and assessments considerably influenced teaching.* Different methods of instruction and learning were likely to be required for different key objectives. For example, in the comprehension exercise some guidance (instruction) in how to read academic articles and respond to short-answer questions about them was desirable. Exemplars could be undertaken in examination conditions or for homework. In either case, alterations in the time available for covering the syllabus would be required. Similarly, some form of instruction and response would be desirable for the written planning exercise, that is the teaching of design, but there was no syllabus requirement for this (see Section 3.2).

Again in Ireland, when this approach was used to design examinations for secondary school mathematics, the research workers were not allowed to consider the instruction and learning necessary to obtain the key objectives. Thus, students performed badly in exercises designed to test problem-solving abilities because the teachers had not provided such instruction. The state public examination syllabus did not envisage activities of the kind designed by the teachers. Even so they were taught the requirements of the public examinations rather than those of the experimental examinations they had designed.

Similarly, different objectives are likely to be obtained from problem-based learning courses than traditional courses. The design of assessment in problem-based learning has been found to be problematic (Shelton and Smith, 1998; Vu and Black, 1997) particularly when the objectives are not clear to the students (Sadlo, 1997). The point to be made is that assessment cannot be the afterthought of the process because it dictates what will be, and how and what should be taught. Therefore the learning objectives and the assessment objectives have to be the same and in some way this has to be demonstrated in curriculum diagrams, and this is what makes Exhibit 3.9 different to most other illustrations. Assessment is integral to the curriculum process that has to take into account the entering characteristics of the students. If any of the links in the model are broken, it is not possible to be certain that the objectives will be obtained. Recent publications in engineering suggest that this problem is beginning to be understood and the term "alignment" is used to describe it (Streveler, Smith, and Pilotte, 2012).

Similarly, Felder and Brent's diagram (Exhibit 3.10) makes clear the differences between program outcomes and learning outcomes but as Abro and Cuper (2013) show, it is possible to design learning objectives that are "a reliable source of program learning outcomes evaluation." They distinguish between direct and indirect assessments. The former is when the instructor sets the level of achievement that is designed for each objective, and uses tests, assignments, and projects to assess that level. Indirect assessment is the student's ranking of achievement of the same learning objectives from his or her perspective. The model is designed to evaluate the level of consistency. They argue that this approach can lead to a process of continuous improvement.

Stice (1976) reported that when he adopted an objectives approach to the design of curriculum, it made him ask questions about what was essential and what was not essential. He found that some of what he had been teaching was inessential. This suggests that a really thorough ongoing analysis of a curriculum by objectives might lead to a reduction in content and, therefore, reduce the load on students. This means that the syllabus should not be fitted to the objectives but derived from an analysis of the time required to learn key objectives.

Engineering educators have become very familiar with concept inventories and their significance (Streveler et al., 2008, 2014). While there has been much attention to the teaching of concepts to avoid misperceptions, there is little in the literature that concerns itself with the time students of varying learning profiles required to learn concepts. The experience of middle and high school teaching (ages 12–18 years) in Ireland suggests that students require more time to learn concepts than is currently allowed. Apart from the need for investigations in the time required for deep as opposed to surface learning of concepts and principles, one approach to curriculum design that takes this into account is by mapping the key concepts (e.g., Donald, 1982; Fordyce, 1992; Streveler, Smith, and Pilotte, 2012). They can be an invaluable aid in the integration of subjects (Blyth et al., 1973). Integration within the disciplines of engineering is likely to be necessary at some levels of the curriculum simply because of the explosion of engineering knowledge, hence the importance of the epistemological analysis of the curriculum. There is no excuse for rejecting new knowledge areas because they will overload the curriculum without such analyses.

Reductionist approaches to curriculum design are equally prone to continually increasing the number of outcomes required to the extent that assessment becomes a checklist exercise. This is why the engineering science assessment rubrics were limited to a number of key domains. It made reasonable as opposed to unreasonable demands on assessor workloads. But, it was understood much else would be tested in the responses to those domains believed to be essential to producing a rounded student. The central question that should determine the content of the curriculum is: "What are the necessary key concepts that should be learnt?" Key concepts are as much objectives as those that describe the cognitive skills required. It is important to remember that there are many sources of objectives especially those derived from the work that engineers do.

Perhaps the most important idea that was learnt in these studies is just how difficult it is to design tests that align with the objectives and that question design is not, or should not be an off-the-cuff exercise (Edwards, 1984; Laurillard, 1984). This problem continues to be a challenge (Abro and Cuper, 2013).

3.5 Multidimensional Assessment and Evaluation: A Case Study

It will be noticed that in Exhibit 3.9, "evaluation" is separated from "examinations and assessment." Its task is to establish if a course is achieving its aims and objectives. It is about questions such as "what factors cause some students to fail and others to succeed?" or "are the learning strategies the most suitable for obtaining the objectives that have to be achieved?" It may be argued that the two should be fused and in some American studies they are (e.g., Farnsworth et al., 2013). This chapter is now completed with a summary of a British multidimensional study of assessment for general practice in medicine that was considerably influenced by the outcomes of research in the United States.

In England and Wales, in the 1950s, medical students received 5 years of basic training, the last two of which were in a clinical setting. At the end they could choose to remain in hospital and begin a path to consultancy or enter general practice. Patients registered with a general practice or practitioner (GP) who became their first port of call in the event of illness. If necessary, the GP referred them to a consultant. In the 1960s, a Royal Commission on Medical Education recommended that all intending general practitioners should receive vocational training and programs were set up nationwide, and university departments of general practice were set up in some medical schools. A research program to investigate the assessment of trainee performance at four universities and three postgraduate centers was established at the University of Manchester's Department of General Practice by Professor P. S. Byrne. Just as Furneaux had indicated the need for reform of university examinations in engineering, so too had McGuire (1963) made a similar criticism of medical examining in the United States. Also, in the United States, the opening chapter of Miller's (1962) seminal work on teaching medical students was devoted to the problem of assessing medical students.

Byrne was joined by a psychologist James Freeman to conduct this study. A governing principle of Freeman's approach came from the field of aptitude and intelligence testing. Namely, to understand a student, you had to make as many measurements of different characteristics as possible in order to construct a psychometric profile of the student. It is a principle that is as relevant now as it was then. Furthermore, following Furneaux, Freeman and Byrne held that the primary purpose of assessment is formative, that is, to "monitor the student's progress so that the processes of teaching and learning can be matched to his needs at every stage" (Freeman and Byrne, 1976, p 4). They also accepted that there was a causal link between teaching and assessment so they attempted to develop a system of assessment procedures in conjunction with defined educational and professional aims and objectives directly related to the teaching methods used. Support for this approach was found in McGuire's (1967) report on evaluation in medical education in the United States. In this section, the concern is primarily (1) with the procedures used for deriving the specific objectives of the program and (2) with the range of tests used. The reader is referred to the second edition of their report, which contains a critical review of assessment procedures used in higher education at the time (Freeman and Byrne, 1976). The philosophy is contained in a statement of aims published by the Royal College of General Practitioners from which the operational philosophy of the department was drawn. The aims represent what was held to be the basic knowledge required by every future general practitioner. The areas were (1a) clinical knowledge, (1b) human development, (1c) human behavior, (2) society and medicine, (3) the practice, (4) research, and (5) continuing education. In terms of educating a person for professional practice in engineering, the parallels are profound. During the 60 or more years covered by this text, industry has continually complained that engineers are weak in areas that they commonly call "soft skills." But general practitioners are required to possess knowledge of human development and human behavior in order to be able to respond accordingly. There is no requirement for such understanding from engineers, yet much of their work involves working with other people. Although it is with the first three areas listed above that Freeman and Byrne's investigation was concerned, it will be seen that it also embraced an area of "attitudes" to which the College had drawn attention. They summarized these as attitudes toward (1) patients, (2) colleagues, and (3) special aspects of professional life, including special disciplines and techniques. Again, industry often comments on the poor attitudes of newly graduated engineers.

Freeman and Byrne (1976) turned the statements into behavioral objectives from which they devised a scheme of assessment. They achieved this by modifying nine criteria that had been developed by McGuire from a study of critical incidents applied to orthopedic surgery. The modified criteria distinguished between behaviors that were acceptable and those that were unacceptable. Two of the criteria are shown in Exhibit 3.11. Information gathering is chosen because it is a key skill that is generic and is often underrated in other academic studies. It is also a contributory skill in problem finding. Relationship (professional and collegial interaction) with colleagues is chosen because this is an area about which industry complains, and as Trevelyan (see Chapter 4.8) reports technical coordination is a key skill required by engineers (see Section 10.5). The nine criteria were (1) Information Gathering, (2) Problem solving, (3) Clinical Judgment, (4) Relationship to Patients, (5) Continuing Responsibility, (6) Emergency Care,

Criterion 1: Information Gathering

This criterion is concerned with the trainee's willingness, ability, and skill in gathering information necessary for diagnosis and/or decisions.

Behavioral objectives

The unacceptable trainee:	The acceptable trainee:
1. Follows no routine of history taking.	1. Takes a comprehensive history, when appropriate, including clinical, psychological, and social factors.
2. Fails to identify or does not follow salient leads.	2 Records information accurately.
3. Does not pursue alternative hypotheses.	3 Uses previous and continuing records intelligently.
4. Does not seek information on clinical, psychological, and social factors.	4 Plans investigations and uses diagnostic services intelligently.
5. Recording is sketchy and not systematic.	
6. Tends to use investigations in a "blunderbuss" fashion.	

Criterion 7: Relationship with Colleagues

This criterion is concerned with the trainee's ability to work effectively with his colleagues and members of the health team.

Behavioral objectives

The unacceptable trainee:	The acceptable trainee:
1. Has difficulty in personal relationships and lacks the ability to give and take instructions gracefully.	1. Gets on well with other people. Is conscious of the need for teamwork and fits in well as a member or, on occasion as a leader of a team.
2. Tends to be tactless or inconsiderate.	2. Seeks consultation when appropriate and respects the views of others.
3. Is unable to inspire the confidence or cooperation of those with whom he or she works.	3. Acknowledges the contributions of others.
4. Is unwilling to make referrals or seek consultation. Does not support colleagues in their contacts with patients.	4. Creates an atmosphere of "working with" not "working for" in other people. Demonstrates self-control

EXHIBIT 3.11. Two of the nine criteria for the assessment of performance in general practice developed by Freeman, J., and Byrne, P. (1976 edition). *The Assessment of Post-Graduate Training in General Practice.* Guildford, Society for Research into Higher Education. (Adapted and reproduced by permission of the authors). The assessors were asked to rate the candidate for each criteria on a 12-point scale, where 1–3 = poor; 4–6 = marginal; 7–8 = Good; and 10–12 = Excellent.

(7) Relationship With Colleagues, (8) Professional Values, and (9) Overall Competence Which Takes Into Account Criteria 1–8.

This scheme was adapted and extended by D. E. Murphy for use in schools (Murphy cited in Heywood, 1974). His scheme like Freeman and Byrne's included criteria within the affective domain, some of which were controversial. Scoring a scheme like this requires a lot of hard work and this is major impediment to adopting it with large numbers of students. The consultants and practitioners in the medical investigation would only be completing it for one or two students during a 3-year period. Freeman and Byrne developed a 15-scale rubric for information gathering as a result of their evaluation. It would have been interesting to know how a holistic judgment would have correlated with this type of reductionist data.

Pre- and posttest data were also obtained from a battery of tests. At entry to the course tests were administered for (A) Knowledge, (B) Skills, (C) Intelligence, Aptitude, and Personality,[5] and (D) Self-rating by students of their attitudes. The investigators also interviewed the students and conducted group discussions. Four consultants and two general practitioners rated (assessed) coursework during the 3-year period of postgraduate study.

Two approaches to testing knowledge and skill were used. These were multiple-choice questions and modified essay questions. The modified essay question is essentially a case history that requires short answers to a series of questions that indicate a candidate's skill in diagnosis at various stages of the case. Twelve such questions were undertaken in a period of 75 minutes. They are difficult to design. They are a reminder that "diagnosis" is an important skill in other areas of life. Yet in engineering, it scarcely gets a mention (see Section 5.1). Similar tests were conducted at the end of the course. The results showed a significant improvement in the posttest scores "particularly among the poorer students and particularly in the crucial area of problem solving skills. This was related to organizational structure of the courses, the poorer students gained greatest benefit from the courses which featured the regular and frequent use of seminars and group teaching methods" (p. 80). Many reports testify to the value of group work, particularly cooperative learning in the United States, yet it is still resisted by many faculty. Freeman and Byrne (1976) showed that when trainees rated at the top of the distribution were in the same training scheme as those at the lower end, the improvement in performance of those at the lower end of the scale was much greater than when this was not the case (p. 89). Remarkable changes were found among the poorer trainees in their personality profiles that were also remarked upon by the course organizers. Scores for "rigidity" and "authoritarianism" were greatly reduced and since high scores on these dimensions are known to impede problem solving, this goes some way to explaining the improvement in performance reported. Freeman and Byrne suggest that any system of remedial teaching that is introduced has to be comprehensive. It should be noted that the sample of trainees assessed was above the average caliber in intellectual ability of other groups of medical students. In the United Kingdom, medical students are selected from the top end of the intellectual ability range.

Engineering educators may be surprised to know that by 1983, a well-developed outcomes approach is to be found in medical education worldwide. A major Australian study for the World Health Organization details outcomes for the assessment of clinical competence and gives examples of a variety of assessment techniques. It remains a useful source of ideas (Fabb and Marshall, 1983).

3.6 Discussion

The engineering science examination that has been described was founded on the philosophy that the ways of thinking of engineers were different from the ways of thinking of scientists. Nevertheless, it was constrained by the public notion that engineering is the application of science. Therefore, if it was to be accepted by engineering departments in the universities for the selection of students, it would have to be seen as the equivalent of physics if it were to acquire status. The proponents attempted to persuade the university professors of engineering that they should accept the products of the subject in preference to those with physics. In spite of the fact that the professors complained that they could not get a sufficient number of students with high quality "A" levels in physics they were not, with one or two exceptions, prepared to look at alternatives that might have remediated this situation (Carter, Heywood, and Kelly, 1986). In consequence, although the subject ran for another 15 or so years, it eventually became an alternative to physics. It undoubtedly influenced the development of assessment techniques in physics syllabus B, and it was that model that is cited by the National Research Council in its 2000 report.

At the time the engineering professors supported the development of a nonexamined national "Project Technology" in schools in the belief that if substantial extra-curriculum project work was done in schools it would be sufficient to encourage students to become engineering students. "Project Technology" had an influential role in the development of engineering science through D. T. Kelly.

In the 1960s, whether or not design could be taught was hotly debated. This was the case in engineering science where the Board took the view that design could be learnt through project work and the planning and evaluation aspects assessed in a written examination without the need for a specific syllabus or instruction (Carter, Heywood, and Kelly, 1986). It seems that the written examination assessed something quite different from that which was being assessed in coursework. What that might be was the subject of some theoretical speculation.

A new approach to the assessment of course work was taken, which involved criterion (competency)-based assessments. They were not found to be easy to design. One lesson that continues to apply is the need to establish that faculty and students each believe that the same thing is being assessed (see also Section 6.3). The examiners had the prior knowledge that enabled them to develop the system of assessment and the curriculum. Similarly, much attention was paid to the design of questions for the written problem solving and analysis paper. However, they were not analyzed in the detail they should have been. The value of doing this is demonstrated by Lorimer and Davis (2014) in the longitudinal study referred to in Chapter 1 for question-by-question analysis showed that preengineering students were collectively competent in some areas (algebra) but lacked competence in others (e.g., trigonometry). They also showed up common misconceptions. One of the advantages of multiple-choice tests is that item analysis reveals this information instantly and recent developments in the design of concept inventories in engineering and science greatly facilitate teacher understanding of student needs as well as helping students understand their weaknesses.

In sum, the development of engineering science was impeded by the same factors in the British culture that were outlined in Chapter 2 and although it was offered for 20 years it did not acquire the status necessary for it to be attractive to large numbers of students and teachers. However, arguments about its philosophy have resurfaced during the last decade and the procedures and techniques of assessment it developed continue to be of interest since those for coursework correspond with the approach to Standards Based Assessment in the United States. It was probably the first course in the United Kingdom to state precisely what attitudes and interests a course in engineering science should foster but it did not evaluate whether or not they were developed as was the case with the evaluation of a postgraduate vocational training course for general practitioners. The medical case study illustrates the need for course evaluations to be multidimensional if an understanding is to be obtained of the factors that enhance and impede learning. Some engineering educators have recognized this need (e.g., Husted et al., 2013). The coursework assessment profile developed by Freeman and Byrne may be adapted for use in other subjects including engineering. Given the criticisms of industry about the quality of new graduates particularly as they relate to the so-called soft skills, (professional skills) the emphasis in the profile on inter- and intrapersonal skills is of value to those concerned with the development of these skills whether in college or in industry.[6]

No attempt was made to take a criterion-referenced approach to the written examinations in engineering science. In recent years, however, in the United Kingdom, examination boards with responsibility for setting assessments for the Higher National Diplomas in engineering subjects have developed strict criterion measures of the academic components of the courses. According to some teachers, problems arise because while the Examination Boards seem to advocate a multiple strategy approach to assessment the external assessors are perceived to restrict the assessments to assignments in which the students have to demonstrate that each criterion has been covered, irrespective of the quality of the work submitted. In their view, some creative students are penalized. This was the experience of teachers with the first criterion-referenced scheme developed for the assessment of coursework in engineering science.[7] Since ABET have recently complained about the lack of innovation and creativity in the implementation of EC 2000, is it possible that schemes of this kind cause a checklist mentality, or is the failure to innovate and be creative due to a lack of knowledge among assessors and teachers of the potential that assessments of this kind can offer? How many, for example, have read *Learning That Lasts*?

In contrast to engineering science where the objectives were derived from a model of what it was thought to be the work of professional engineers was, the objectives of the general practitioner training program and its assessment were based on a substantial knowledge of the work undertaken by GPs. The next chapter begins with a description of an attempt to analyze the work by done by engineers in an innovative firm with a view to developing a statement of objectives for training technicians and technologists.

Notes

1. In the 1960s and 1970s, the General Certificate of Education was set at two levels—Ordinary and Advanced or "O" and "A." These examinations acted as both leaving certificates and

examinations taken as an alternative to matriculation for entry to University. The minimum requirements for such entry were three "A" levels and one "O" level or, two "A" levels and three "O" levels. The "O" levels were taken in the age range 15–16 and the "A" levels 2 years later. From "O" level the student undertook 2 years in the "sixth form" and studied for "A" levels in three and sometimes four subjects. University entrance requirements dictated what subjects a student took. Engineering Departments would normally require good grades in mathematics and physics and one other science subject (commonly chemistry). This specialization and the high standard of the "O" level meant that "A" levels in science subjects were considered to be at a level equivalent to at least that of completing freshman students in the United States. Subsequently the "O" levels became the General Certificate of Secondary Education (GCSE) and were intended for the majority of the school population. A subject is a field of study—a particular department of art or science studied or taught (e.g., History, Physics, and Engineering Science). *The New Shorter Oxford English Dictionary).*

2. All the difficulties discussed in Section 1.2 about ambiguities in the language of assessment are present in these paragraphs. In papers published about the coursework assessment rubrics for engineering science the terms "criterion" and "semi-criterion" referenced were used, which gets over the difficulties in the use of competency, performance, and competence. As with assessment, it helps if the context in which they are used is understood.

3. The complete rubrics for both the experimental investigations and project will be found in Heywood, J. (1989) *Assessment in Higher Education,* 2nd edition. Chichester, UK: Wiley, pp. 260–263.

4. By *comprehensive* is meant that a variety of techniques would be used because no one form of assessment can serve all the purposes an examination is intended to serve. One dimension of *coherence* is that the "conceptual base or models of student learning underlying various external and classroom assessments within a system should be compatible" (Pellegrino et al., 2001). Another was that curriculum, instruction, and assessment should be aligned. Continuity requires that student performance should be measured over time "akin to a videotape record than to snapshots" (Pellegrino et al., 2001), Hence the importance the assessors attached to the keeping of a journal in the JMB course for Engineering Science.

5. The principle tests used were the AH5 test of high-grade intelligence. The Wechsler Adult Intelligence (Vocabulary Scale), Abstract-Concrete Performance Test giving relative scores for convergent and divergent thinking. Also included were adaptations of Guilford's tests of divergent/convergent thinking and measures of rigidity, personality and vocational interest blanks the items being recast from American tests for use in Britain. The testees were also interviewed from the point of view of their attitudes and opinions on various aspects of their training (Freeman and Byrne, 1976 pp. 33–46).

6. Recently some writers have begun to refer to the "soft-skills" as "professional skills." The latter is to be preferred because "soft-skills" implies a value judgment and infers they may not be as important as the hard skills of engineering science. The use of the term "professional skills" implies that they are equally important.

7. Through the courtesy of Professor John Sharp, I was able to attend a 2-day workshop with teachers on problems of assessing criterion-referenced schemes.

References

Abro, S., and Cuper, J. (2013). Constant course assessment model. In: Proceedings of Annual Conference of American Society for Engineering Education, June 2013. Paper 5886.

Angelo, T. A., and Cross, P. K. (1993). *Classroom Assessment Techniques.* San Francisco, CA: Jossey-Bass.

Bloom, B. S. (ed.) (1956). *The Taxonomy of Educational Objectives. Handbook 1. Cognitive Domain*. New York David Mackay. (1964) London: Longmans Green.

Blyth, W. A. L., Derricott, R., Elliot, G. F., Sumner, H. M., and A. Waplington (1973). *History, Geography and Social Studies 8–13*. Liverpool: Schools Council Project, School of Education, University of Liverpool.

Brown, A. O., Crawford, R. H., Jensen, D. D., Rencis, P. E., Liu, J., Watson, K. A., Jackson, R. S., Hackett, R. A., Schimpf, P. H., Chen, C.-C., Akasheh, F., Wood, J. J., Dunlop, B. U., and Sargent, E. R. (2013). Assessment of active learning modules: an update of research findings. In: Proceedings of Annual Conference of American Society for Engineering Education, June 2013. Paper 7462.

Carberry, A. R., Siniawski, M. T., and Dionisio, D. N. (2012). Standards based grading. Preliminary studies to quantify changes in affective and cognitive student behaviour. In: ASEEE/IEEE Proceedings of Frontiers in Education Conference, pp. 947–951.

Carter, G., Armour, D. G., Lee, L. S., and Sharples, R. (1980). Assessment of undergraduate electrical engineering laboratory studies. *Institution of Electrical Engineers Proceedings*, 127, A(7), 460–474.

Carter, G., Heywood, J., and Kelly, D. T. (1986). *A Case Study in Curriculum Assessment. GCE Engineering Science (Advanced)*. Manchester: Roundthorn Publishing.

Davis, D. C., LeBeau, J. E., Trevisan, M. S., Davis, H. P., Brown, S. A., and French, B. F. (2013). Growing assessment capacity of engineering educators through ASSESS. In: Proceedings of Annual Conference of American Society for Engineering Education, June 2013. Paper 5948.

Donald, J. G. (1982). Knowledge structures: methods for exploring course content. *Journal of Higher Education*, 54(10), 31–41.

Edels, H. (1968). Technology in the sixth-form. *Trends in Education*. No 10. April. London: Ministry of Education.

Edwards, R. M. (1984). A case study in the examination of systems analysis. *Assessment and Evaluation in Higher Education*, 9(1), 31–39.

Eggleston, J. (1977). *The Sociology of the School Curriculum*. London: Routledge.

Elliott, R. (2014). Do students like the flipped classroom? An investigation of student reaction to a flipped undergraduate IT course. In: ASEE/IEEE Proceedings of Frontiers in Education Conference, pp. 492–498.

Fabb, W. E., and Marshall, J. R. (1983). *The Assessment of Clinical Competence in General Family Practice*. Lancaster: MTP Press.

Farnsworth, C. R., Welch, R. W., McGinnis, N. J., and Wright, G. (2013). Bringing creativity into the lab environment. In: Proceedings of Annual Conference of American Society for Engineering Education, June 2013. Paper 6691.

Felder, R. M., and Brent, R. (2003). Designing and teaching course to satisfy the ABET engineering criteria. *Journal of Engineering Education*, 92(1), 7–25.

Fordyce, D. (1992). The nature of student learning in engineering education. *International Journal of Technology and Design Education*, 2(3), 23–40.

Freeman, J., and Byrne, P. S. (1976). *The Assessment of Post-Graduate Training in General Practice*. 2nd ed. Guilford. Society for Research into Higher Education.

Furneaux, W. D. (1962). The psychologist and the university. *Universities Quarterly*, 17, 33–47.

Goldman, S. L. (2004). Why we need a philosophy of engineering. A work in progress. *Interdisciplinary Science Review*, 29(2), 163–176.

Head, M. H., Owalbi, O. A., and James, P. A. (2013). Comparative assessment of student performance on exams when using on-line homework tools in an undergraduate engineering mechanics course. In: Proceedings of Annual Conference of American Society for Engineering Education, June 2013. Paper 7957.

Heywood, J. (1977). *Assessment in Higher Education*. Chichester, UK: John Wiley & Sons, Ltd.

Heywood, J. (2014). The evolution of a criterion referenced system of grading for engineering science coursework. In: ASEEE/IEEE Proceedings Frontiers in Education Conference, 1514–1519,

Heywood, J., Carter, G., and Kelly, D. T. (2007). Engineering science A level in the UK. A case study in the balanced assessment of student learning. Educational policies and educational scholarship. In: ASEE/IEEE Proceedings of Frontiers in Education Conference, S4F-9 to 13.

Heywood, J., and Kelly, D. T. (1973). The evaluation of course work—a study of engineering science among schools in England and Wales. In: ASEE/IEEE Proceedings of Frontiers in Education Conference, pp. 269–276.

Husted, S., Gutierrez, J. V., Ramirez-Corona, N., Lopez-Malo, A., and Palou, E. (2013). Multidimensional assessment of creativity in an introduction to engineering design course. In: Proceedings of Annual Conference of American Society for Engineering Education, June 2013. Paper 8789.

Intermediate Certificate Examination. (1975). *The ICE Report*. Report of the Committee on the Form and Function of the Intermediate Certificate Examination. Dublin: Government Publications.

Joint Matriculation Board. (1972). *GCE A Engineering Science. Notes for the Guidance of Schools*. Manchester: Joint Matriculation Board.

Laurillard, D. M. (1984). Learning from problem solving In: F. Marton, D. Hounsell, and N. J.Entwistle (eds.), *The Experience of Learning*. Edinburgh: Scottish Academic Press.

Lorimer, S. (2013). Concept inventories as predictors of changing pre-engineering skills. In: Proceedings of Annual Conference of American Society for Engineering Education, June 2013. Paper 6215.

Lorimer, S., and Davis, J. A. (2014). Consistency in assessment of pre-engineering skills. In: Proceedings of Annual Conference of American Society for Engineering Education, June 2014. Paper 9596.

McCormack, J. (2011). Assessing professional skill development. *International Journal of Engineering Education*, 27(6), 1308–1323.

McDonald, F. J. (1968). *Educational Psychology*. Belmont, CA: Wadsworth.

McGuire, C. H. (1963). A process approach to the construction of medical examinations. *Journal of Medical Education*, 38, 556.

McGuire, C. H. (1967). *An Evaluation Model for Professional Education: Medical Education*. Chicago, IL: Chicago Medical School.

Mentkowski, M., and Associates (2000). *Learning That Lasts. Integrating Learning, Development and Performance in College and Beyond*. San Francisco, CA: Jossey-Bass.

Miller, G. E. (1962). *Teaching and Learning in Medical School*. Cambridge, MA: Harvard University Press.

Miskioghi, E. E., and Wood, D. W. (2014). That's not my style. Understanding the correlation of learning style, preference, self-efficacy and student performance in an introductory chemical engineering course. In: ASEE/IEEE Proceedings of Frontiers in Education Conference, 1988–1995.

Murphy, D. E. (1975). Experimental Pupil Evaluation Form for Teachers. Appendix A. In: J.Heywood (ed.), *Assessment in History: Twelve to Fifteen*. Dublin: Public examinations Evaluation Project, School of Education, University of Dublin.

Narayanan, M. (2013). Assessment of learning based on the principles of learning and meta-cognition. In: Proceedings of Annual Conference of American Society for Engineering Education, June 2013. Paper 6562.

Pellegrino, J. W., Chudowsky, N., and Glaser, R. (eds.) (2001). *Knowing What Students Know? The Science and Design of Educational Assessment*. National Research Council. Washington, DC: National Academies Press.

Prince, M. J., Vigeant, M. A., and Nottis, K. E. K. (2013). Assessment and repair of critical misconceptions in engineering heat transfer and thermodynamics. In: Proceedings of Annual Conference of American Society for Engineering Education, June 2013. Paper 6584.

Sadlo, G. (1997). Problem based learning enhances the educational experiences of occupational therapy students. *Education for Health*, 10(10), 101–114.

Saupé, J. (1961). Learning In: P. L.Dressel (ed.), *Evaluation in Higher Education*. Boston, MA: Houghton Mifflin.

Shelton, J. B., and Smith, R. F. (1998). Problem based learning in analytical science undergraduate teaching. *Research in Science and Technological Education*, 16(1), 19–31.

Sternberg, R. S. (1985). *Beyond IQ. A Triarchic Theory of Intelligence*. Cambridge University Press.

Stice, J. E. (1976). A first step toward teaching. *Engineering Education*, 67, 394–398.

Stickel, M., Hari, S., and Liu, Q. (2014). The effect of the inverted classroom teaching approach on student/faculty interaction and students' self-efficacy. In: Proceedings of Annual Conference of American Society for Engineering Education, June 2014. Paper 10492.

Streveler, R. A., Brown, S., Herman, G. L., and Montfort, D. (2014). Conceptual change and misconceptions in engineering education: curriculum, measurement, and theory focused approaches. In: A.Johri and B. M.Olds (eds.), *Cambridge Handbook of Engineering Education Research*. New York, NY: Cambridge University Press.

Streveler, R. A., Litzinger, T. A., Miller, R. L., and Steif, P. S. (2008). Learning conceptual knowledge in engineering. Overview and future research directions, *Journal of Engineering Education*, 97(3), 279–294.

Streveler, R. A., Smith, K. A., and Pilotte, M. (2012). Aligning course content, assessment and delivery. Creating context for outcomes based education In: K. M.Yusof et al. (eds.), *Outcomes-Based Science, Technology, Engineering and Mathematics Innovative Practice*. Hersey, PA: IGI Global.

Thomas, R., and Izatt, J. (2003). A taxonomy of engineering design tasks and its applicability to university engineering education. *European Journal of Engineering Education*, 28(4), 535–547.

Tyler, R. W. (1949). Achievement testing and curriculum construction. In: E. G.Williamson (ed.), *Trends in Student Personnel Work*. Minneapolis: University of Minnesota.

van de Pol, I., and Goldberg, D. E. (eds.) (2011). *Philosophy and Engineering. An Emerging Agenda*. Dordrecht, the Netherlands: Springer.

Vu, N. V., and Black, R. (1997). Problem analysis questions for assessment in problem based learning. Development and difficulties. *Education for Health*, 10(1), 79–89.

Widmann, J. M., Self, B. P., and Prince, M. J. (2014). Development and assessment of inquiry based learning: a case study in identifying source and repairing student misperceptions. In: Proceedings of Annual Conference of American Society for Engineering Education, June 2014. Paper 9969.

4

Categorizing the Work Done by Engineers: Implications for Assessment and Training

Given the integrated nature of the curriculum process, it is inevitable that learning will be driven by assessment. What is assessed is determined by objectives that are, in turn, interpretations of what the aims of education are perceived to be. There are many sources of objectives: the key concepts of the curriculum; the ways in which we learn and develop, and lists or classifications of objectives such as The Taxonomy of Educational Objectives, as well as the literature of the subject that is taught.

The objectives derived for new curricular in engineering (Chapters 2 and 3) were criticized because they were based on models (beliefs) of what engineers did rather than actual studies of what they do. Three approaches to the study of what engineers do in practice are described. In the first, on the basis of a task analysis, objectives for training engineers and technicians were classified by operational groupings and work types. The study was unique in that it took into account the attitudes of the respondents, and the techniques used were able to show strengths and weaknesses in organizational structure as well as point to training needs. They were generalizable and suitable for use by any organization. The second approach was undertaken by W. Humble, who analyzed the tasks of middle managers in a steel-manufacturing company by reclassifying them in terms of the two domains of the Taxonomy of Educational Objectives. He highlighted the importance of the affective domain in understanding work. The third investigation

The Assessment of Learning in Engineering Education: Practice and Policy, First Edition. John Heywood.
© 2016 The Institute of Electrical and Electronics Engineers, Inc. Published 2016 by John Wiley & Sons, Inc.

reported a comparative study of the work done by project engineers in Germany and the United Kingdom. It highlighted the fact that while overall the jobs done had coherence they derived from many activities that gave the appearance of fragmentation.

Taken together, these studies showed engineering to be a complex activity that the objectives of assessment should reflect. But how far should the preparation of skill development in areas of expertise other than engineering science be the province of the university, and to what extent should it be a shared enterprise with industry? The concept of a labor arena is advanced. Given changing patterns of work and lengths of employment, if employees are to widen the span of job search which is likely to be the case, employers will not be able to escape the responsibility for proper performance-based assessments that can contribute to an employee's profile and future career. Clearly there is a role for a well-thought-out portfolio approach to assessment in which continuing professional development becomes a normal part of one's career. In these circumstances, the ability to self-assess becomes important. Changes in the structure of the workforce may dictate a change in the structure of higher education and the way students are credentialed.

4.1 Introduction

The engineering science project was criticized because it was based on models held by its academic sponsors of what it was thought an engineer did and not on what was actually accomplished. In spite of the fact that the 1966 Lancaster project was a semi-Delphi derived development from a group whose members came primarily from industry, the same criticism was also leveled at that project. The same criticism may be made of Meuwese's (1969) attempt to derive objectives from among the academics of the industrial engineering department at Eindhoven Technological University. No debate ensued about these criticisms that came from an engineer who was also Personnel and Training Director of Lucas AeroSpace Albert Hirst, since he did not publicize them. He also criticized reports by the Engineering Industries Training Board (EITB, 1968a, 1968b) and the Council of Engineering Institutions (1968a, 1968b). Nevertheless, one of the participants in both the Lancaster and Engineering Science projects, and the Department of Employment, was sufficiently impressed to sponsor a project in which every person defined as an engineer in Lucas Aerospace would be task analyzed (subject to their permission) with a view to determining the objectives of training technologists and technicians (Youngman et al., 1978).

The assumption was that an engineering department had an obligation to prepare students for industry although it was not assumed, as seems to be the case with some present-day employers, that university programs would make their products immediately employable. It is an issue of current concern.

However, in the 1950s and early 1960s, large employers in the United Kingdom continued to provide training programs for graduates and in some cases supported sandwich (cooperative) courses (Heywood, 1969). In the absence of any thorough ongoing discussion of the aims of higher education, a utilitarian model was adopted and industry took the view, as it does now, that it had a right to comment on university programs and expect a response (see Section 5.6). The cooperative attitude engendered by the German dual system of education and vocational training was not present. Nevertheless, given the requirements for chartered membership of the professional institutions for experience of practical work in employment, in principle if not in practice, that experience should be rigorously assessed. A taxonomy or classification of training objectives for technologists and technicians would therefore be valuable. It was assumed that this study would be specific to the type of engineering undertaken in the organization but that the methods developed would be generalizable to other engineering activities.

4.2   A Study of Engineers at Work in a Firm in the Aircraft Industry

D. L. Marples, an engineering educator at Cambridge, reported in a private communication to Monk (Gregory and Monk, 1972, p. 66) "that an analysis of the roles filled in the daily work of a project design engineer has suggested that he has to undertake activities as an individual, as a designer, as a manager, and as a service provider." Monk concluded that "what this leads to is the conclusion that a professional engineer, through his need to fulfill many roles in his job, has to be able to deploy problem solving skills which cover the range of requirements." He found support for this thesis in the work of Turner (1969). At the time that he wrote, Monk had just joined Heywood, Oxtoby, and Youngman to participate in their study of engineers at work, which would inevitably determine the extent to which his thesis held in this particular organization. The purpose was not only to describe the work actually done but also to account for key factors that contributed to satisfactory performance on the job, such as attitudes to work and self, and significantly, organizational structure because the project was about structuring work for training purposes, that is, the development of competence at work (Youngman, 1975). The studies reported in Chapter 2 by Barnes (1960) and Burns and Stalker (1961) had shown the importance of the impact of organizational structure on performance and the values, attitudes, and beliefs that employees bring to their work. However, no attempt would be made to measure the competence of the individuals contributing to the enquiry: rather, the organization with this methodology would be able to analyze the jobs for the knowledge, skills, and attributes required and then provide tailored training for particular jobs, or bring suitably qualified personnel to fill those positions. Clearly, if competences are expected to predict potential performance, knowledge of this kind is essential for the design of the curriculum and the assessment of student learning.

Just prior to the study, Meuwese (1969) had asked members of the Department of Industrial Engineering at Eindhoven Technological University to rate some 300 objectives stated in *Taxonomy* like terms for their importance. Using factorial analysis and a type of cluster analysis, Meuwese derived six factors (groupings of objectives). These were objectives related to (1) the social system, (2) machine shop technology, (3) systems analysis, (4) critical analysis and synthesis in industrial situations, (5) organization and planning, and (6) management of mechanical systems. He followed this up with a similar study in mechanical engineering and, together with faculty, designed a course of study, each unit of which consisted of stated objectives, a list of references to specific pages in books that could be used, supplementary texts, a series of study questions, and six diagnostic multiple-choice tests of approximately 12 items each (Meuwese, 1971).

Since Meuwese's objectives were generated by faculty, the study was open to Hirst's criticism that they were not based on what actually happened in industry. Moreover, some of the objectives were very broad and not highly focused. If, however, engineers in industry could be persuaded to develop objectives based on their work, the criticism would disappear. The task analysis carried out in Lucas Aerospace did not go that far but went some way to achieve that goal.

In the pilot study, a series of 39 interviews of engineers' representative of the engineering functions in the organization information was obtained using a modified personal construct approach during the interview (Bannister and Mair, 1968; Kelly, 1955) about what the engineers actually did in comparison with those with whom they worked (Youngman et al., 1978). Each engineer determined the content of his interview: to a large extent, no two interviews resulted in the same information. The result was the generation of a 434-item checklist that was administered to 208 persons in engineering functions.[1] It is of interest in the context of this text to note that the activities described in the checklist were at first called "abilities" but later "operations" (see Section 5.5).

The cluster analysis produced 14 groupings of engineering activities (Youngman, 1975) that, given Anastasi's later expressed view of the nature of ability, might well have continued to have been called abilities, but they were not and neither were they intended to be a taxonomy (see Section 5.5). Their titles are given in Exhibit 4.1 together with a more detailed description of one of the activities.[2] They might equally be called competencies, a competence in this case being the ability to perform one of these tasks. They do give clear guidance on training needs that is accentuated when they are related to particular individuals. For this reason since the organizational structure was known person by person, it was also possible to see if there were groupings of engineers using the same operations and to examine the extent of organizational fit. The second analysis generated 12 work types (see Exhibit 4.2). A comparison of the work types with the job functions listed by Woodson (1966), an American engineering educator, in his widely used textbook on engineering design revealed little correspondence between the two classifications. However, while this might support Hirst's argument, it might also have been a function of American usage.

It was understood that job descriptions that rely on a limited range of worker behaviors are barely adequate in most circumstances, a point that is reinforced by Barnes (see Chapter 2, 1960). For example, the way in which individuals of differing personalities interact with organization is an important factor in the work that they do and the competence they develop (see Section 3.5 and Chapter 11). For this reason, the investigation evaluated the merits of determining the attitudes that these engineers had to their work and themselves by the use of two semantic differentials (Oxtoby, 1973). These instruments were administered in face-to-face interviews during which additional questions were asked about attitudes to training (Youngman et al., 1978). Apart from the more obvious point that such measures provide information about job satisfaction, they can be important in vocational guidance and professional development. Of significance

The 14 Engineering Activities
1. Quality monitoring
2. Customer liaison
3. Testing
4. Organize materials and methods
5. Contract supervision
6. Drafting
7. Project supervision
8. Design
9. Production scheduling
10. Facilitate manufacturing
11. Organize testing
12. Produce specifications
13. Organize information
14. Long-term planning

Example of an Activity—Activity 10 Facilitate manufacture.
Nucleus operations
Examine design schemes for manufacturing problems
Identify possible production difficulties
Examine designs for possible assembly difficulties
Examine manufacturing implications of a new technique
Anticipate possible side effects of machining processes
Consult design engineers regarding alterations for the manufacturer
Recommend design alternatives

EXHIBIT 4.1. The engineering activity groupings together with the nucleus operations for one activity from Youngman, M. B., Oxtoby, R., Monk, J. D, and Heywood, J. (1978). *Analysing Jobs*. Aldershot, England: Gower Press.

Management and liaison   Specification development   Project engineering
Draftsmen   Methods and planning   Design
Technicians   Quality support
Service functions   Manufacture and production   Standards and materials   Contracts and sales

EXHIBIT 4.2. The 12 work types shown in the four broad groupings that emerged from the cluster analysis from Youngman, M. B., Oxtoby, R., Monk, J. D, and Heywood, J. (1978). *Analysing Jobs*. Aldershot, England: Gower Press.

to this discussion is the finding that age and job level were more significant variables than educational qualifications in the explanation of differences in job descriptions.

One finding that may continue to be the case, given recent reports of unemployment among middle-aged engineers, arises from the attempt to test Hesseling's (1966) view that "specialism (experience) fosters autistic tendencies because one tends to define each situation as fitting in one's own schemata." The analysis tended to support this view that as engineers grow older they tend to place increasing reliance on experience and reject the notion that training can be beneficial. Excessive reliance on experience may prevent an individual from looking outside of the box and may, in the end, be destructive of innovation.

The general case has been argued elsewhere (Heywood, 1989) and, if correct, has implications for the credentialing of performance/competence, since to justify what a person is good at might be to support his or her autistic tendencies. It is a major reason for continuing professional development and associated assessment (see Chapter 12). It also has implications for the design of experiential education.

Of significance is the fact that no traditional manager work type emerged, nor do the so-called management abilities appear in any recognizable form within the 14 activities. Broader generalizations inherent in the function structure did not suggest a general management role that seems to suggest that some definitions of management could be unrealistic. Although one of the work types is termed "Management and Liaison," not every person within this category was a manager *per se* or had senior status. They undertook a variety of investigatory and product support roles. One member of the team followed this up with an illuminative evaluation of the data and came to the conclusion that more or less everyone in the organization exhibited direction and control in one way or another and was, therefore, managing to a greater or lesser degree in order that he or she could take the actions required of them (Heywood, 1976). "[…] it seemed that persons were appointed to roles which they have to change in order to communicate. The organization was rather more a system of persons in relations than a hierarchical structure. It is in such situations that feelings of responsibility are acquired" (Youngman et al., 1978). This feature of the organization illustrates the significance of the firms "informal organization." In that period, in firms such as this design and designers had lowly status but Monk (1972) noted that "it was not so much status that was sought by the designers but responsibility […] it is almost as if they have to justify themselves that they are doing something worthwhile by measuring it in terms of responsibility."

Everyone had to cope. Thirty or more years later, Mintzberg (2009) citing Charles Handy (1994) wrote "Manage did mean 'coping with' until we purloined the words to mean planning and control," or as in the *Pocket Oxford Dictionary* "direction and control." Mintzberg also cited Chester Barnard (1938), who said that the function of the executive is "to reconcile conflicting forces, instincts, interests, conditions, positions and ideals" which is one that requires considerable competence! Moreover, in terms of the personal (self), it is as much about the utilization of the affective domain as it is the practical. Heywood (1976) also tried to organize the elements that made up one of the activity groups into Bloom like educational categories and showed they would be different to *The Taxonomy*. He demonstrated categories of application, communication,

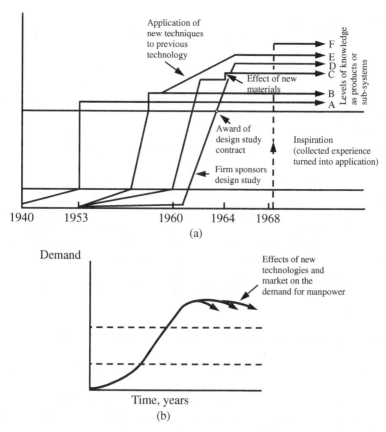

EXHIBIT 4.3. (a) The pattern of innovation in the firm and (b) Illustration of the rate of change of demand for manpower as a function of the learning/innovation curve.

diagnosis, evaluation, and management (direction and control). They were not hierarchically organized, but as he had shown elsewhere, application of the Bloom *Taxonomy* to the engineering curriculum places severe restraints on what is likely to be achieved in terms of engineering practice, a point that has been reiterated recently by McCahan and Romkey (2014).

Finally, he was able to show from the organization's history that it had demonstrated all the characteristics of a learning organization when learning is defined as a process of problem finding and problem solving (Exhibit 4.3; Heywood, 1974).[3]

Changes in the recruitment policy of the organization that reflected the increasing stratification of the education system impacted on the negative attitudes of draftsmen and technicians who saw their career route to the assistant engineer level cutoff. In this respect, a case study reported in Section 11.5 that shows how in France some technicians were helped to gain technologist status is of considerable interest.

4.3   The Application of *The Taxonomy of Educational Objectives* to the Task Analysis of Managers in a Steel Plant

Notwithstanding Heywood's criticisms of *The Taxonomy*, Bill Humble, a senior training officer with the British Steel Corporation, showed that it was possible to analyze the role of a unit manager in a steel plant using both the cognitive domain and affective domains (Krathwohl, Bloom, and Mersia, 1964). Humble's analysis of one of the objectives is shown in Exhibit 4.4. In the original there were eight different objectives and he also suggested which of the domains of the affective taxonomy applied (Exhibit 4.5; and Heywood, 1970).

It is evident that the manager required communicative, instructional, and negotiating skills in large measure. Humble argued that such managers required abilities to adapt, control, and deal with people. He drew up a taxonomy for these domains (Exhibit 4.6). Humble's work was supported by evaluations of the in-company training programs for which he was responsible. One of these brought together a group of participants with this writer for a week for purpose of reflecting on the training they had received. The skill that these managers thought was most important was "communication." To this day, industrialists complain that new graduates lack skill in communication.

4.4   The Significance of Interpersonal Competence

At about the same time as the Youngman study, the EITB deduced the classification of knowledge and skill areas shown in Exhibit 4.7, from a survey of some 1095 mostly high-level technicians. Later from a study of 1295 professional engineers and scientists, they obtained the average work load (in time) spent on a number of different activities (see Exhibit 4.7B and also Exhibits 4.8 and 4.9 given later).

From the perspective of competence, perhaps the most important finding relates to interpersonal competence for on average 60% of an engineer's time involved him in interpersonal transactions. That this is a characteristic of behavior in organizations employing engineers is shown by Trevelyan (2010) who reported that not only senior engineers but novice engineers working on technical things were engaged in some form of social interaction for 60% of their time. Notwithstanding problems in the naming of categories in Exhibits 4.1 and 4.2, whichever way they are interpreted, interpersonal contact is large, as is the need to demonstrate skills frequently associated with management.

While the investigation was being conducted, arising from the cancellation by the government of projects in the aircraft industry (not in this firm) two sociologists Thomas and Madigan (1974) tracked what had happened to those who had been made redundant in order to establish what employment they had obtained. This highlighted the difficulties that individuals had in demonstrating to future employers that they had transferable skills (if they were aware of them). Thomas and Madigan suggested that a "theory of Labour Arenas which reflects the 'political' nature of job search might provide a more adequate basis for the analysis of job search and job change." Youngman et al. (1978) used this concept to describe an *arena* as a group of skills that is already

Objective: to plan and maintain work schedules to secure the required production of goods and services						
Activities	Knowledge	Comprehension	Application	Analysis	Synthesis	Evaluation
Planning for and causing the required quantity and output to be maintained	Targets, tolerances, customer preferences etc.			Ability to recognize when plan is not being met. Skill in causes of disruption		
Assigning employees to meet work schedules	Schedules, policies methods, limits of authority	Skill in interpreting and translating policies and producing outcomes	Skill in predicting probable effect of changes			
Obtaining and/or checking the availability of the necessary materials, tools, machines and services in accordance with policies and procedures	Procedures, sources, policies					
The proper care and use of materials, tools, machines and equipment within his unit	Familiarity with equipment			Ability to instruct in proper use and maintenance of equipment		
Recommending and controlling overtime	Criteria, agreements, limits of authority	Skill in interpreting and translating agreements and predicting outcomes				
Providing adequate materials and tools to meet the work program of the following shift	Work program. Sources of tools and materials, Methods					
Recording status of work and general conditions at the end of each shift	Criteria	Skill in communicating				
Checking performance against standards and schedules and taking corrective action	Criteria, standards, schedules	Skill in interpreting standards		Identifying causes of nonachievement	Ability to take corrective action	

EXHIBIT 4.4. Example from one of eight schedules that role analyzed the work of a unit manager in a steel plant due to W. Humble, Llangattock Training Centre, British Steel Corporation (reproduced in Heywood, J. (1970). Qualities and their assessment in the education of technologists. *International Bulletin of Mechanical Engineering Education*, 9, 15–29).

Principal categories of the Taxonomy of Educational Objectives for the Affective domain		
1. **Receiving attending** 1.1. Awareness. 1.2. Willingness to receive. 1.3. Controlled or selected attention	**Awareness**	Listens to others, receives others as coworkers, and listens to advice
		Verbally pays attention to alternative points of view on a given issue.
		Refers to subgroups (social, intellectual, sex, race, etc.)
		Acknowledges aesthetic factors. Aware of feelings of others. Recognizes own bias as bias. Recognizes other bias as bias.
	Willingness to receive	Seeks agreement from another. Seeks responsibility. Seeks information from another. Pursues another way of doing something. Seeks materials. Asks another to examine aesthetic factors. Inquires how another felt about an event or subject.
2. **Responding** 1.1. Acquiescence in responding 2.2. Willingness to respond	**Acquiescence**	Complies with existing regulations (rules). Complies to a suggestion or directive. Offers materials on request. Gives opinion when requested.Responds to a question. Takes responsibility when offered. Remains passive when a response is indicated. Actively rejects directions (or suggestions).
2. **Valuing** 2.1 Acceptance of a value 1.2. Preference for a value 1.3. Commitment (conviction)	**Preference**	Seeks the value of another. Defends the value of another. Clearly Openly defends the right of another to possess a value. Tries to convince another to accept a value. Agrees with value of another. Disagrees with the value of another.
2. **Organization** 2.2. Conceptualization of value 2.3. Organization of a value system	**Characterization**	Makes deductions from abstractions. Makes judgments (implies compares own value to that of another). Attempts to identify the characteristics of a value or value system.
3. **Characterization by a value or value complex** 3.2. Generalized set 3.3. Characterization		Revises judgments based on evidence. Bases judgments on consideration of more than one proposal. Makes judgment in the light of situational context.

EXHIBIT 4.5. The principal categories of the *Taxonomy of Educational Objectives for the Affective Domain* (Krathwohl, Bloom, and Mersia, 1964). The illustrations of some of the categories are due to Leonard Kaplan, who developed the Taxonomy for use in schools. (Kaplan, L., (1978). *Developing Objectives in the Affective Domain.* San Diego, CA: Collegiate Publishing, also reproduced in Heywood, 2000).

The ability to adapt involves:

The ability to perceive

1. That organizational structure and formal/informal relationships, value systems and language, and therefore, its needs, knowledge of the technical, and human and financial aspects of the system or situation
2. The different thought processes involved in the solution of human or technological problems
3. Our own (self) attitudes and needs

The ability to control involves:

1. Knowledge of

   (i) How the skills of those who have to be controlled should be used.
   (ii) His or her requirements in relation to needs for communication, competence, and excellence.
   (iii) What people ought to be doing.
   (iv) Whether or not they are doing it effectively.
   (v) How to create a climate in which jobs will be done effectively.

2. The ability to make things happen.
3. The ability to discriminate between relevant and irrelevant information etc.

The ability to relate with people involves

1. Knowledge of rights, responsibilities, and obligations.
2. Knowledge of the ways of thinking (determinants of attitudes and values) of people in all parts of the organization.
3. Ability to understand when action in the key environment is right and acceptable in those circumstances (i.e., to understand the effect of his or her behavior on a situation).
4. Ability to be able to predict the effects of his or her behavior and that of others on a situation.
5. Ability to create the feeling that the job is important, etc.

EXHIBIT 4.6. W. Humble's partial derivation of a taxonomy of industrial objectives derived from a typical work situation in which managers and workmen were in confrontation to some degree (Heywood, J. (1970). Qualities and their assessment in the education of technologists. *International Bulletin of Mechanical Engineering Education, 9*, 15–29).

possessed or that may be readily acquired that crosses the divide of job perceptions derived from job titles. They argued that it would be relatively easy for the employer or employee to check whether or not they can cope with the operations in other jobs by comparing the nucleus operations necessary for the performance of their present job as derived from task analyses, with those presented by the job they are seeking. Such an approach avoids judgments made on the basis of a job title that may well be flawed. Given changing patterns of work and lengths of employment, this idea assumes some importance if employees are to widen the span of job search, and employers cannot escape the responsibility for proper performance-based assessments that can contribute to an employee's profile and future career (see Chapter 12). Clearly there is a role for a well-thought-out career-based portfolio approach to assessment.

A. EITB Technician Survey. Knowledge/skill categories (EITB, 1970)	B. EITB Survey of professional engineers and scientists. Work distribution: average time spent on different activities (%) (Venning, 1975)	
1. Communication for technical information (including drawing) 2. Design (of components, circuits, tests, etc.) 3. Organizing work (planning and control of work) 4. Diagnosis (measuring, inspecting, fault finding, etc.) 5. Practical work (servicing, maintenance, assembly, manufacturing, operation) 6. Supervision of other staff	Written communication Verbal communication Technical activities (including design) Management services Organization of production Commercial activities and finance Supervisory duties Other activities	25% 27% 24% 4% 5% 1% 7% 7%

EXHIBIT 4.7. Data from surveys of the Engineering Industries Training Board (EITB, Watford, UK) reported in 1970 in *The Technician in Engineering*, EITB Research Report No. 1 and Venning, M. (1975). *Professional Engineers, Scientists and Technologists in the Engineering Industry*. EITB Research Report No. 6.

Type of Activity	German Production Managers (N = 16), % of Managers' Time Spent on:	British Production Managers (N = 25), % of Managers' Time Spent on:
Formal scheduled recurrent meetings	9.78	15.50
Convened special purpose meetings	12.62	14.46
Ad hoc discussions	20.07	17.93
Tours of works	16.87	17.35
Telephoning	10.66	7.23
Desk work	11.56	11.16
Explanations to observer	10.45	12.08
Total time accounted for under these headings (%)	91.98	95.92

EXHIBIT 4.8. Activities that accounted for the time worked by production managers in Germany and Britain. (Hutton, S. P, and Lawrence, P. (1982) *The Work of Production Managers: Case Studies at Manufacturing Companies in the United Kingdom*. Southampton: University of Southampton (p. 121). Study carried out between 1975 and 1977 for the UK Department of Industry. See also Hutton, S. P., and Lawrence, P. A. (1981). *German Engineers: The Anatomy of a Profession*. Oxford University Press.

Activity	
Supervision/management	26.3
Conferences	14.0
Routine technical work	12.5
Nonroutine technical work	10.9
Report writing	10.3
Nonroutine designing	8.6
Routine designing	3.7
Drafting	3.3
Personal matters	2.0
Teaching	1.8
Miscellaneous nontechnical skills	1.5
Routine laboratory work	0.9
Data searching	0.5
Unclassified	3.7
	100.0

EXHIBIT 4.9. Percentage of engineer's total time devoted to various activities. Survey of engineers who attended engineering management courses at the English Electric Staff Training Centre 1960–1966 (number not given). Turner comments on this group of engineers in mid-course positions that the quantity of routine technical work might be due to insufficient technical support. That said, it is possible to infer that there was significant human interaction (Turner, B. T. (1969). *Management Training for Engineers*. London: Business Books, p. 25).

4.5 A Comparative Study of British and German Production Engineers (Managers)

During the 1970s, a major study of the recruitment, employment, and status of the mechanical engineer in the German Federal Republic was carried out for the UK Department of Industry by Hutton, Lawrence, and Smith (1977)[4] One of the things that they did was to repeat a slightly adapted version of Hutton and Gerstl's (1965) survey of British mechanical engineers among a 1000 German engineers, making it a truly comparative study (Hutton and Lawrence, 1981).

Among the findings of the comparative study of production managers of relevance to this discussion was that German "companies are generally satisfied with the engineers produced by the educational system, except that they are critical of recent changes de-emphasizing practical experience, in the way in which students are recruited to the Ing.Grad course." They used questionnaires to establish the training and experience of production engineers in both countries together with observations of production engineers at work (Hutton and Lawrence, 1979). They give a detailed account of the daily work of 16 German and 25 British production managers, which is summarized in

Exhibit 4.8. There is not much difference between the two groups except that German managers attend far fewer formal scheduled meetings than British managers. There is also a difference in the level of usage of the telephone. One wonders if the advent of the mobile (cell phone) has changed that situation in either country!

They found, as did Turner (Exhibit 4.9), and others who had made similar studies that the engineering manager's working day is fragmented although Hutton and Lawrence thought the brevity and fragmentation, part of a commonly held thesis, were exaggerated. "Most of the managers chaired or attended formal meetings with only occasional interruptions, and all of them were able to conduct (frequently protracted) man to man discussions with colleagues. Again making a tour of the works, daily is regarded by the managers themselves as desirable and there was one instance in which this did not happen. Further to this point of self-control of time and action we have already shown that the bulk of telephoning was outgoing, and thus largely a manifestation of the manager's own initiative" (p. 67). It is evident from Hutton and Lawrence's study that these managers had a high level of interpersonal contact. This may also be inferred from Turner's study, which included engineers from all branches of the company from which they came. Clearly, some of the skills required may not be accomplished effectively without some training, as for example chairing a meeting. Probably, the most difficult thing for many managers is how to deal with critical incidents (p. 79). Perhaps the most important comments relate to the predilection for *Technik* among the German engineers… "All the non-commercial functions and departments… are part of Technik. And so are all ranks and grades on this side of the firm from the skilled worker to the most senior technical manager" (p. 74).[5] It is part of the cultural bias in Germany.

As indicated, a similar study of production managers was undertaken in Britain (Hutton and Lawrence, 1982). Once again the case studies showed what these managers did during the day. They wrote that "the pace was fast in most cases, with a sense of a lot of things happening quickly, often under pressure, and this holds more for the British sample than for the German." In Britain, there was some friction between production, and other departments—engineering, quality control, design, maintenance, purchasing, sales […]. The source of much of this frustration was the availability, for one reason or another of parts and materials. The German firms "never seemed to run out of anything." But the authors commended the British managers for their "general resourcefulness, their ability to get things done, to find ways out, to adapt, to work out spot alternatives, and to generally win through in spite of circumstantial difficulties" (p. 120). In this respect, unlike Youngman et al. (1978), who had some evidence of a relationship between status and morale, despite the perceived lower status of the British production managers, there was no evidence of low morale among the participants in their case studies.

Hutton and Lawrence's study showed that Germans emphasized the technical whereas their British counterparts emphasized the managerial. The German managers were also better qualified technically than the British; moreover, the Germans seemed to have better relations with persons in other technical functions, and "they were more likely to have patents to their (personal) credit, more likely to make direct contributions to the production technology or to experiment with machinery layouts" (p. 123). In Britain, "people tell you that management and human problems is what it is all about, not engineering, methods and product refinement" […] and there is a particular emphasis on man management.

Just as in the Youngman et al. study, the persons investigated by Hutton and Lawrence had a range of qualifications ranging from having served an apprenticeship to one with a doctorate. In both samples, there was an element of heterogeneity. Similar enquiries conducted today would come up with a very different pattern since the changes in the 1960s to raise the level of qualification of engineers will by now be embodied in the system. At the same time, it is highly unlikely that the value systems of the two countries production management systems will have changed. The cultures in which the engineering systems are embedded are very different and highlight the need for competencies that enable mobile engineers to understand new cultural requirements very quickly. High priority will be still given to interpersonal relationships in both systems.

It is clear from the two studies that engineering is something much more than the application of science. It is less clear if either study contributes to our understanding of what has come to be called engineering knowledge. It is doubtful if Youngman et al. had such a concept in mind yet it is clear that the performance of professional tasks depends on professional knowledge often, as many authorities have shown, acquired tacitly. Youngman et al. established what engineers did, not how they did it. While there have been one or two major studies of what engineers know and how they know it, to quote the title of Vincenti's (1990) seminal study of aeronautical engineers at work, we need to unify these two dimensions so that it provides a focus that differentiates engineering from science on the one hand, and on the other hand places it in its economic, human, social (cultural), and psychological contexts. With that knowledge, we can develop an appropriate framework for assessment. How then is engineering knowledge generated?

4.6 Engineering Knowledge

Vincenti (1990) writes, "The generation of engineering knowledge involves an astonishing range of individuals. The cases of this book display people from the following (not mutually exclusive) groups: aerodynamicists, design engineers, research engineers, applied mathematicians, instrumentation engineers, production and process engineers, aircraft manufacturing executives, inventors, pilots (airline, military, and research), academics, military service monitors, and airline executives. Cases from other branches of engineering would lengthen the list. This remarkable diversity of skills and training is perhaps greater than that for any other branch of knowledge. It reflects the fact that engineering is an extraordinarily broad and ramified area of human endeavor. Engineers do a multitude of things in a modern society, and representatives from a large number of areas contribute to the vast knowledge this complex-activity requires" (pp. 237–238).[6] He goes on to argue that this is achieved by people working in informal communities that are the central agencies for the accumulation of knowledge in their areas. He agrees with Constant (1984) that the central locus of technological cognition is the community. The question for this text is whether or not the contribution that a person can make to a community is sufficient justification for the continuation of educational programs as presently structured? At its simplest, is teamwork (and its formative assessment) sufficient preparation for life in one of these communities? It is not much use requiring students to participate in teams if they are not subject to the views of other members and do not reflect (self-assess) on their own performance? Such reflection would seem to be pointless if they are not given a chance to turn the reflection into action in other team exercises? But there is another problem, and that is, much of the research that is done is in large companies whereas many engineers work in much smaller organizations, and some are self-employed. Indeed, many large firms now farm out their engineering work.[7] Hirst's criticism continues to be relevant because very little is known about what these engineers actually do particularly in terms of the epistemologies they use and how they differ or are similar to those used in science on the one hand and on the other hand business. In the aircraft component firm, discussions with management yielded the fact that the firm functioned at the frontiers of manufacturing technology not the frontiers of pure knowledge that was the prerogative of science. For this reason, the kind of postgraduate and postexperience courses valued by the company were those in allied or peripheral techniques. At the time they had made use of courses in computer-aided design and transistors as they changed over from thermionic valves (vacuum tubes).

Currently, much attention is being paid to the epistemologies used by engineers through the study of the philosophy of engineering and its implications for engineering education (Heywood, Carberry, and Grimson, 2011; Michelfelder, McCarthy, and Goldberger, 2013; Heywood and Cheville, 2014). At the operational level, practitioners are beginning to appreciate the different kinds of knowledge that come into play in any activity, as for example, tacit versus procedural knowledge, or "know how" versus "know what" (e.g., McCahan and Gomkey, 2014). Trevelyan (2014) lays great stress on the concept of distributed knowledge. He found that the knowledge map of an organization was much more complex than he had imagined in spite of 30 years of engineering experience. Moreover, "every aspect of technical knowledge is required somewhere along the way in nearly every enterprise" (p. 133). But this knowledge is not vested in any one individual rather each individual in the organization possesses technical knowledge that is largely unwritten and related to the particular task he or she performs. The knowledge required to complete the overall objective of the organization is distributed among its members who necessarily have to "share it through skilled and knowledgeable performances."

That sharing creates its own social system. For example, Youngman et al. (Section 4.2) found that engineers had to create and work within an informal organization to get things done. "A person is a psychosocial system. Within the boundaries of this system most individuals wish to be 'organic' to use a term first suggested by Burns and Stalker (1961; see Section 2.1). They wish to be able to take actions and decisions as well as mature. The boundaries of these psychosocial systems arise as a function of the needs of the job and the needs of the person. When these are matched for each person in the organization, a hierarchic organization becomes structured by individuals who are organic within their own system. The system itself becomes organic if it can respond to the needs of individuals. Both systems have to be self-adjusting."

Fast forward 36 years and Trevelyan (2014) as a result of his studies of engineers at work concluded that "Technical coordination" which seems to be close or the same as Youngman et al.'s, liaison, "seems to be a major part of day-to-day practice for practically all engineers, and it is an intrinsic requirement in any engineering enterprise that requires coordinated performances by different people." He notes that "technical coordination is all about getting things done and informal leadership is an important part of that." It is not about management which he saw as part of the formal organizational structure "rather it was like knot-working, mostly involves short-term relationships in which people collaborate without organizational authority." Youngman et al. would, in contrast, have found some longer-term relationships.

Trevelyan's thesis is supported by a Portuguese study that showed that at least 35% of the time spent by engineers was on technical liaison, the understanding of which is clearly explained in their essay (Williams and Figueiredo, 2013). From the perspective of the curriculum and therefore assessment, "Technical Coordination" is a generic competence or personal transferable skill (see Sections 5.4 and 5.6). Trevelyan points out that its aim is to get action.

More generally, Williams (2015) has argued that engineering practice is a legitimate area for study in the same way that it has been argued that engineering education is a distinct area of scientific inquiry (Froyd and Lohmann, 2013). It is likely to blossom in the future.

4.7 Discussion

These studies focused on what people did at work and not on the knowledge they used or how they used it. This remains an important issue because first, curricular tend to overload students, and second, technology continually changes the knowledge requirements. The activity of engineering is shown to be very much more than the application of science to solving problems. The engineering task requires a high level of interpersonal competence that in turn requires high-level skill in communication. By implication, these studies support the view expressed in earlier chapters that engineers have to speak many "languages."

Engineering is a complex activity that requires both collaboration and technical coordination. But how far should the development of abilities in areas other than engineering science be the province of the university, and how much a shared enterprise? Given changing patterns of work and lengths of employment, if employees are to widen the span of job search that is likely to be the case, employers will not be able to escape the responsibility for proper performance-based assessments that can contribute to an employee's profile and future career.

Clearly there is a role for a well-thought-out portfolio approach to assessment in which continuing professional development becomes a normal part of one's career. In these circumstances, the ability to self-assess becomes important (see Section 8.3), and thought will have to be given to the way students are credentialed (see Chapter 12). Various approaches to competency-based assessment are considered in the next chapter.

Notes

1. The study was initiated in the special products group of the English Electric Co., which was sold shortly afterward to Lucas Aerospace. The firm was located on four sites. Most of the activities of interest to the investigators took place on two sites. It had around 1000 personnel.

About one-third were employed on the shop floor, another third in clerical and administrative roles, and the remainder in engineering functions. The inquiry was focused on the latter group who had a variety of qualifications ranging from a degree or degree equivalent to lower-level qualifications used to qualify persons in production and manufacturing (see Youngman et al., 1978, p. 21).

2. Each activity comprised between 20 and 35 operations. The cluster analysis can be continued. For example, activities 4 and 10 can be combined into one activity and these can be brought together with design and drafting. Four large groupings were found in this way with information (activity 13) standing alone among the four. The nucleus activities are central to the task. In the case of activity 10, the average activity score overall was 42, which means that on average each engineer uses 42% of the 14 operations that constitute the activity.

3. Published information shows that during the Second World War this company was involved in the manufacturing of a very limited range of equipment for aircraft. Early in the 1950s, a decision was made to expand activities in this field. There was increasing demand for greater capacity from generating systems as well as for a reduction in the weight of components, and a new approach was required because existing methods could not cope. A decision was made to embark on the manufacturing of equipment using a different technology. This stage can be described as the problem formulation stage of a learning curve. However, the problem is not solved merely by correctly formulating it: knowledge has to be acquired and structured. Knowledge may be acquired by a research and development team, or it can be purchased. Within the firm in question, knowledge was initially structured by the purchase of information in the term of a license to manufacture a component from a firm in America. With the information, the firm was in a position to design and manufacture systems. Improvements in the system involved the engineers in restructuring knowledge that was already well-organized, such restructuring being necessitated by the unreliability of the earlier devices. Thus, in the mid-1950s the company set out to change the technology yet again. In this, they were aided by developments in the semiconductor industry. Inspection of the papers written about the developments in the aircraft industry (e.g., Gledhill, 1966) shows that each new problem causes the structuring of relatively disorganized knowledge. If learning is the process by which experience develops new responses and reorganizes old ones, the process of bringing a product into regular manufacture may be regarded as a process in which the organization proceeds from a relatively disorganized state of knowledge to a relatively organized one and in so doing considers alternative solutions. Exhibit 4.3 attempts to show the innovation process in terms of learning (Heywood, 1974). To survive, the first must maintain the ability to learn. It is this wish for survival that should inspire the desire to maintain effective skills in learning.

4. They wrote together or separately articles in the British engineering press and journals of the Institution of Mechanical Engineers. The reports were very influential in changing the attitudes of the authorities toward the need for 4-year courses to qualify professional engineers. They raised many issues about the status of engineers in the United Kingdom when compared with that of Germany. At the time the two Germanys had not been reunited after the Second World War. These reports are about engineers in the then West Germany.

5. "The German term *Wissenschaft* covers all formal knowledge subjects, whether arts, science, or social science in our terms. This explains the rather casual use of the word "scientific" by Germans who speak English: in their view, it can be applied as readily to historical scholarship as to nuclear physics. In the German scheme Kunst denotes art—not the "the arts" in the Anglo-Saxon sense, but the products of the arts. The criterion for inclusion is aesthetic not critical-intellectual. And the "third culture" in the German scheme of things is Technik. Technik is for the Germans an independent domain, embracing knowledge and skills relevant to manufacturing. Thus, it is an autonomous cultural rubric tending to dignify engineering,

and certainly serving to differentiate it from Natural science. (Hutton, S. P., and Lawrence, P. (1981). *German Engineers. The Anatomy of a Profession.* Oxford: The Clarendon Press, pp. 107–108).

6. A countervailing thesis to some of Vincenti's views will be found in the equally seminal work by M. Davis (1998), *Thinking like an Engineer. The Ethics of Profession.* Oxford University. Davis writes, "although Vincenti's work contributed to mine his has a narrower focus. He tries to understand engineering as a developing body of technical knowledge, a discipline; I, on the other hand, try to understand engineering as a profession. Knowledge though of course a part of what makes the engineer, is only a part. At least as important is the way the knower moves (or, at least is supposed to move) from knowledge to action. That movement from knowledge to action is the 'thinking' of my title" (p viii).

7. To be fair, 12 of the production managers in Hutton and Lawrence's British study came from firms where the numbers of employees ranged from 150 to 1000.

References

Barnard, C. (1938). *The Functions of the Executive.* Cambridge, MA: Harvard University Press.

Bannister, D., and Mair, J. M. M. (1968). *The Evaluation of Personal Constructs.* London: Academic Press.

Barnes, L. B. (1960). *Organizational Systems and Engineering Groups. A Comparative Study of Two Technical Groups in Industry.* Boston, MA: Harvard University, Graduate School of Business Administration.

Burns, T., and Stalker, G. (1961). *The Management of Innovation.* London: Tavistock.

Council of Engineering Institutions. (1968a). *Guidelines for Training Professional Engineers.* No. 6. London: Council of Engineering Institutions.

Council of Engineering Institutions. (1968b). *Guidelines on Education and Training for Management.* No. 9. London: Council of Engineering Institutions.

Constant, E. W. (1984). Communities and hierarchies: structure in the practice of science and technology. In: Laudan, I. R. (ed.), *The Nature of Technological Knowledge. Are Models of Scientific Change Relevant?* Dordrecht, the Netherlands: Springer.

Engineering Industries Training Board. (1968a). *The Training of Engineers.* Booklet No. 5. Watford: Engineering Industries Training Board.

Engineering Industries Training Board. (1968b). *The Training of Managers.* Booklet No. 6. Watford: Engineering Industries Training Board.

Engineering Industries Training Board. (1970). *The Technician in Engineering.* Watford: Engineering Industries Training Board.

Froyd, J. E., and Lohmann, J. R. (2013). Chronological and ontological development of engineering education as field of scientific in inquiry. In: Johri, A., and Olds, B. M. (eds.), *Cambridge Handbook of Engineering Education Research.* New York: Cambridge University Press.

Gledhill, J. (1966). Recent developments in electric power generating equipment for modern aircraft. *The English Electric Journal,* 21(6), 35.

Gregory, S. A., and Monk, J. D. (1972). Creativity: definitions and models. In: Gregory, S. A. (ed.), *Creativity and Innovation in Engineering.* London: Butterworths.

Handy, C. (1994). *The Age of Paradox.* Boston, MA: Harvard Business School Press.

Hesseling, P. (1966). *A Strategy for Evaluation Research.* Assen, the Netherlands: Van Gorcum.

Heywood, J. (1969). An Evaluation of Certain Post-War Developments in Higher Technological Education. Thesis. University of Lancaster, Lancaster. (Two Volumes).

Heywood, J. (1970). Qualities and their assessment in the education of technologists. *International Bulletin of Mechanical Engineering Education*, 9, 15–29.

Heywood, J. (1974). *Assessment in History (Twelve to Fifteen)*. 1st Report of the Public Examinations Evaluation Committee. Dublin, School of Education, University of Dublin. pp. 61–67.

Heywood, J. (1975). Toward the classification of objectives in training technicians and technologists. *International Journal of Electrical Engineering Education*, 13, 217–233.

Heywood, J. (1976). Engineers at work. An "illuminative" evaluation. *The Vocational Aspect of Education*, 28, 25–38.

Heywood, J. (1989). *Learning, Adaptability and Change. The Challenge for Education and Industry*. London: Paul Chapman/Sage.

Heywood, J., Carberry, A., and Grimson, W. (2011). A Select and Annotated Bibliography of Philosophy in Engineering Education. In: ASEE/IEEE Proceedings Frontiers in Education Conference, PEEE pp. 1–26.

Heywood, J., and Cheville, A. (eds.) (2014). Philosophical Perspectives in Engineering and Technological Literacy *1*. Washington, DC: Division of Technological literacy of the American Society for Engineering Education.

Hutton, S. P., and Gerstl, J. E. (1964–65). Engineering education and careers. Proceedings of a Symposium on Education and Careers. *Proceedings of Institution of Mechanical Engineers*, 178, Part 3F. (later published as *The Anatomy of a Profession*. London: Tavistock).

Hutton, S. P., and Lawrence, P. A. (1979). The work of Production Managers. Case Studies of Manufacturing Companies in West Germany. Interim report to the Department of Industry, Southampton. Department of Mechanical Engineering, University of Southampton

Hutton, S. P., and Lawrence, P. A. (1982). The work of production managers: Case Studies of Manufacturing Companies in the United Kingdom, Southampton. Department of Mechanical Engineering, University of Southampton.

Hutton, S. P., Lawrence, P. A., and Smith, J. H. (1977). The Recruitment, Deployment, and Status of the Mechanical Engineer in the German Federal Republic. Report to the UK Department of Industry. Department of Mechanical Engineering, University of Southampton.

Kelly, G. A. (1955). *The Psychology of Personal Constructs*. Vols. 1 and 2. New York: Norton.

Krathwohl, D., Bloom, B., and Mersia, B. B. (1964). *Taxonomy of Educational Objectives. The Classification of Goals. Vol. 2 Affective Domain*. London: Longmans Green.

McCahan, S., and Romkey, L. (2014). Beyond Bloom's. A taxonomy for teaching engineering practice. *International Journal of Engineering Education*, 30(5), 1176–1189.

Meuwese, W. (1969). Preview. Measurement of industrial engineering objective. In: Proceedings of the Sixteenth International Congress on Applied Psychology, Amsterdam, pp. 159–160.

Meuwese, W. (1971). Construction and evaluation of a course in technical mechanics. Report CCC/ESR (71), 14. Strasbourg, Council of Europe.

Michelfelder, D. P., McCarthy, N., and Goldberger, D. E. (eds.) (2012). *Philosophy and Engineering: Reflections on Practice, Principles and Process*. Dordrecht, the Netherlands: Springer.

Mintzberg, H. (2009). *Managing*. Harlow: Pearson.

Monk, J. D. (1972). An investigation into the role of the design function in the education, training and career patterns of professional engineers. Unpublished thesis, University of Lancaster, Lancaster, UK.

Oxtoby, R. (1973). Engineers, their jobs and training needs. *The Vocational Aspect of Education*, 25, 49–59.

Thomas, B., and Madigan, C. (1974). Strategy and job choice after redundancy: a case study in the aircraft industry. *Sociological Review*, 22, 83–102.

Trevelyan, J. (2010). Restructuring engineering from practice. *Engineering Studies*, 2(3), 175–195.

Trevelyan, J. (2014). *The Making of an Expert Engineer*. London: Taylor and Francis Group.

Turner, B. T. (1969). *Management Training for Engineers*. London: Business Books.

Vincenti, W. G. (1990). *What Engineers Know and How They Know It. Analytical Studies From Aeronautical History*. Baltimore, MD: The Johns Hopkins University Press.

Williams, B., and Figueiredo, J. (2013). Finding workable solutions: Portuguese engineering experience. In: B. Williams, J. Figueiredo, and J. P. Trevelyan (eds.), *Engineering Practice in a Global Context. Understanding the Technical and the Social*. London: CRC Press/Taylor and Francis.

Williams, B. (2015). Doctoral Thesis. Engineering and Management Department of IST University of Lisbon.

Woodson, T. T. (1966). *An Introduction to Engineering Design*. New York: McGraw Hill.

Youngman, M. B. (1975). Structuring work for training purposes. *The Vocational Aspect of Education*, 27(68), 77–86.

Youngman, M. B., Oxtoby, R., Monk, J. D., and Heywood, J. (1978). *Analysing Jobs*. Aldershot: Gower Press.

5

Competency-Based Qualifications in the United Kingdom and United States and Other Developments

This chapter describes some of the attempts made in the United Kingdom and United States to develop precisely defined competency-based qualifications. Many educators would not assume that only those things that are defined are the only elements of competence. A system of General National Vocational Qualifications was introduced to try and obtain parity of esteem between vocational and academic qualifications for the 16- to 19-year-old group. In 1991, the UK Employment Department attempted to introduce a National Record of Achievement. These developments were the subject of much misunderstanding and prejudice.

A brief description of the development and purposes of "Standards" in schools in the United States is given. It would be impossible to teach all that is required by the long lists of bench marks. Attention is drawn to the standards for technological literacy, and the distinction that is now being made between engineering and technological literacy.

An alternative to the competency approach for higher education was advocated by the Royal Society for the Encouragement of the Arts, Manufactures and Commerce (RSA) under the title "Education for Capability." Capability embraces competence but is also forward looking since it is concerned with the negotiation of potential.

The Assessment of Learning in Engineering Education: Practice and Policy, First Edition. John Heywood.
© 2016 The Institute of Electrical and Electronics Engineers, Inc. Published 2016 by John Wiley & Sons, Inc.

The ability-led curricula of Alverno College is described. The college's definition of assessment has many similarities with that of the NVQs viz., "A multidimensional process of observing and judging an individual in action on the basis of public, developmental criteria." The curriculum is assessment led by eight ability categories. Each category comprises six levels of conceptual development, each of which is more comprehensive than the previous level. Much attention is paid to learning, and because learning is self-awareness then students should be trained to self-assess. Learning has to be integrative and experiential. Integration is commensurate with synthesis, which is a skill required by engineering designers. The development of innovative curricula depends on it having a firm philosophical foundation and Alverno faculty provided that foundation for their most innovative curriculum.

An initiative of the UK Employment Department to develop "Personal Transferable Skills" among students from all subject areas in universities is described. This project was initiated because of complaints by industry that new graduates were not prepared for employment. The sponsors argued that these skills could be developed within the programs offered by any university department, be it history or engineering or whatever. A similar approach recommended by the SCANS Committee of the US Department of Labor for the high school curriculum is described. A description of the American College Testing College Outcome Measures Program is included. It is evident that a multiple strategy approach to the assessment of these "skills" is required. Taken together, there are remarkable similarities between the objectives of these different programs and the concept of intelligence as derived from the views of experts and laypeople by Sternberg.

5.1 The Development of Competency-Based Vocational Qualifications in the United Kingdom

In 1991, a seminal work in the development of competency-based testing in the United Kingdom was published by Gilbert Jessup. During the preceding 20 or so years, various governments had messed about with the system of vocational education and training and had made many interventions that were by no means successful. Among other things, they did not know how to deal with entrants from the school system with few if any qualifications, many of whom were likely to join the ranks of the unemployed. For those with some ability, the range of vocational qualifications was very large ranging from those that certificated a limited skill to those that were recognized by a professional organization. The Confederation of British Industry (1989), an employers' organization, said that the interventions that had been made focused on structure and delivery at the expense of content and outcomes. Prior to that, the Government of the day wanted to provide a common structure for all these qualifications that embraced content and skill. To this end in 1986, it established a National Council for Vocational Qualifications (NCVQ), which would award national qualifications commonly known as "NVQs." An NVQ is defined by competence. It "is a statement of competence clearly relevant to work and intended to facilitate entry into, or progress in, employment and further learning, issued to an individual by a recognized awarding body" (Jessup, 1991, p. 15). Clearly, the coursework of the engineering science syllabus (Section 3.1) met the criteria that NCVQ sought because the assessment requirements incorporated specified standards that showed the ability to perform a number of practical related activities that related to performance in some engineering activities. The reports submitted by the students had to show that they had the knowledge to perform these tasks, both practical and theoretical.

The NCVQ required that "the statement of competence should incorporate speci-fied standards in—the ability to perform in a range of work related activities and—the underpinning skills, knowledge and understanding required for performance in employ-ment" (NCVQ, 1989, cited by Jessup, p. 15). Assessment in NVQs is based directly on statements of competences and Jessup shows how five levels can be made to cor-respond with the examination levels in the general education system of England and Wales, with level 5 being that of a university degree. It is important to note that it was recognized that there was a substantial distinction to be made between job competence and occupational/professional competence (Jessup, p. 26).

The NVQ system of assessment is based on a collection of evidence. Jessup (1991) wrote, "The two aspects of the statement of competence which the definition (above) goes on to add are significant, as is their relationship to each other." The statement leads on "performance," which is of course central to the concept of competence, and places "skills, knowledge and understanding" as underpinning requirements of such performance. This does not deny the need for knowledge and understanding, but does make clear that they, however necessary, are not the same thing as competence. "This position has considerable implications for assessment" (p. 16). The assessors of engi-neering science paid great attention to the student's theoretical understanding, although some teachers would have said too much. This remains a critical issue especially when successful projects seem to be based on inadequate understanding of their theoretical rationale.

Jessup (1991) writes: "Assessment may be regarded as the process of collecting evidence and making judgments on whether performance criteria have demonstrated he or she can meet performance criteria for each element of the competence specified" (p. 18). Thus assessment can take place within a formal education setting or be formal-ized at work. How one acquires the knowledge, skills, and understanding required of performance is of less importance than securing situations in industry that will enable that performance to be demonstrated.

Jessup (1991) gives examples that show how a vocational qualification is con-structed. The components out of which it is constructed are similar to the components that contribute to a domain in the *Bloom Taxonomy*. However, the earlier outcome approaches of persons such as Tyler and Bloom are rejected by NCVQ because they measured only outcomes that could easily be assessed. The assessors of engineering science would probably have contradicted that view. For example, the assessors had difficulty in dealing with creativity. In the statement of objectives for coursework they preferred the term "originality," and in the assessment they avoided the use of the term "creativity" preferring a rubric with the title "design activity," although it is clear from the domain objective in the *Notes for Guidance* shown in Exhibit 3.4 that the criteria would be met in the other domains of assessment (see also Section 2.4). The earlier approaches were also rejected by NCVQ because they considered that they were limited to existing psychometric and examination-based approaches too which the assessors of engineering science would have to plead guilty.

Jessup (1991) argued that the outcomes-based approach that he and others had developed was applicable to all forms of learning. In this context, it should be noted that an outcomes approach was being developed in higher education but without the sophistication of the NVQs. A report from the Department of Employment (Otter, 1992) included a *Taxonomy of Outcomes for Engineering Education* by R. G. Carter (1984, 1985) that he had previously developed in 1981. A significant feature of his taxonomy is that it combines both cognitive and affective dimensions. Arising from the Bologna agreement (Bologna, 1999), course outcomes are like the NCVQ specified for the different levels of educational attainment in European universities with the exception that there are nine levels.

Two objections have been made to the NCVQ approach. The first is that the focus is on discrete behaviors that ignore the complexity of performance in the real world where judgments have to be made about intelligent performance (Gonczi, 1994). The second is that much work is done collectively so that it is not always possible to separate out an individual's contribution (Ashworth, 1992). Competence is developed collectively and better explained by the "outside" theory than the "inside" (see Section 1.3 and Chapter 10). Ashworth argues that an individual would not have to be totally competent when working together with others.

Smith (2010), who reviewed the experience of the Australian system of competency-based Vocational Education and Training, which has many similarities with the British system, used a framework developed by J. West (2004) to analyze that system. Philo-sophically, these systems are based on the needs of industry and prevent Vocational Education and Training systems from providing a more liberal education. Technically "competency standards cannot create a sensible and sound syllabus that teachers can deliver [...] it is difficult for a competency based system, despite its focus on criteria. To enable effective communication or desired performance standards" (Smith, 2010). In Australia, where the implementation of the system is through construction of the syllabus and the standards of assessment by the teachers, there can be no assurance of student outcomes. Educationally because the focus is on meeting the performance requirements of industry, teachers often place much less emphasis on the knowledge that underpins the competence. Finally, it is argued that too much centralization may lead to inflexibility. The relevance of these criticisms to engineering education is briefly considered in Chapter 12.

Engineering educators are likely to side with medical educators who do not assume that only those things that are defined are the only elements of competence. "The competencies are many and multi-faceted. They may also be ambiguous and tied to local custom and constraints of time, finance, and human resources. Nevertheless, a competency based curriculum in any setting assumes that the many roles and functions in a doctor's (engineer's) work can be clearly defined and expressed. It does not imply that the things defined are the only elements of competence, but rather that those which can be defined represent the critical point of departure in curriculum development. Careful delineation of these components of medical practice is the first and most critical step in designing a competency based curriculum" (Griffin, 2012). It is a view that those who designed the Alverno ability-based curriculum for general education would concur (see Section 5.5).

5.2 Outcomes Approaches in High Schools in the United Kingdom

At the time of the publication of Jessup's (1991) book, the government of the day had accepted his principle and in 1991 extended the work of the Council to introduce a system of General National Vocational Qualifications (GNVQs) that would seek to obtain parity of esteem between academic and vocational qualifications (DES (Department of Education and Science), 1991). To achieve such parity, the concept of "core," "common," or "key" skills was introduced: skills that everyone would need to have, irrespective of whether it was a program in a university or a technical college (Dodderidge, 1999; Oates and Harkin, 1994). Some universities took seriously the need to ensure that "key skills" were developed. All students in possession of a GNVQ would have to possess a range of core skills.

The GNVQs were targeted at the 16- to 19-year-old range. One of their objectives was to encourage young people to remain in education beyond the school leaving age (16 years). The advanced GNVQs which had the intention of providing access to higher education would be aligned with GCE "A" levels and NVQ level 3. Each vocational unit would be equal to 1/6th of an "A" level. Advanced GNVQs would require satisfactory completion of 12 units or two "A" levels. Knight, Helsby, and Saunders (1998) were of the opinion that advanced GNVQs were hybrids that fell between the two stools of the academic and the vocational. They were too educationally led to offer vocational qualifications and too vocationally led to allow students to be introduced to critical academic engagement. During the last decade, Lord (Kenneth) Baker, a former Secretary of State for Education, has sought to reconcile these differences in university technical colleges established for that purpose.

In 1991, the Employment Department introduced a National Record of Achievement as part of its Technical Vocational Education Initiative. The idea was that it would encourage the recording and provision of evidence of formal and informal achievements and was based on work that had already been done in schools in the previous decade. It was hoped that it would be of use to employers. Jessup (1994) pointed out that the Royal Society (1993) had recommended an expanded use of Records of Achievement in undergraduate education and it foresaw replacement of the current "honors" classification by them. In 2013, the Higher Education Achievement Report (HEAR) was introduced after much experimentation that dates back to the early 1990s. It is resourced by the Higher Education Academy. If the structure of the work force changes as is suggested that it might in Chapters 4 and 12 and many individuals find they have to embark on a second career, records of achievement, and/or portfolio assessments are likely to become significant instruments for both employees and employers.

At the end of his 1994 introduction to NVQs, Burke (1994) confides in the reader that the development of GNVQs would experience much misunderstanding and prejudice and he gives several examples to support his thesis. It is apparent from discussions in the press and parliament toward the end of 2013 that parity of esteem between the systems of vocational and general education has not been achieved. There are many critics who believed that vocational education continues to be undervalued by many politicians, and in spite of the efforts of Lord Baker to restore the good image of the technical colleges, debate centers on the restoration of the school examination system to its former glories. Such is the problem of the English class structure.

In Scotland, the school examinations are now based on learning outcomes that the students are given. The most interesting feature of this development is the attempt to get assessors to use the full range of marks.

5.3 Standards in Schools in the United States

In the United States, the National Association of Teachers of Mathematics set in trail a "standards" movement when it listed the standards that schools should obtain in mathematics. In 1996, the National Academies published *National Science Education Standards*, and in 2000, the International Technology Education Association (ITEA) published *Standards for Technological Literacy: Content for the Study of Technology.* Standards were published in all the subjects of the curriculum: essentially they were long lists of outcomes aimed at setting bench marks for the curriculum. They are not a national curriculum in the sense of the national curriculum for schools in England and Wales. The following points have been made about standards:

- Standards are an attempt to define what students should be able to know and do.
- The standards are informed by the latest theory and research regarding the written curricula.
- Standards are field based: They build on past successes of teachers and students.
- Standards are met through a variety of teaching styles and strategies.
- The standards project emphasizes that all students can learn and achieve at high levels if their background needs and interests are considered.
- Standards should be a source of professional conversation and critique about what to do and how to do it.
- Teachers are members of a professional community and a variety of professional organizations are available to support teacher growth.
- The literacy demands of the twenty-first century will require students to construct meaning with a variety of tools and texts.

This list was constructed by Glatthorn, Boscher, and Whitehead (2006), who cite Marzano to the effect that it would take 23 years of schooling to cover all the benchmarks. "Teachers can't possibly teach it all… and kids couldn't possibly learn it all."

When they were created, there was no attempt to relate them to higher education. A working group of the Technological Literacy Division of the American Society for Engineering Education (ASEE) examined the standards for technological literacy (K-12) (Krupczak et al., 2012). They noted that there were five main categories used to define technological literacy. These were:

1. Understanding the nature of technology
2. Understanding of technology and society
3. Understanding of design
4. Abilities for a technological world
5. Understanding of the designed world

"The ITEA Standards represented a significant advance and elaboration of the parameters defining the technological world, and the recognition that, given the importance, all students should begin to develop an increasingly sophisticated understanding of technology starting at the earliest years of school."

More than a decade ago, ASEE decided to set up a division for K-12 technology education. In recent years, it has been a vehicle for the promotion of engineering in schools. Accompanying this and in the Technological Literacy Division has been discussion about the need to draw up a statement of standards for engineering. It should be noted that the working party drew a distinction between engineering and technological literacy (Blake and Krupczak, 2012).

The idea that standards-based grading should be adopted in engineering has been put forward by Carberry, Siniawaski, and Dionisio (2012; see Section 10.5).

5.4 Education for Capability: Capability vs. Competence

An alternative to the competency-based developments in the 1980s is the capability approach that was sponsored by the Royal Society for the Encouragement of Arts, Manufacturers and Commerce (RSA) that it described in the *Education for Capability Manifesto* published in 1979 (Manifesto, 1979; Stephenson and Weil, 1992; Stephenson and Yorke, 1998). The capability thesis began with a criticism of British education that it saw as the education of a "scholarly individual who has neither been educated or trained to exercise useful skills; who is able to understand not to act" (Manifesto, 1979). This is rather a weak criticism, given that no one really understands what is learnt in the humanities by way of understanding and skill. Similarly, Furneaux's study (1962, chapter 2.5) showed that at the time probably little was known about what was and was not achieved in science and technology courses. Nevertheless, the argument that education is unbalanced can be supported in many ways. For example, it is clear that the curriculum did not support the development of spatial ability that is essential in the learning of mathematics and technological subjects (MacFarlane-Smith, 1964). Equally it may be more to do with culture than the courses taken. The manifesto admits as such when it defends the culture of doing, making, and organizing and the creative arts in its own right. Stephenson and Weil (1992) argue that "higher education should be judged by the extent to which it: (1) gives students the confidence and ability to take responsibility for their own continuing personal and professional development; (2) prepares students to be personally effective within the circumstances of their lives and work; and (3) promotes the pursuit of excellence in the development, acquisition and application of knowledge and skills." Detailed analysis of these goals suggests that the movement is not, on the contrary, at odds with those, such as John Henry Newman, who promoted the cause of liberal education (Heywood, 1994).

It is evident from Stephenson and Weil's collected reports of initiatives that the capability movement, together with other endeavors such as the Enterprise in Higher Education Initiative (EHEI; see Section 5.6), had considerable impact on thinking in educational institutions. Like the EHEI, it emphasized the development of what came to be called "personal transferable skills." It advocated action learning as a means of helping students learn how to apply their knowledge and skills. It also argued that students should be able to negotiate their programs of study and should learn through collaborative learning. Capability may be assessed by observing if students are able to take effective action, explain what they are doing, live and work effectively with others, show that they learn from experience, be it their own or with others (Stephenson and Yorke, 1998).

In contrast to competency, capability is holistic. Stephenson argues that "capability is a broader concept than competence. Competence is primarily about the ability to perform effectively, concerned largely with the here and now. Capability embraces competence but is also forward looking concerned with the negotiation of potential. A capability approach focuses on the capacity of individuals to participate in the formulation of their own developmental needs and those of the context in which they work and live." By 1992, Weil, Lines, and Williams were able to report a number of initiatives in higher education, including engineering to bring about capability. They noted that while this work was not widely publicized "implicit in the examples is recognition of changing priorities, in terms of skills and qualities that a capable engineer in future will require and how this might be best developed" (Stephenson and Weil, 1992, p. 102).

Cowan (2006, p. 35), who participated in the early development of the capability project, notes that the manifesto and the criteria did not explicitly call for reflection on learning that became a distinctive feature of progressive education and his own work in the 1990s. It seems that there must have been some reflective learning, given that students were asked to negotiate their programs of study and work with others. With the aid of an RAS award, Cowan (1980) was able to offer a first-year course in Properties and the use of Materials in which the students chose their individual syllabuses. Third-year students who heard about the success of this scheme asked if the same thing could be done with their third-year courses in Design. While Cowan agreed, he argued that if they were to become professional people in a year or so's time, they ought to formatively assess their capabilities, needs, and achievements, and thus began Cowan's work on self and peer assessment for which he is rightly well known (Cowan, 2006, pp. 88–92). One of the important tools that he uses as an aid to learning and self-assessment is the learning journal (see Section 8.4).

5.5   Ability (Assessment)-Led Curricula: The Alverno College Model

The most surprising development in higher education, certainly the most innovative I have seen, seems to be a development that derives from the view that assessment leads learning; therefore, assessments have to be designed to have a positive impact on learning. This cannot be achieved unless the whole curriculum is directed to the attainment of that goal. In this case, the total college curriculum was devoted to achieving the goals of liberal education through the enactment of this principle. It began in 1973 (circa) at Alverno College in Milwaukee and was the brainchild of a Religious Order of Franciscan Sisters who wanted to rejuvenate a small liberal arts college for women. Their program, which has been thoroughly evaluated (Mentkowski and Associates, 2000), has brought them world-wide renown (Cowan, 2006). Within the United States, the college has directed faculty teams in 23 institutions in the "field testing of performance assessment instruments that directly assessed students' progress in developing the abilities inherent in their institution's stated outcomes" (Schulte and Loacker, 1994). Its approach has been of interest to engineering educators in the United States.[1]

Alverno College defines assessment as "a multidimensional process of observing and judging an individual in action on the basis of public, developmental criteria" (Schulte and Loacker, 1994, p. 6). This process was brought to bear on the assessment of eight domains of competence, each of which was described in four levels of achievement in general education and two advanced levels specific to the student's selected major and minor subjects (Alverno, 1994). It is important to emphasize that these are levels of conceptual development and not a checklist of items that contribute to level 6 (Exhibit 5.1). Each level represents a greater degree of comprehensiveness. By 1999, the term "competence" had been replaced by "ability" although the domains remained the same (see Exhibit 5.2). Although this is not the reason given by Mentkowski, it is this writer's impression that one of the reasons why competency was dropped was that

---

An example of the six levels of an Alverno generic ability domain: The development of analytic capabilities

1. Show observational skills.
2. Draw reasonable inferences from observations.
3. Perceive and make relationships.
4. Analyze structure and organization.
5. Establish ability to employ networks from area of concentration or support area discipline in order to analyze.
6. Master the ability to employ independently the frameworks from area of concentration or support area discipline to analyze.

EXHIBIT 5.1. The internal construction of the abilities is regularly reviewed. This is taken from a list supplied by Dr. Georgine Loacker of Alverno College for reproduction in Heywood, J. (2000). *Assessment in Higher Education. Student Learning, Teaching, Programmes and Institutions*. London: Jessica Kingsley, p. 52.

The Alverno College Ability Led Curriculum—Ability Domains to be developed

1. Develop communication ability (effectively send and respond to varied audiences and purposes).
2. Develop analytical capabilities.
3. Develop workable problem-solving skill.
4. Develop facility in making value judgments and independent decisions.
5. Develop facility for social interaction.
6. Develop responsibility for the environment.
7. Develop awareness and understanding of the world in which the individual lives.
8. Develop aesthetic responsiveness in the arts.

EXHIBIT 5.2. The Alverno College eight abilities (the term *domain* is this writer's).

some judgments of competency-based teacher education programs were to the effect that they de-professionalized teaching because they were too reductionist. This is the last thing that the Alverno curriculum is. However, the scholarly interpretation must be that Alverno learned more about what they were doing in terms of developments in psychological thinking. Thus, in 1983, Anastasi (1983), a distinguished psychometrician, had conceived the idea of "developed abilities," and in 1992, Alexander (1992) had discussed the concept of domain knowledge so "what can be learned encompasses developed attitudes and domain knowledge." Alverno preferred the term "developed abilities" to aptitudes "because we have found the language of abilities communicates broadly across educators, employers and professionals [...] moreover, developing abilities is a broader term than acquisition of skill, competence, or expertise. Other terms than ability, such as skill can serve if there is no dichotomy between knowledge and skill, and skills are defined and discussed in the context of learning in the disciplines" (Mentkowski and Associates, 2000, p. 10).[2]

Alverno believed that learning had to be integrative and experiential, and that if this was the case assessment had to judge performance. "If learning is to be characterized by self-awareness, assessment must include self-assessment as well as expected outcomes and developmental criteria that are public. If learning is to be active/interactive, assessment must include feedback and elements of externality as well as performance. If learning is to be developmental assessment must be cumulative and expansive. Finally if learning is to be transferable, assessment must be multiple in mode and context" (Alverno, 1994).

These principles sit well alongside those of comprehensiveness, coherence, and continuity stated in the National Research Council's report on *Knowing How They Know* (see Section 3.3).

By integration is meant the ability "to continually create new wholes out of multiple parts." It would differ from synthesis in the Bloom *Taxonomy* in so far as synthesis in the *Taxonomy* is confined to a single subject. Integration brings knowledge from a variety of different perspectives, and it may be argued that it is the basic skill that is required for coping with the ever-increasing knowledge that individuals face. It is this skill that engineers need to solve the multiple-faceted problems they have to face. It is particularly

---

Synthesis

Whether one is "designer" or not, as an engineer one is involved in a main stream that synthesizes, the requirement therefore is for **sympathy** toward design. The engineer must be aware of the broad features of design, even if his personal creativity is minimal. The features are:

(a) Specification of the **need** in detail.

(b) **Feasibility** demanding the creation of several solutions, and the selection of one (or two) for further study. A corollary is that the quality of analysis must match the quality of the required answer. The student must be able to spot cases where a quick sum will suffice for the moment and those where greater detail is required.

(c) The value of **rigorous** methods in the overall design function.

(d) **Optimization**.

(e) **Implementation** of the chosen solution. The subfeatures here are that this often involves large-scale management—leading to knowledge of management techniques—and the reconciliation of material supply and production resources with the chosen solution in detail.

(f) At all points in the process of synthesis, the engineer, of whatever "kind," must be prepared to specify **criteria** by which decisions—often several interdependent decisions—can be measured.

(g) The decisions noted above (f) to a large extent also depend on **judgment** and **opinion**. (Surely, any university must set out to cultivate judgment, quite overtly, whatever else it does?)

EXHIBIT 5.3. Deere, M. (1968). Communication to a working party on professional examinations of the Society for Research into Higher Education (unpublished but cited in Heywood, J. (1969). Vol. 1, p. 333).

---

required by engineering designers and in this respect Deere's view of synthesis shown in Exhibit 5.3 is of more than passing interest. However, both Alverno and Deere would, in today's world, be criticized for not encouraging the ability to think in terms of "systems."

Integration as defined by Alverno is the antidote to specialization. *Feedback* is considered to be important and provides the matter for reflection and growth. It simply does not happen in summative examination systems that are terminal where there is no provision for formative assessment as was the case with the examinations analyzed by Furneaux (1962, chapter 2.5), and the written component of engineering science (Section 3.1). *Feedback* is related to *externality* in that if the abilities are to be *transferable* some assessment must be made at a distance from the classroom. Hence, the requirement for an "assessment-Centre." Within the concept of *externality* they attach considerable importance to *self-assessment*, as they do to the use of external assessors from the world-at-large. Cumulative assessment shows how the students progress or regress and student portfolios are used to plot this journey. More generally, portfolios are a key instrument in the assessment of competence and performance.

There is one aspect of Alverno that has been noted by engineering educators and that is the educational climate of the institution. Alverno both created and required a climate in which this approach to assessment could flower (George, 1992). Borrego (2008) points out that if assessment in engineering is to be taken seriously a culture has to be created if that goal is to be met, and she cites Alverno as a place where that has been achieved. Having stayed at the college on several occasions, I can testify that it is an experience that visitors feel. Tonso (2006) has shown how campus culture influences the interactions between engineering teammates. More generally, if education has to change, as some argue, then it will require a most substantial change in culture (Godfrey and Parker, 2010; Felder, Brent, and Prince, 2011).

The objections that are brought by those who do not wish to experiment with an Alverno-type curriculum model are that Alverno is a small private liberal arts college offering a 4-year program where teachers are freed from the obligation of research. But that is a nonsense any department can create such a curriculum and approach to assessment provided that it has the will. It is the will to try alternatives that is missing from so much teaching and assessing in engineering. That is a problem for engineering education that continues to have to demonstrate that it is a valid field of inquiry.

From an engineering perspective, the idea of bringing in external assessors from industry might go some way to resolving the tensions between education and industry, particularly if they can contribute to curriculum discussions. In this respect, cooperative education programs that require inputs from employers about student abilities are a way forward (Plouff, 2013).

Recently, some engineering educators have begun to talk about reconceptualizing engineering education. Any reconceptualization has to be grounded in a firm philosophy. The Alverno curriculum grew out of such a philosophy. One way forward for engineering educators might be to try and establish ability domains from a philosophy of engineering education (see Chapter 12). It is remarkable how the particular abilities that Alverno selected would seem to suit the developing field of engineering and technological literacy.

5.6 The Enterprise in Higher Education Initiative in the United Kingdom and the SCANS Report in the United States

An initiative by the Employment Department in the United Kingdom in the late 1980s, the EHEI is of relevance. The government of the day in response to complaints from industry that graduates *per se*, that is, irrespective of subject, were ill-suited to the needs of industry offered a £1,000,000 to universities willing to develop the skills of enterprise across all subject areas. As part of this endeavor, a unit was sponsored by the UK Employment Department at Sheffield University to derive "personal transferable skills" and show how they might be taught. Today, they are often simply called "transferable skills." At that time many of these skills were either not picked up or if they were, it was haphazardly. Yet they were common to most adult situations. Nearly 20 years earlier, G. Holroyde of the English Electric Co. had drawn up a list of skills shown in Exhibit 5.4 that he thought were essential for living. Confirmation of many of the items in his list is to be found in one of the Sheffield studies that involved the analysis of 10,000 advertisements for graduates in British quality newspapers collected over a short period. Fifty-nine percent made an explicit reference to personal transferable skills that would be required for performance on the job; a further 15% could be inferred to require such skills. They isolated 32 significant characteristics of which 20 were thought to be genuine transferable skills. They were collated into four generic categories of

- Perception, analysis, diagnosis
- Unstructured problem solving, judgment
- Communication
- Understanding, sympathy, tolerance
- Management of oneself and resources
- Management of other people.
- Working cooperatively in groups
- Coping with frustration
- Responsible attitude to life and work
- Self-confidence through self-knowledge
- Innovation and calculated risk taking
- Emotional resilience
- Physical dexterity and endurance

EXHIBIT 5.4. Holroyde's list of skills that are picked up haphazardly, or in a few cases taught formally, which are common to most adult situations. Unpublished circa 1972, cited in Heywood, J. (1977). *Assessment in Higher Education*. Chichester, UK: John Wiley & Sons, Ltd., p. 218.

communication, teamwork, problem solving (creativity), and management and organizing, all of which involve competences that utilize both cognitive and affective domains. They worked with university teachers to provide examples of how the different skills might be developed and were able to draw up a framework that showed how a variety of action learning strategies could be used for that development. (A detailed summary of the work of this unit is found in Heywood, 2005, pp. 39 ff.) The project affirmed the view of the EHEI sponsors that these skills could be learned within normal subject teaching as a function of the strategies used, and that bolt on subjects to achieve the goals of enterprise learning should not be developed, although strong arguments were presented against this view (Heywood, 1994). Much depends on the learning strategy that is used. For example, in an investigation of the development of transferable skills in Slovenia the research workers were able to randomly divide engineering students into two groups of 38 each (Vidic, 2008). The control group received traditional instruction and the experimental group undertook problem-based learning (PBL) for a semester. Comparisons of the results of the projects showed that those who solved problems cooperatively in the PBL group performed better than those in the control group. Self-report questionnaires that analyzed attitudes to teamwork skills showed that the PBL group rated their progress higher than those in the control group. Given that the control group worked individually, this result is not surprising. They did not get a comparative measure of progress in independent learning although the PBL group reported significant improvement in this dimension. The assessment of the influence of the development of team skills without some independent observation is problematic. But this does not take away from this study Vidic's comment that PBL enhanced the motivation to learn scientific content.

At the time of EHEI, a Management Charter Initiative was also sponsored by the Employment Department. It developed the personal competence model for managers

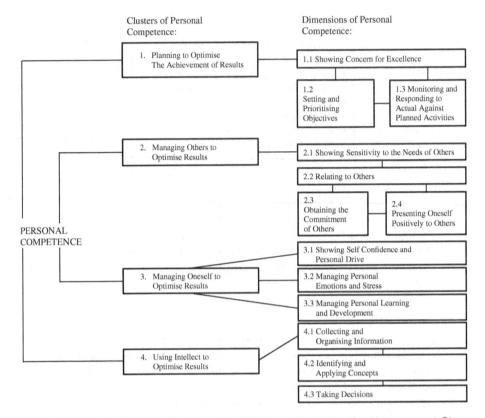

EXHIBIT 5.5. The Personal Competence Model developed by the Management Charter Initiative (1990) funded by the UK Employment Department. *Personal Competence Project*. Summary Report, 1990. Standards Methodology Branch, Employment Department, Sheffield cited in full in Heywood, J. (1994). *Enterprise Learning and Its Assessment in Higher Education*. Technical report No. 20. Learning Methods Branch, Employment Department, Sheffield, p. 15.

shown in Exhibit 5.5 on the basis of managers' perceptions of effective and ineffective behavior and research on management development. Any engineer in a management position would need to have these skills as the commentary on teamwork shows (see Section 9.5).

The recommendations of the SCANS (1992) report for the US Secretary of Labor are directly relevant to this view. The commission argued that the high school curriculum was not equipping students for the world of work. To achieve this goal, students required to develop five workplace competencies: handling of resources; interpersonal skills; handling information; thinking in terms of systems; and technology. These were to be supported by basic skills, higher-order thinking skills, and the development of personal qualities. All of these would seem to meet the requirements of an Alverno-type curriculum or one designed to meet the objectives of the Sheffield Personal Transferable

Skills model. The exception in both cases would seem to be the inclusion of the need to be able to think in terms of systems. Like the EHEI, the SCANS Committee believed that their goals could be achieved through the ordinary subjects of the curriculum and gave examples of how this might be accomplished. Surprisingly, a weakness of the model was that it paid little attention to the "practical" areas of the curriculum like the arts and crafts and music. However, the SCANS report was the starting point for the development of a community college program that integrated four key skills—communication, critical thinking, technical skills, and interpersonal skills—into the whole of its curriculum. By technical skills is meant the demonstration of knowledge and competence in academic and technical fields of study. The other skills were considered to lie within the modern definition of liberal studies. Integration also took place at the assignment level.

Bailey (2013) describes how the integration with assessment was made with liberal studies in an engineering program. Included in his description was the following, which related to a particular course. "Students were asked to research a topic of their choosing, relevant to the course, and write short paper about it. Students would submit a topic, a first draft, and a final draft at various points throughout the semester. Only the final draft was graded. The topic and first draft provided a mechanism for the instructor to give feedback and guide the research and writing. A design project had been part of the course and was expanded. Students were required to submit a short paper that justifies their design decisions including citations of research sources. Both of these assignments reinforce critical thinking and reflection, and promote student directed learning." The Scans program attracted the interest of some engineering educators (Christiano and Ramirez, 1993) and continues to be referenced by instructors.

5.7 The College Outcome Measures Program

Much earlier ACT developed a College Outcome Measures Program (COMP), which was designed as a measure of general education. It was designed to help institutions evaluate their curricular and/or design learning activities that would help students obtain the knowledge, skills, and attitudes necessary for functioning in adult roles after graduation. These aims go beyond education for work and take into account more general aspects of living. A distinction is made between *content areas* (functioning within social institutions and using science and technology) and *process areas* (communicating, solving problems, and clarifying values). The assessments covered the six outcomes expected of general education shown in Exhibit 5.6 (Forrest, 1985).

Two kinds of assessment are used. The first set of measures comprises an objective test and an activity inventory. The former has as its goal the estimation of the group's ability to apply general education skills and knowledge to problems and issues commonly confronted by adults. The latter assesses the quality and quantity of involvement in key out-of-class activities in the six outcome areas. It is a self-report inventory in multiple-choice format. It may be completed at home or in groups. It asks students what they do, not what they think. ACT believes it can help determine if a nontraditional student could earn credits through portfolio assessment.

1. Communicating	Can send and receive information in a variety of modes (written graphic, oral, numeric, and symbolic), within a variety of settings (one-to-one, in small and large groups), and for a variety of purposes (to inform, to persuade, and to analyze).
2. Solving problems	Can analyze a variety of problems (scientific, social, and personal) and can select or create and implement solutions to problems.
3. Clarifying values	Can identify personal values and the values of other individuals, can understand how values develop, and can analyze the implications of decisions made on the basis of personally held values.
4. Functioning within social institutions	Can identify social institutions (religious, marital, and familial institutions: employment and civic, volunteer and recreational organizations) and can analyze own and others functioning within social institutions.
5. Using science and technology	Can identify those activities and products that constitute the scientific/technological aspects of culture (transportation, housing, clothing, health maintenance, entertainment and recreation, communication, health, and data processing), can understand the impact of such entities on the individuals and the physical environment in a culture, and can analyze the uses of technological products in culture and personal use of such products.
6. Using the Arts	Can identify those activities and products that constitute the artistic aspects of a culture (music, drama, literature, dance film, and architecture) and can analyze uses of works of art within a culture and personal uses of art.

EXHIBIT 5.6. Areas of evaluation in the ACT College Outcome Measures Program (COMP). American College Testing Program, Iowa City, Iowa. See Forrest, A., and Steele, J. M. (1982). *Defining and Measuring General Education Knowledge and Skills*. Technical Report. Iowa City: American College Testing Program.

The second set of measures is called "authentic." There are four that may be measured individually or in groups. They are a composite examination, assessment in reasoning and communication, speaking assessment, and writing assessment. Questions in the composite examination are based on television documentaries, recent magazine articles, advertisements, short stories, art prints, music, discussions, and newscasts that are presented in a variety of formats. The responses range through short and long answers to audiotaped response, and what ACT perceived to be innovative multiple-choice questions.

Forrest and Steele (1982) claim that the scores correlate strongly with instruments such as indicators of effective functioning as a job supervisor and social economic status of job function. This suggests that the tests may be valid insofar as they are representative of potential performance in adult functions. Score gains between entrance to college and graduation were measured at 44 institutions. In 4-year colleges, the most gains were found in the first 2 years. It is argued that this was to be expected. These institutions varied considerably in their approaches to general education. At one end of the spectrum, there was a group of 19 characterized by an individualized, practical instructional style. In these institutions, the persistence to graduation rate was much higher than that in the others. Forrest concludes that it seems likely that persistence is related to the perceived relevance and practical skill building within the structure of the education program. Comp Tests provide a profile that could be used as an alternative to grade point averages. They have the potential to define thresholds and in consequence benchmarks.

There are many teachers, particularly outside the United States, who would not accept that this is the best way of measuring college outcomes. Some 17 institutions working with Alverno College reported 65 approaches to assessing general education outcomes that are of general interest (Schulte and Loacker, 1994). In any event, it is unlikely to be possible to measure the outcomes of general education without a multiple strategy approach to assessment and there is evidence of this in the ACT approach. Taken together, the different approaches to assessment have relevance for the evaluation and assessment of courses in engineering and technological literacy as well as the general education component of engineering and programs.

5.8 Discussion

The EHEI committee responsible for advising on assessment described four areas of learning that equip students for their working lives. They were knowledge/skills, social skills, managing one's self, and learning how to learning (Exhibit 5.7). These are reflected in the SCANS proposals (Exhibit 5.8). They are compared with a description

---

Cognitive knowledge and skills

Knowledge: Key concepts of enterprise learning (accounting, economics, organizational behavior, inter- and intrapersonal behavior).

Skills: The ability to handle information, evaluate evidence, think critically, think systematically (in terms of systems), solve problems, argue rationally, and think creatively.

Social skills: For example, the ability to communicate, and to work with others in a variety of roles as both a leader and a team leader.

Managing one's self: For example, to be able to take initiative, to act independently, to take reasoned risks, to want to achieve, to be willing to change, to be able to adapt, to know one's self and one's values, and to able to assess one's actions.

Learning to learn: To understand how one learns and solves problems in different contexts and to be able to apply the styles learnt appropriately to the solution of problems.

EXHIBIT 5.7. Five broad areas of learning that are important for equipping students for their working lives, as defined by the REAL working group of the UK Employment Department, 1991 (cited in Heywood, J. (2005) *Engineering Education. Research and Development in Curriculum and Instruction.* Hoboken, NJ: IEEE/ John Wiley & Sons, Inc.

Workplace competencies	Effective workers can productively use
1. Resources	They know how to allocate, time, money, materials, space, and staff.
2. Interpersonal skills	They can work in teams, teach others, serve customers, lead, negotiate, and work well with people from culturally diverse backgrounds.
3. Information	They can acquire and evaluate data, organize and maintain files, interpret and communicate, and use computers to process information.
4. Systems	They understand social, technological, and organizational systems, they can monitor and correct performance, and they can design or improve systems.
5. Technology	They can select equipment and tools, apply technology to specific tasks, and maintain and troubleshoot equipment.
Foundation skills	**Competent workers in a high performance workplace need**
1. Basic skills	Reading, writing, arithmetic, and mathematics Speaking and listening
2. Thinking skills	The ability to learn, reason, think creatively, make decisions, and solve problems
3. Personal qualities	Individual responsibility, self-esteem and self-management, sociability, and integrity

EXHIBIT 5.8. The SCANS competencies. (SCANS Secretary's Commission on Achieving Necessary Skills—Competencies and Foundation Skills for Workplace Know-How). *Learning a Living: A Blueprint for High Performance.* Washington, DC: US Department of Labor, 1992.

of intelligence derived by Sternberg (1985) from the views of academics and laypeople (Exhibit 5.9). The similarities between them are profound, and in this writer's view, it is easy to see how they can be turned into statements of performance/competence for the purposes of assessment. It can be argued that the Alverno abilities cover much of what would be deduced from such an analysis. However, both lists draw attention to the need to think in terms of systems (systems thinking).

Overall, given Sternberg's description of intelligence, it would seem that all these curricular are trying to develop "intelligent behavior," or in Sternberg's terms "a mental activity directed toward purposive adaptation to, selection and shaping of, real world environments relevant to one's life."

Both the NVQ and Alverno College developments place assessment in a whole new context. At Alverno, "it is a multi-dimensional process of observing and judging an individual in action on the basis of public developmental criteria." In Alverno's case, these criteria are expressed through eight ability groupings, each of which comprises six levels of conceptual development with each level being more comprehensive than the previous. It is a system that avoids the long lists generated by "Standards" and is based on firm philosophical foundations (see Chapter 12). Much attention is given to

---

Practical problem-solving ability: reasons logically and well, identifies connections among ideas, sees all aspects of a problem, keeps an open mind, responds to other's ideas, sizes up situations well, gets to the heart of the problem, interprets information accurately, makes good decisions, goes to original sources of basic information, poses problems in an optimal way, is a good source of ideas, perceives implied assumptions and conclusions, listens to all sides of an argument, and deals with problems resourcefully.

Verbal ability: speaks clearly and articulately, is verbally fluent, converses well, is knowledgeable about a particular field, studies hard, reads with high comprehension, reads widely, deals effectively with people, writes without difficulty, sets times aside for reading, displays a good vocabulary, accepts norms, and tries new things.

Social competence: accepts others for what they are, admits mistakes, displays interest in the world at large, is on time for appointments, has social conscience, thinks before speaking and doing, displays curiosity, does not make snap judgments, assesses well the relevance of information to a problem at hand, is sensitive to other people's needs and desires, is frank and honest with self and others, and displays interest in the immediate environment.

EXHIBIT 5.9. Abilities that are thought to contribute to intelligence. Obtained from questions about the nature of intelligence, academic intelligence, and unintelligence put to experts in research on intelligence and laypersons by R. H. Sternberg and his colleagues. Among the findings was the fact that research workers considered motivation to be an important function of intelligence whereas laypersons stressed interpersonal competence in a social context. In Sternberg, R. H. (1985). *Beyond IQ. A Triarchic View of Intelligence.* Cambridge University Press. The comparability between this view of intelligence and various statement so the aims of education is discussed in Heywood (1994).

learning, and since learning is self-awareness, to self-assessment. The idea of capability has significance for higher education and more particularly engineering because while it embraces competence capability is forward looking as it is concerned with the negotiation of potential. That potential is partly to continue with learning because technology requires constant renewal.

It is clear that those responsible for the Alverno Curriculum and the EHEI believed that much more could be done in the university curriculum to prepare students for work. The essential difference between them was that the Alverno curriculum derived from the need to prepare students for life of which work is a component and was informed by a substantial belief in liberal education. American engineering educators may not have responded to all of these ideas as well as they might, but students are required to undertake some form of liberal study where they might be expected to begin to develop some of these so-called soft skills. Within engineering subjects, the debate continues about creativity and design, the need to think in terms of systems, and learning how to solve complex problems. Team projects have come to be seen as one resolution of the problem as have cooperative and problem-based learning. The danger is that shifting the burden to colleges will lead industry to ignore the contribution that it can make to a person's development (see Chapters 10 and 11). There is also a danger that educators become obsessed with particular activities at the expense of other equally valuable activities such as the replacement of individual projects by team work projects rather than providing a combination. To a large extent the direction that engineering education takes is in the hands of the accrediting authorities, but there is one big question. It is—why has nothing happened or if it did happen why did it not become embedded in the system? In the United Kingdom, in the 1950s the Colleges of Advanced Technology were established; in the 1980s the Capability movement, and the EHEI were created, each with the purpose of ensuring that higher education produced graduates of the quality required by industry. Yet the Royal Academy of Engineering continues to exhort colleges to produce graduates suitably trained to meet the needs of industry (Spinks, Silburn, and Birchall, 2006; Royal Academy of Engineering, 2010). Much the same can be said about the major reports on engineering education published in the United States.

Notes

1. Ruth Streveler organized a special session on Alverno's ability-based curriculum at the 2007 Frontiers in Education Conference (Proceedings, T3F-1).
2. The two lists may be compared in Heywood (1989) pp. 336–337 (competencies), and Heywood (2000) p. 52 (abilities).

References

Alverno (1994). *Student-Assessment-as-Learning at Alverno College*. Milwaukee, WI: Alverno College Institute.

Alexander, P. A. (1992). Domain knowledge: evolving themes and emerging concerns. *Educational Psychologist*, 17(1), 33–51.

American College Testing. (1981). *Defining and Measuring General Education, Knowledge and Skills*. COMP technical report 1976–1981. Iowa City, IA: American College Testing Program.

Anastasi, A. (1983). Evolving trait concepts. *American Psychologist*, 38(2), 175–184.

Ashworth, P. (1992). Being competent and having competencies. *Journal of Further and Higher Education*, 16(3), 8–17.

Bailey, B. D. (2013). Integrating liberal studies at the assignment level—a case study. In: Proceedings of Annual Conference of American Society for Engineering Education, June 2013. Paper 5902.

Blake, J., and Krupczak, J. (2012). Defining engineering and technological literacy. In: Proceedings of Annual Conference of American Society for Engineering Education. Paper 5100.

Borrego, M. (2008). Creating a culture of assessment within engineering education. In: ASEE/IEEE Proceedings of Frontiers in Education Conference S2B-1 to 6.

Burke, J. (ed.) (1994). *Outcomes Learning and the Curriculum. Implications for NVQ's GNVQ's and Other Qualifications*. London: Falmer Press.

Carberry, A. R., Siniawski, M. T., and Dionisio, D. N. (2012). Standards based grading. Preliminary studies to quantify changes in affective and cognitive student behavior. In: ASEEE/IEEE Proceedings of Frontiers in Education Conference, pp. 947–951.

Carter, R. G. (1984). Engineering curriculum design. *Institution of Electrical Engineers Proceedings*, 131, Part A. 678.

Carter, R. G. (1985). Taxonomy of objectives for professional education. *Studies in Higher Education*, 10(2), 135–149.

Christiano, S. J. E., and Ramirez, M. (1993). Creativity in the classroom. Special concerns and insights. In: ASEE/IEEE Proceedings of Frontiers in Education Conference, pp. 209–212.

Cowan, J. (1980). Freedom in the selection of course content—a case study of a course without a syllabus. *Studies in Higher Education*, 3(20), 139–148.

Cowan, J. (2006). *On Becoming an Innovative University Teacher*, 2nd edition. Buckingham: Open University Press.

DES. (1991). *Education and Training in the 21st Century*. Vol. 1. London: HMSO.

Dodderidge, M. (1999). Generic skill requirements for engineers in the 21st century. In: ASEE/IEEE Proceedings of Frontiers in Education Conference, 3 13a-9 to 14.

Felder, R. M., Brent, R., and Prince, M. J. (2011). Effective instructional development strategies. *Journal of Engineering Education*, 100(10), 89–122.

Forrest, A. (1985). Creating conditions for student institutional success In: Nole et al. (eds.), *Increasing Student Retention*. San Francisco, CA: Jossey-Bass.

Forrest, A., and Steele, J. M. (1982). *Defining and Measuring General education Knowledge and Skills*. Technical report. Iowa City, IA: American College Testing Program.

Furneaux, W. D. (1962). The psychologist and the university. *Universities Quarterly*, 17, 33–47.

George, J. (1992). Alverno College. A learner's perspective. *British Journal of Educational Psychology*, 22(3), 194–198.

Glatthorn, A. A., Boscher, F., and Whitehead, B. M. (2006). *Curriculum Leadership, Development and Implementation*. Thousand Oaks, CA: Sage.

Godfrey, E., and L. Parker, L. (2010). Mapping the cultural landscape in education. *Journal of Engineering Education*, 99(1), 5–122.

Gonczi, A. (1994). Competency based assessment in the professions in Australia. *Assessment in Education. Principles, Policy and Practice*, 1(1), 27–44.

Griffin, C. (2012). A longitudinal study of portfolio use to assess the competence of undergraduate student nurses. Doctoral dissertation, University of Dublin, Dublin.

Heywood, J. (1994). *Enterprise Learning and Its Assessment in Higher Education*. Report No. 20. Sheffield: Employment Department.

Jessup, G. (1991). *Outcomes. NVQ's and the Emerging Model of Education and Training*. London: Falmer.

Jessup, G. (1994). Outcome based qualifications and the implications for learning. In: Burke, J. (ed.), *Outcomes, Learning and the Curriculum. Implications for NVQ's, GNVQ's and Other Qualifications*. London: Falmer.

Knight, P., Helsby, G., and Saunders, M. (1998). Independence and prescription in learning. Researching the paradox of advanced GNVQs. *British Journal of Educational Studies*, 46(10), 54–67.

Krupczak, J., Blake, J. W., Disney, K. A., Hilgarth, C. O., Libros, R., Mina, M., and Walk, S. R. (2012). Defining engineering and technological literacy. In: Proceedings of Annual Conference of the American Society for Engineering Education. Paper 5100.

MacFarlane-Smith, I. (1964). *Spatial Ability*. London: University of London Press.

McGahie, W., Miller, G. E., Sajid, A. W., and Telder, T. V. (1978). *Competency-Based Curriculum Development in Medical Education. An Introduction*. Geneva, Switzerland: World Health Organization.

Manifesto (1979). *Education for Capability Manifesto*. London: The Royal Society for the Encouragement of the Arts, Manufactures and Commerce.

Mentkowski, M., and Associates (2000). *Learning That Lasts. Integrating Learning, Development, and Performance in College and Beyond*. San Francisco, CA: Jossey-Bass.

Oates, T., and Harkin, J. (1994). From design to delivery. The implementation of the NCVQ core skills units. In: Burke, J. (ed.), *Outcomes, Learning and the Curriculum. Implications for NVQ's, GNVQ's and Other Qualifications*. London: Falmer Press.

Otter, S. (1992). *Learning Outcomes in Higher Education*. London: HMSO for the Employment Department.

Plouff, C. (2013). Cooperative education as the catalyst for effective and efficient assessment of ABET student learning outcomes for an engineering program. In: Proceedings of Annual Conference of American Society for Engineering Education, June 2013. Paper 8033.

Royal Academy of Engineering. (2010). *Engineering Graduates for Industry*. London: Royal Academy of Engineering.

Royal Society (The) (1993). *Higher Education Futures*. London: The Royal Society.

Schulte, J., and Loacker, G. (1994). *Assessing General Education Outcomes for the Individual Student: Performance Assessment as Learning. Part 1. Designing and Implementing Performance and Assessment Instruments*. Milwaukee, WI: Alverno College Institute.

Smith, E. (2010). A review of twenty years of competency-base training in the Australian vocational education and training system. *International Journal of Training and Development*, 14(1), 54–64.

Spinks, N., Silburn, N., and Birchall, D. (2006). *Educating Engineers for the 21st Century. The Industry View*. Henley-on-Thames, UK: Henley Management College.

Stephenson, J., and Weil, S. (1992). *Quality in Learning. A Capability Approach in Higher Education*. London: Kogan Page.

Stephenson, J., and Yorke, M. (1998). *Capability and Quality in Higher Education*. London: Kogan Page.

Sternberg, R. S. (1985). *Beyond IQ. A Triarchic Theory of Intelligence*. Cambridge University Press.

Tonso, K. L. (2006). Teams that work: campus culture, engineer identity and social interaction. *Journal of Engineering Education*, 9591, 25–38.

Vidic, A. D. (2008). Development of transferable skills within an engineering science context using problem-based-learning. *International Journal of Engineering Education*, 24(6), 1671–1677.

Weil, S., Lines, R., and Williams, J. (1992). Capability through engineering higher education is Stepenson, J., and Weil, S. (eds.). *Quality in Learning*. London: Kogan Page.

West, J. (2004). *Dreams and Nightmares. The NVQ Experience*. Working paper No. 45. Leicester: Centre for Labour Market Studies.

6

The Impact of Accreditation

The chapter begins with a brief introduction to present developments in the regulation (accreditation) of the curriculum. The review of the literature in this and other chapters shows that these regulations dominate thinking about the curriculum. They also show the lasting influence of the Taxonomy of Educational Objectives that, despite criticisms of its use in engineering, continues to be attractive to some educators. A brief description of the revised taxonomy and an example of its use is given. Of considerable importance are attempts to analyze questions set in examinations to evaluate the extent to which critical thinking and problems-solving skills were being tested. But the judgments were based on face validity. Some educators have looked to other taxonomies for the design of their courses. This point is illustrated by the design of a service course using Fink's taxonomy, which at first sight shows as much concern for the affective domain as it does for the cognitive.

Another attempt to reflect the variability of student performance that indicates what objectives students should attain at different levels is reported. Performance on the objectives is a measure of student achievement. One of very few attempts to check the validity of outcome statements is reported. The perceptions of outcomes of students and their instructors were compared with not very promising results. Outcomes assessment surveys are often used.

The Assessment of Learning in Engineering Education: Practice and Policy, First Edition. John Heywood.
© 2016 The Institute of Electrical and Electronics Engineers, Inc. Published 2016 by John Wiley & Sons, Inc.

One study is reported in which the Myers-Briggs Type Indicator (personality indicator) is used to assign students to teams. A substantial study of entrepreneurially minded engineers that embraced the affective domain is reported. It compared practicing engineers with students. The practicing engineers demonstrated a different "footprint." Suggestions are made for schools offering entrepreneurship education. A major question is the extent to which a college can help develop personal and professional competencies, an issue that is followed up in Chapters 10 and 11.

A section on mastery learning in which interest has resurfaced concludes the chapter. Throughout the chapter, attention is drawn to the difficulty of writing valid examination and test questions.

6.1 ABET, European Higher Education Area (Bologna Process), and the Regulation of the Curriculum

In the late 1990s ABET, the US accreditation agency reviewed its approach to accreditation (ABET). From 2000, the emphasis would focus on preparing graduates for industry and programs were required to prepare graduates for professional practice. The third criterion, which stated the program outcomes and assessments listed in Exhibit 6.1, has had a profound influence in many parts of the world. Seven domains are called abilities. They are necessarily composed of many subabilities. Some commentators (investigators) call them competencies, but they are domains of competencies, each of which needs to be specified. Sometimes they are called "skills." It is by no means clear what "understanding" in (f) means, or for that matter "contemporary issues." Clearly the approach is in the tradition of Tyler and Bloom (Chapters 2 and 3), and Besterfield-Sacre et al. (2000) adapted *The Taxonomy of Educational Objectives* for the purpose of measuring

---

Criterion 3. Program Outcomes and Assessment
Engineering programs must demonstrate that their graduates have

a. an ability to apply knowledge of mathematics, science, and engineering;
b. an ability to design, conduct experiments, as well as analyze and interpret data;
c. an ability to design a system, component, or process to meet desired needs;
d. an ability to function in multidisciplinary teams;
e. an ability to identify, formulate, and solve engineering problems;
f. an understanding of professional and ethical responsibility;
g. an ability to communicate effectively;
h. the broad education necessary to understand the impact of engineering solutions in a global and social context;
i. a recognition of the need for, and an ability to engage in lifelong learning;
j. a knowledge of contemporary issues; and
k. an ability to use techniques, skills, and modern engineering tools necessary for engineering practice.

EXHIBIT 6.1. ABET EC 2000 Criterion 3. Program outcomes and assessment.

the ABET outcomes. Cheville et al. (2008) wanted to develop a simplified taxonomy, which could be used for program and self-course assessment. For this purpose, they too adapted the Bloom *Taxonomy*. A major revision of EC2000 has been proposed by ABET (see Chapter 12).

The outcomes approach also informs the European Higher Education Area (EHEA-Bologna process) that describes the outcomes required for the different levels of higher education in a similar way to those defined by the NVQ system in England (see Section 5.1). EHEA standards have influenced methods of teaching in some countries (e.g., Oliveira and André, 2014). Such statements, and many countries have made such statements, enable the mutual recognition of qualifications between countries, for example, the Washington Accord (1989; Oladiran, Pezzotta, and Uzjak, 2013). These developments give rise to the following questions: first, do these criteria meet the present and perceived future needs of industry? Do programs that say they do actually do? And, would the language of educators give the list more life (such as critical thinking, synthesizing, creating, problem solving to be found in the EHEI model [...])? Problem solving is implicit in the list but so are other terms, especially, creativity.

It is clear that the regulators have been influenced by the opinions of industry expressed in earlier years and that that this is not just an American phenomenon. These complaints do not differ in many respects from those made in the 1950s and 1960s in the United Kingdom in that the demand is for what came to be known as "soft skills," for which the term "professional" is now preferred. Pistrui, Layer, and Dietrich (2012) remind us that these are right hemisphere skills in common place thinking about the brain. Early in that tradition in the United Kingdom, the psychologist MacFarlane Smith (1964, chapter 3) pointed out that the neglect of the practical (let alone the personal) in the curriculum in high schools was to the detriment of engineering education since it was here that spatial ability was developed. There is no doubt that advances in neuropsychology will have a great bearing on our understanding of learning within the disciplines, but that is not a matter for discussion here (see Section 7.5). It is, however, important to note that faculty, worldwide, are responding to these demands. There is no suggestion, as there was in the United Kingdom in the 1960s that new curricula should be developed to meet the needs of industry (Chapter 2), rather present curricula should be designed to ensure that these outcomes are obtained.

The review of the literature that is contained in this chapter and those that follow shows that these regulations dominate thinking about the curriculum.

6.2 Taxonomies

Four papers from the FIE conferences (2006–2012) had the word "taxonomy" in their title and one from ASEE. The four FIE papers all referred to the Bloom's *Taxonomy*. One was based on the original version of *The Taxonomy*, and the others on the 2001 revision (Anderson et al., 2001). A proposed application of the revised Taxonomy to final year projects at the University of Auckland has been described by Thambyah (2011). This paper includes the statement of graduate competencies and attributes made by the Institution of Professional Engineers of New Zealand, and which includes decomposed definitions of "complex engineering problems," "broadly defined engineering problems," and "well-defined engineering problems."

Spivey's (2007) paper explains the differences between the two versions of *The Taxonomy* although it does not deal with the criticisms of the original and how much the authors of the revised edition took on board these criticisms (Anderson and Sosniak, 1994). Some of the difficulties of using it in engineering were summarized by Heywood (2005). It is fairly clear that the criticism of taxonomies of this kind that they are not hierarchical has been taken on board. The fact that creativity was not mentioned in the original is partially recognized but in the form of "create." Analyze, Evaluate, and Create are at the same level. Some may regret the passing of "synthesis" and its relation to the ability to be able to think in term of systems, but the authors have incorporated it to some extent within "create." All of the cognitive process dimension categories are now expressed as verbs. Exhibit 6.2 shows that the knowledge process dimension is

---

1. **Remember**

    1.1 Recognize (identify)
    1.2 Recalling (retrieving)

2. **Understand**

    1.3 Interpreting (clarifying, paraphrasing, representing, translating)
    1.4 Exemplifying (illustrating, instantiating)
    1.5 Classifying (categorizing, subsuming)
    1.6 Summarizing (abstracting, generalizing)
    1.7 Inferring (concluding, extrapolating, interpolating, predicting)
    1.8 Comparing (contrasting, mapping, matching)
    1.9 Explaining (constructing models)

3. **Apply**

    3.1 Executing (carrying out)
    3.2 Implementing (using)

4. **Analyze**

    4.1 Differentiating (discriminating, distinguishing, focusing, selecting)
    4.2 Organizing (finding coherence, integrating, outline, parsing, structuring)
    4.3 Attributing (deconstructing)

5. **Evaluate**

    5.1 Checking (coordinating, detecting, monitoring, testing)
    5.2 Critiquing (judging)

6. **Create**

    6.1 Generating (hypothesizing)
    6.2 Planning (designing)
    6.3 Producing (constructing)

EXHIBIT 6.2. The cognitive process dimensions of the revised taxonomy with alternative names in brackets. Anderson, L. W., Krathwohl, D. R., et al. (eds.) (2001). *A Taxonomy for Learning, Teaching and Assessing. A Revision of Bloom's Taxonomy of Educational Objectives.* New York: Addison-Wesley/Longman. (Inside back-cover).

Objective	Cognitive Process Dimension	Knowledge Dimension
1. Understand how logic expressions express an output as a function of various inputs.	Understand	Conceptual
2. Represent basic logic expressions (AND, OR, NOT) using series and parallel combinations and switches.	Understand	Conceptual
3. Represent a logic expression with a truth table.	Understand	Conceptual
4. Identify the logic functions of AND/OR/NOT gates.	Remember	Factual
5. Implement circuits from logic expressions using gates.	Apply	Procedural
6. Interpret the logical function in a schematic.	Understand	Conceptual
7. Draw a timing diagram for a logic circuit.	Apply	Procedural

EXHIBIT 6.3. Reorganization of Spivey's text to show the matrix consequent on the revised 2001 taxonomy functions. Spivey also relates the objectives to activities that should obtain the objective, homework, and quizzes. Spivey, G. (2007). A Taxonomy for learning, teaching and assess digital logic design. In: IEEE/ASEE Proceedings of Frontiers in Education Conference, F4G-9 to 14.

changed to "Remember" with subcategories of recognizing and recalling. Knowledge is divided into four major types expressed as nouns: factual, conceptual, procedural, and meta-cognitive. The two dimensions can be arranged to form a matrix, and Spivey gives an example of its use in respect of a digital logic taxonomy (Exhibit 6.3). Spivey presents the idea of "the taxon," which is "the bundling of lecture notes, handouts, activities, homework assignments, solutions, quiz questions, assessment techniques, all tied together around the unit of a class lesson, aligned and advertised using the taxonomy table" for sharing among the digital logic design community. Clearly, the same idea could be applied to other courses that are generally required. It has some similarity with the view put forward by Carter and Lee (1974) in the United Kingdom that all universities could set a common first-year examination in electrical engineering since they were all offering much the same course.

Even if not specifically in the title, there are numerous references to *The Taxonomy* in the literature often in relation to the achievement of ABET outcomes (e.g., Almarshoud, 2011).

Related to Spivey's analysis is the query by Johnson and Fuller (cited by Highley and Edlin, 2009) as to whether *The Taxonomy* is really suitable for computer science since application is the primary aim of computer science, and higher-order skills are reached only in the final year. The starting point of Highley and Edlin's (2009) assessment of a two-course Discrete Mathematics program is the principle that the grades students receive should correspond to their levels of understanding. From a qualitative analysis of what the students could do for each of the grades, they defined their learning objectives. Exhibit 6.4 is a précis of their text.

They implemented the system with the expectation that students had passed all of the objectives that preceded a given grade level. Candidates for the grade of B would

Objective	Self-assessment Questions	Cognitive Process Dimension
A. Shows mastery of all course material and can formulate proofs on even difficult questions and apply their knowledge to new questions.	Given the description of a method of numeric representation (e.g., one's complement, two's complement, binary code decimal, Cantor expansion) I can use it effectively.	Analyze Evaluate Create
B. Shows mastery of the basic skills and proficiency with the advanced skills in the course, including completing proofs.	I can convert a number between any two bases. I can describe how to convert between different bases.	Apply
C. Shows proficiency with a core set of skills	I can convert a number from its decimal expansion to its binary or hexadecimal expansion. I can convert a number from its binary or hexadecimal expansion to its decimal expansion.	Understand
D. Should understand the language of discrete mathematics.	I know these terms: base b expansion of n, binary expansion, hexadecimal expansion and octal expansion.	Remember (tested by vocabulary comprehension)

EXHIBIT 6.4. Précis of Highley and Edlin's text. The illustrations they give are clearly questions that students can use for self-assessment, hence the title of the second column. Highley, T., and Edlin, A. E. (2009). Discrete mathematics assessment using learning objectives based on Bloom's Taxonomy. In: IEEE/ASEE Proceedings of Frontiers in Education Conference, M2J-1 to 6.

have to have passed grades C and D. Students received credit for the preceding grades. While it caused students to focus on the material they did not know well, which is a purpose of positive assessment (i.e., the enhancement of learning), it also caused an "undue amount of anxiety." In another implementation of the course, they based the grades on the number of learning objectives completed so deficiency in one area could be made up by stronger performance in another. They were surprised by students who gave up working when they had acquired the number of objectives for the grade required. They then gave a course where 60% of the grade was based on progress with learning objectives and 40% divided equally. This caused students to focus on the material they did not know well, which, as indicated, is the purpose of positive assessment (i.e., the enhancement of learning), but it also caused an "undue amount of anxiety."

Given the problems that students have with mathematics, the following quote from Highley and Edlin is given in full: "We were surprised at how challenging some students found the vocabulary sections on tests. As these were tested with either multiple choice or by matching the appropriate word from a given list to the correct definition we expected that students would find this quite straightforward. However, there were cases where the students were unable to meet this objective but were able to obtain the C and B level objectives for the same topic. This does perhaps demonstrate the challenge of learning to speak mathematics. Understanding the difference between 'at least one' and 'at most one' within a definition can be harder than visually recognizing a one-to-one or onto function."

This point is also a reminder that question setting is often trivialized when in fact it is much more difficult than it seems. In the United Kingdom, where the approach to examining and testing has been different from that in the United States, Jones et al. (2009) were caused to ask "are student's learning what we think they are learning?" which is to repeat the question in another form that Furneaux (1962) asked in 1962 (see Section 2.5). They, like the engineering scientists, tried to answer his question what objectives we should be testing, and like that group they chose Bloom's original taxonomy to analyze what they were doing, but without the consideration the engineering scientists gave to problems of utilizing the categories of the *Taxonomy* directly. However, their statements of learning outcomes followed the same pattern, namely a statement of what is was that a student should be able to do.

Jones et al. analyzed the questions in examination papers given to second- and final-year students in eight programs. Their purpose was to distinguish between lower-order, intermediate-, and higher-order cognitive questions in the light of Bloom's *Taxonomy* in order to ascertain whether critical thinking and problem-solving skills were being tested by effective questions. While the "verb" in a question indicates the cognitive level sought, any analysis has to take into account the focus and perspective of the question, so the whole statement had to be reviewed. They give the following example: "*State Ohms Law*" is a lower-order question whereas "State reasons why feedforward control is preferable to feedback control" is an intermediate-order question. It is not possible to do justice to the study because the work had to take into account numerous variables that the authors describe in detail. They confined themselves to questions in examination papers only in order to establish relationships between expected cognitive levels of a course and the questions asked. It was not simply a face validity exercise on the questions but took into account the examinee's marks per question.

Just how difficult such analyses can be, and just how important they are in setting questions, are shown by the fact that no direct relation was found between question level and mean mark. Neither was there any obvious link between student choice and the cognitive level of the question. However, they did find "plenty of examples of good practice in aligning the examination question levels to the required learning outcomes. But there were also a significant number of instances where there is little association matching taking place." More significantly, they found few learning outcomes in the higher-order area. They thought that this might be appropriate because while the skills of evaluation and creativity are required of engineering graduates, they are difficult to assess in traditional examinations and might, therefore, be better assessed by some form of coursework or project. One might suggest nontraditional type of examinations such as extended prior notice examinations (Cameron and Heywood, 1995). It would be interesting to know what a factorial analysis of their results would have yielded.

The assumption with these three papers is that the Bloom *Taxonomy* either in its original or revised forms necessarily embraces the skill sets that an engineering education should provide. But as the previous sections show there are things that such an education should seek to obtain that are not adequately provided for (or perceived to be) by *The Taxonomy*, for example, communication. A number of other criticisms of *The Taxonomy* have been made (summarized in Heywood, 2005). Eisner's comments on the limitations of the objectives approach are given in Section 10.9. The papers summarized above do, however, highlight the link between objectives, assessment, and teaching, and Castles, Lohani, and Kachroo (2009) indicate how hands-on learning activities in which concepts are developed through a spiral approach might obtain the goals of the original *Taxonomy*.

Poyner et al. (2011) have suggested that *The Taxonomy* is inadequate for the task of modeling or visualizing complex learning or delivery systems like just-in-time teaching. Their focus is on the utilization of simulation by industrial engineering students. They modify *The Taxonomy* by reducing the number of levels to 4. Knowledge and comprehension form one level. Analysis and synthesis are placed at the same level. Because students can memorize the applications of knowledge without fully comprehending the knowledge they are utilizing their model allows the learner to "skip" a level "For example, if asked to calculate the average time-in-queue for an M/M/1 queuing system students are quite capable of memorizing the formula, selecting the appropriate data to 'plug' into the formula, and solving for the unknown parameter—but they are often not capable of 'stating in their own words' what the average time in queue means for a steady-state queuing system. So, in contrast to what Bloom supports, we do allow for learning to 'skip'—a student may not master comprehension but will (i) go from knowledge to application, or (ii) from application to knowledge (the 'learn-by-doing' style)." Coincidentally, and once again, this example illustrates the weakness of *The Taxonomy* in not having a category for communication. What happens to the student who comprehends but cannot communicate that understanding?

Poyner et al.'s model supports learning between and within levels in any direction which would seem to be an important development, irrespective of the other dimensions of the model. They show the relationship of their version of *The Taxonomy* to concurrent learning by the use of a pyramid/prism (3D Knowledge Pyramid/Prism Model) in which the second face is "concurrent learning." They write "the student simultaneously requires prior (old) knowledge (e.g., statistics and queuing theory). The just-in-time delivery and learn-by-doing teaching and learning styles may be viewed as 'concurrent learning'; where separate learning objectives (knowledge) are being achieved (built) in a synchronous or asynchronous manner." Their hope is that the model will assist instructors to view the relationships between and among knowledge for complex learning topics. The kind of situation they envisage is where in new degree structures students have to obtain pre-requisite knowledge for themselves where once it was a formal requirement particularly in accelerated programs.

There are other taxonomies and one due to Fink has been used by Ferro (2011) to establish course objectives for a redesigned materials engineering course. Fink's (2003) taxonomy has six categories. These are Foundational knowledge, Application, Integration, Human Dimensions, Caring and Learning-how-to-Learn. These categories emphasize the importance of the names (titles) given to the domains. Apparently (at first sight), these categories show as much concern for the affective domain as they do for the

Learning Objectives	Fink Taxonomy Category
1. Identify important sources of materials information.	
2. Be more interested in materials.	Caring
3. Respect their own abilities to make informed engineering decisions about materials.	Human dimension
4. Describe the effect of atomic bonding on properties for a solid material.	Integration
5. Select the appropriate class of material for general application	Integration
6. Predict the effect of increased C on the mechanical properties of steel	Application
7. Calculate weight fractions of phases in a microstructure from an equilibrium phase diagram, given an alloy composition and temperature	Application
8. Determine phase compositions for all phases in a microstructure from an equilibrium phase diagram, given an alloy composition and temperature.	Application
9. Determine the ultimate tensile strength, yield strength, elastic modulus, and ductility of a material, given a plot of stress as a function of strain from a tensile test failure.	Application
10. Identify and classify the major types of solid materials	Foundational knowledge
11. List the three crystal structures	Foundational knowledge

EXHIBIT 6.5. Course objectives for MENG221, 2010. Showing Fink Taxonomy category. The Podcast project is for learning objectives 1, 2, and 3 in Ferro, P. (2011). Use of Fink's Taxonomy in establishing course objectives for a resigned materials engineering course. In: Proceedings of Annual Conference of the American Society for Engineering Education, Paper 309.

cognitive, and may be more attractive to some engineering educators than the categories of the Bloom *Taxonomy.*

Fink argues that a redesigned course should contain a "rich learning experience" that is designed to obtain more than one objective. The example cited by Ferro is of a service learning course where a community service project involving recycling would involve Fink's dimensions of "Caring" and "Application." Ferro's paper focuses on the design of the rich learning experience that involves foundational knowledge, application, caring, and learning-how-to-learn objectives. These are set out in Exhibit 6.5. Teams of students are required to produce an audio podcast on "materials" that is suitable for middle or high school science classes. It had to be between 1.5 and 3 minutes length. In order to understand what was required of them, the students had to watch some award-winning "materials" podcasts and evaluate them as part of a listening assignment. (This is a good educational practice as Cowan (2006) and Heywood (2008) show.) When a team had produced their own podcast, its members were required to listen individually to the podcasts of four other teams and evaluate them against specified criteria—for "overall quality," "entertainment," "good for kids," and "educational." Ferro found that student evaluations revealed a "relatively higher level of enthusiasm for the Podcast project compared with a written report even though a research paper can provide a rich learning experience." Ferro concluded that taking together all the things students had to do, "the podcast project may be more effective in increasing interest in 'materials,' which may be a necessary step to learning about 'materials." Put in another way, it may be part of the first stage of learning that Whitehead calls "romance" (Whitehead, 1932). Although learning objectives 4 through 9 are accomplished by active learning exercises and assessed by a combination of homework, quizzes, and final examination, no information is given about these so it is not possible to tell how well the questions asked obtained the required learning outcomes. Neither is it possible to say how near it approaches the model of multiple-objective (strategy) examining described in Chapter 3.

McCahan and Romkey (2014) have also published a taxonomy that spans the cognitive, affective, and psychomotor domains that is intended for teaching engineering practice. It is based on Miller's (1962) pyramid (see Section 1.3). Engineering practice levels are described for each level of the pyramid, that is, knows how, shows how, and does. While it makes substantial reference to the literature on professional practice *per se*, there are no references to the major studies of engineers at work and the bearing they might have on the taxonomy (see Chapters 2, 4, 10, and 11). Nevertheless, it shows a way forward in understanding how the knowledge of engineering science is utilized in practice.

6.3 Outcomes-Based Engineering Education

The Bloom *Taxonomy* is the earliest expression of part of Tyler's theory of curriculum put into practice (Tyler, 1949), although the idea of giving precision to the objectives of education can be traced back to Bobbitt in the early years of the twentieth century (Flinders and Thornton, 2009). It is not possible to say that the term "objectives" has been substituted with the term "outcomes" because some writers continue to use the term *objectives*. ABET use both to the confusion of some engineering educators (see below). Declaring objectives seems to imply that what we do with students will obtain those objectives: outcomes seem to imply an unwarranted certainty that that is what will happen. Be that as it may, "outcomes" has been adopted by accrediting agencies and those who dictate policy worldwide. In the United Kingdom, the first reported studies of outcomes in engineering were in 1991 (Otter, 1991, 1992).

A study that has some similarity with Highley and Edlin's study (above) has been reported by Slamovich and Bowman (2009). It is of developments in the Purdue School of Materials Engineering to integrate course objectives with student outcomes so as to provide a process for continuous improvement. In order to replace the hypothetical "average" student with something that reflected the variability of student performance, three categories of performance were introduced and related to course objectives. The first level of the tier expected that *all* students should obtain the objectives set. The second level required that *most* students should achieve its objectives, and the third level was the expectation that *some* students should achieve its objectives. Slamovich and Bowman point out that these levels are related to the original *Taxonomy* (see Exhibit 6.6).

Category	Objectives to Be Achieved by:	Examples as Cited in the Text	Bloom Categories
1.	All students	Information processing with little analysis	I–IV
2.	Most students	Information analysis and some synthesis relating data to physical phenomena, or interrelating physical phenomena	III–V
3.	Some students	Application of information in a different context from where the information was initially gleaned with emphasis on synthesis and evaluation.	V–VI

EXHIBIT 6.6. Categories of students and what is expected of them in the Purdue Materials Engineering tiered assessment Protocol. Slamovich, E. B., and Bowman, K. J. (2009). All, Most or Some: Implementation of Tiered Objectives for ABET assessment in an engineering program. In: IEEE/ASEE Proceedings Frontiers in Education Conference T3C-1 to 6 (tabulation derived from the text by this writer).

Objectives were established for each of the levels, and these, in turn, relate to ABET criteria. Examination questions were designed to obtain the objectives. They reported that "the linking tasks in a laboratory report or a technical paper requires more forethought on the part of the instructor to ensure that the overall score in an assignment can be parsed to reflect the contributions of the components of the assignment and corresponds to different course objectives and ABET outcomes a-k." In this system, performance on course objectives is a measure of student achievement, "Because 'all' students are likely to perform differently on different days and a great deal of coursework and laboratory work involves partial credit, a threshold for performance is used for assessing student performance on 'all' objectives. For any individual to satisfy a given 'all' course objectives, he or she must achieve a score of at least 60% of the possible points on the associated task (e.g. exam question; lab report section) [...]."

The collected data enable the tutors to see at a glance the average score on each question, the percentage of the sample obtaining more than 60%, and whether or not the class goal is obtained in terms of the levels. Together with other data, faculty have been able to consider course improvement and curriculum change and enabled faculty to focus on the relationship between course content and course assignments. Whether or not this influenced instruction is not made clear. The same applies to the description of another assessment process at Washington State University (Lang and Gurocak, 2008). In the extract from the conclusion of their paper that follows, it is not clear if the results impact on teaching even though "faculty may make changes to the curriculum 'to close the loop'." They wrote: "In this paper we presented a 4-level assessment. One of these levels is the performance criteria which ties course outcomes to program outcomes. The performance criteria describe the standards and means by which students demonstrate their achievements of a program outcome and represent faculty's interpretation of the outcomes. In our approach we interpret performance criteria in the context of each course to derive course outcomes. Then, faculty assesses student achievement of the course outcomes by each student through questions and assignments targeting each course outcome. These assessments are direct measures. Data collected

A. Defined Levels	B. Levels of Competency
Basic: You are able to understand a discussion about and follow directions related to the competency.	0—Little to no knowledge 1—Basic level 2—Between basic and intermediate
Intermediate: You are comfortable making decisions about and leading discussions.	3—Intermediate level 4—Between intermediate and expert
Expert: Many others look to you for knowledge about the competency.	5—Expert knowledge

EXHIBIT 6.7. Tabulated by the writer from Squires, A. F. and Cloutier, A. J. (2011). Comparing perceptions of competing knowledge development in systems engineering curriculum. A case study. Proceedings Annual Conference American Society for Engineering Education. Paper 1162.

in each course and through other sources, including exit and student course surveys and focus groups, are assembled to arrive at a percentage score for each program outcome." As to the meaning of the score, they said it represents "the percentage of opportunities given to students to demonstrate the achievement of an outcome where they met or exceeded faculty's expectations." The introduction of their paper is of interest because it points out that the ABET system continues to evolve and they give the criteria for accrediting computer science programs in 2007–2008.

It seems from the papers inspected that there may be a temptation to consider that lists of outcomes and designing criterion referenced assessment are inherently valid; hence, there is apparently little interest in validity. However, validation studies are likely to show, as they did for the engineering scientists that assessment and curriculum design are more difficult than it seems. At the present time this is illustrated by Squires and Cloutier (2011), who compared the perceptions of instructors and students of the competencies addressed in one of the following web-based campus courses in systems engineering (three courses), engineering management (one course), and software engineering (one course). They distinguished between the three levels of competency shown in column A of Exhibit 6.7, scaled as in column B. Thirty-seven competencies were defined.

The instructors were asked for their perceptions of the competencies covered by the course and the expected student competency knowledge growth. The students were asked for biographical data, including education, course completed, delivery, and discussion used, together with a statement of overall satisfaction. About half the students had less than 10 years' professional work experience, but 30% had more than 20 years' experience. The average was 12.5 years' professional work experience and 4.5 years of systems engineering experience which makes it a reasonably mature population. They were then asked for their perception of the competencies covered and their knowledge growth in each of the 37 competencies.

The analysis focused on level of agreement by course and level of agreement in respect of eight competencies. Interestingly, the course that had the highest level of agreement between instructors and students, and between students and students, was a research methods course, but this was due to the fact that the specific competencies were not addressed in the course, specifically, by course content. Otherwise, only 2 of the remaining 11 courses reached 50% or more agreement between instructors and students. When student ratings were compared with each other, the extent of agreement was only slightly higher averaging 45%.

In respect of the eight competencies, the highest level of agreement between instructors and students was for the competency "Define/manager stakeholder expectations," and the lowest was for "perform trade studies."

The authors recognize that much more data could have been collected including biographical data from the instructors to see if experience was an important variable. Apart from recognizing the need to extend the analysis to the other 29 competencies, the authors write "this research, as defined, focused on competency knowledge development, rather than addressing changes in both cognitive competency (knowledge) and affective competency (behavior and attitude) both of which are important to maturing as systems engineers" […]. They might have said "engineers." It is this dimension, particularly as it is extended into the causes of behavior that is missing in so much discussion related to outcomes, and the associated belief that teamwork will solve many problems. This validity study shows that if there is not a common base of understanding between instructors and students, assessment is problematic.

The problem of the perception of assessment instruction is international. A Spanish study reported that 22% of the students participating in a voluntary nongraded Self-Assessment activity said they had completed it to obtain a grade! (Serra-Toro, Travers, and Amengual, 2014).

Given the significance that is currently attached to courses in entrepreneurship and given that there is no definitive definition of entrepreneurship, it is not surprising to find that instructors and student definitions of entrepreneurship differed. The instructor's definitions were broader. Of interest is the fact that students considered that courses in entrepreneurship were a means to becoming their own boss and starting a business. Reeves et al. (2014) pointed out that this perception could put off some students from considering entrepreneurially related careers.

A paper on civil engineering outcomes from the Coast Guard Academy reiterates the need for careful planning (Jackson et al., 2012). To demonstrate the need for planning, the authors compare their current model with their previous model and show how it has been simplified and streamlined. At the beginning of their paper, they rehearse some of ABET's list of shortcomings in their original program, thus:

Inadequate evidence that the process in which objectives are determined and periodically evaluated is based on the needs of the constituents (criterion 2).

Confusion between the definition of program institutional objectives (criterion 2) and program outcomes.

Inadequate evidence of using results of evaluation of objectives and/or assessment of outcomes for improvement.

Inadequate evidence demonstrating achievement of objectives or outcomes. (ABET 2010)

It is argued that ABET visits (and those of other accrediting agencies) are helpful to departmental thinking as well as to getting things "right" for the purpose of accreditation. Similarly, evaluating outcome schedules can show weaknesses and lead to improvements as happened in project work in computer engineering at the University of Technology Mara in Malaysia (Mansor et al., 2008).

Outcomes assessment surveys are often administered. In the case of an Introduction to Mechanical Engineering course at the University of Texas at Austin, students responded to pre-, mid-, and postcourse outcomes assessment surveys (Barr, Krueger, and Aanstoos, 2006). It is interesting to note that the first assignment required the students to complete the Myers-Briggs Type Indicator (MBTI-Personality Inventory; see Section 7.2), and a personal *resumé*. The former is used to assign students to teams and the latter for interpersonal skills development. This is followed by a semester long-team project that is accompanied by six associated assignments. A final project is completed when a team of four students reverse engineer a mechanical assembly. There are other interventions, but the eighth assignment is the completion of a small student portfolio that, it is hoped, the students will update as they matriculate through the curriculum.

In one of the few papers to acknowledge that an increasing number of graduates are finding employment in small firms that require them to have "a broad range of skills and knowledge beyond a science and engineering background" Duval-Couetil, Reed-Rhoads, and Haghighi (2010) describe the use of a survey to help with the design of an assessment instrument to examine the multiple outcomes of entrepreneurship in engineering education (see also Creed et al., 2002). An overview of assessment instruments is provided (see also Shartrand et al., 2008). One of their problems was to identify validated instruments. In the event, six categories were used. They were (1) attitudes, (2) behaviors, (3) knowledge and skills, (4), self-efficacy, (5) perceptions of programs and faculty, and (6) demographic. They note that little research has related the influence of learning outcomes on student attitudes, behaviors, career goals, or professional competence that it is hoped a well-designed entrepreneurship curriculum would develop positively.

The survey that comprised 135 items was administered among senior capstone design course level students at three institutions where entrepreneurship courses were available to engineering students. The preliminary analysis suggested that the scales had high reliability. The evaluation was ongoing.

On an altogether different scale, Pistrui, Layer, and Dietrich (2012) report a study undertaken with the aid of the Kern Entrepreneurship Education Network (KEEN) of a new class of engineers called "Entrepreneurially Minded Engineers" (EME) who are to be found working not only in Fortune 1000 firms but in small- and medium-size enterprises as well. The characteristics of EMEs are shown in Exhibit 6.8. Pistrui, Layer, and Dietrich aimed to study hypothesized relationships between EME behavior, motivation, and exhibited skills (313 individuals), and from that to identify key EME attributes that may be incorporated into current engineering pedagogy. Forty-one practicing engineers took part in the study, together with 2004 students. The assessment dimensions were behavioral (knowledge of self—how one interacts with group in team setting), motivational (what drives actions, the why of your actions, and causes of conflict), and personal and professional competencies (skills learnt through human interaction and practice–communicating–leading–managing–teaming). Each latent variable is

Opportunity orientations	Searching to identify and solve real-world problems that improve people's lives through value creation.
Technical empowerment	View technology as an enabler used to solve problems and create value for customers in a dynamic and changing global market place.
Business fundamentals	Understanding the business and industry the firm is in and support the advancement of the corporate agenda.
Interpersonal dynamics	A clear understanding of given situations and providing projects with leadership and teamwork through good communication.
Forward thinking	Intellectual and personal curiosity in the form of looking for "what's next" and effectively and economically applying new methods.

EXHIBIT 6.8. Some characteristics of EMEs listed by Pistrui, D., Layer, J. K., and Dietrich, S. L. (2012). Mapping the behaviors, motives, and professional competencies of entrepreneurially minded engineers in theory and practice: an empirical investigation. In: Proceedings of Annual Conference of American Society for Engineering Education. Paper 4615.

described by manifest variables that, in the case of behavior, are Dominance, Influence, Steadiness, and Compliance. Those related to Motivation are Theoretical, Aesthetic, Traditional, Individualistic, Social, and Utilitarian. The skills latent variable had 23 manifest variables. This gives some idea of the scale of the analyses undertaken. The important finding was that freshmen ($N = 1717$) and senior students ($N = 287$) exhibit the same "fingerprint" as the EMEs, but it is at a lower level. That "fingerprint" indicated that if an EME's behavior in "high" dominance and "low" steadiness is increased, the associated skills also increase. These attributes do not appear to be affected by time on the course. The authors suggest that more could be done in the curriculum to enhance the attributes of leadership, employee development/coaching, conflict management, flexibility, goal orientation, persuasion, and futuristic thinking. The practicing engineers exhibited a different "fingerprint." The data suggest that they had lower interpersonal skills, lower creativity, lower goal orientation, and lower negotiation skills. This is one of the few studies that link the professional with the personal.

Outside of the ABET remit, Hamid Khan (2009) of Our Lady of the Lake University reported on the redevelopment of a course in Business Ethics, which began with an international survey from which he and his colleagues used the American responses to determine the general outcomes shown in Exhibit 6.9.

6.4 Mastery Learning and Personalized Systems of Instruction

An outcome of competency-based assessment in the United States in the 1970s were two similar approaches to course construction and learning called Mastery Learning

---

1. Demonstrate an understanding of how the social, economic, political, technological, and ecological dimensions of internal and external environments create a moral and social context for business decision making.
2. Demonstrate an ability to apply personal values and ethical principles as a basis of identifying, analyzing, and managing ethical issues in contemporary business settings.
3. Demonstrate the ability to analyze the influence of critical stakeholders on business operations and to apply principles of stakeholder management to contemporary issues in business practice.
4. Demonstrate an understanding of the complex interdependences that exist between business and government, and of their strategic importance to corporate decision making.
5. Demonstrate an understanding of the legal, ethical, and social responsibilities of business toward their members, customer, and natural environment.
6. Demonstrate the ability to recognize and solve contemporary ethical and social issues in the business, economics, or public administration decision-making process.

EXHIBIT 6.9. General outcomes for a business ethics course derived from a survey. Khan, H. (2009). Measuring learning outcomes from the ethics course. In: IEEE/ASEE Proceedings Frontiers in Education Conference M4G- 1 to 6.

and personalized Instruction (PSI), respectively. The former originated with Bloom and Carroll and the latter with Keller—hence, the term "Keller plan" (Stice, 1979). Engineering educators became interested in them and several articles about them appeared in *Engineering Education* (Blair, 1977; Gessner, 1974; Stice, 1979; Koen, 1985).

The principles of learning on which they are based are:

1. Given that aptitude is normally distributed, then variations in the amount and quality of instruction can bring every student to the same level in a particular subject.

2. Anyone learns best when they know what is expected of them.

3. Anyone learns best when they have learnt the procedures that have to be followed in learning new material. For this reason, it is important to relate the new learning to what a student already knows.

4. Programs should be designed to respond to the learning strategies that the students bring to the problem. Students learn in a variety of ways; therefore, a variety of ways should be provided.

5. Learning is most proficient when it is undertaken in relatively small units that are undertaken regularly and for which feedback is given at the time. Regular assessment of small bodies of knowledge is, therefore, a prerequisite of efficient learning.

6. Because learning proceeds in a sequence of logical steps, the units should be arranged to move from the simple to the complex. The learner should complete each step successfully before passing to the next. Therefore, feedback should be provided at each step. Tutoring should be provided to help students over difficult hurdles.

Keller's scheme (PSI) embraces the following features:

1. The student proceeds at his or her own pace.
2. Complete mastery must be obtained.
3. Lectures are used as a means of motivation (students qualify for lectures that are used for a reward: the material in the lectures is not examined).
4. Proctors should be used. They are more senior students to whom students bring their work.

Both systems are criterion referenced. That is, grading is against mastery of the criterion (objective/outcome) and is not related to the performance of the other students. In the early reports of its use in engineering, the 100% criterion was modified to 90% or even 80% because a student could show that he or she knows how to solve the problem yet make a mistake in the arithmetic and get a wrong answer, which would normally lead to a repeat of the unit. This was the same sort of dilemma that the examiners of engineering science coursework faced when they examined course work using a criterion-referenced scheme (see Section 3.2). Bloom's approach used an 80–90% criterion.

In 2014, engineering faculty at Pennsylvania State University, Behrend College, Erie, reported in two papers on the use of mastery learning over a period of 8 years. The former (Sangelcar et al., 2014) describes the system for a statics course and discusses the results of previous and present surveys of student opinion. They show in detail how the open-ended items in their survey were analyzed. The latter (Ashour et al., 2014) gives a similar introduction but reports on interviews with faculty members who have or who are using the mastery system. A modified Bloom approach was adopted for the design of the system and the system was paced by the instructors.

The syllabus was broken down into six modules (force vectors and equilibrium of particles: Force system resultants and moment about an axis: Reduction of distributed loading and equilibrium of a rigid body 2D: Rigid body equilibrium 3D & Method of joints-trusses; Method of sections: Trusses & frame: Internal Forces and friction). Two or three problems were set per test with the intention of testing the ability to solve a particular type of problem. They did not try to test multiple concepts at the same time. The problems were similar to those done in homework and in class. Each problem is graded for whether it is correct, almost correct, retake, and not attempted. One hundred percent is given for a correct first attempt answer. If it is attempted a second time, a correct answer receives only 85%, and on a third attempt only 70%. Similarly with "Almost" correct where the scores are 90%, 75%, and 60%, respectively. The retake scores are 0% in all cases, and not attempted receive −15%, −19%, and −5%, respectively. A final examination for which partial credit is given covers the entire syllabus plus Center of gravity and centroid 2D and 3D. The retake examinations contain similar problems designed to test the same concepts.

If this writer's experience of innovation is anything to go by, the planning for such courses will have taken up much faculty time yet the rewards can be great. In the example above, once the course was going, the method of conduct caused a significant shift in workload, and a greater burden was placed on the students to take responsibility for their own learning. "For the instructors, very little time is spent on agonizing over the partial credit to award for an incorrect solution and more time is spent creating retake problems. Additionally, since detailed comments are not provided on an incorrect solution the time spent grading is markedly reduced for a given exam. For students who were not able to correctly solve a given problem on an exam, more study time is encouraged by not awarding partial credit and the student is required to spend more time attempting to solve problems during the retake exam. By only noting where the solution became incorrect and not giving detailed comments (or providing solutions) the students are encouraged to solve the problems on their own rather than attempting to memorize an instructors solutions. Students are told that help will be given relating to incorrect solutions on an exam problem only if they re-attempt the problem on a separate sheet of paper to the point where they do not understand and bring this to office hours [...]" (Sangelkar et al., 2014).

Comparing previous students with present students led the team to conclude that students recognize the benefits of the mastery approach after the course but not necessarily while they are taking it. One study found that students were against the principles of PSI even though there were fewer dropouts (Hereford, 1979). Hereford thought that the type of pacing might influence the dropout rate. Those on the course felt that they might have obtained better grades had a partial grading method been used. The authors suggest that grading inconsistencies between professors could possibly affect student opinion about mastery learning for which reason a more consistent grading rubric could be designed. In my department, we had the greatest of difficulty persuading our students that standardizing the marks of the different examiners was much fairer than simply adding raw scores (Cameron, 1984; Heywood, 2000). It also reduced the hassle of adjusting marks at the examiner's meetings.

In their other report, the six instructors who were interviewed thought that mastery learning helped them understand better student needs and achievements (Ashour et al., 2014). They also thought that the learning experience was improved. However, they felt that the flexibility in writing examination questions was reduced. Academics like to be flexible and many of my colleagues believe that students should be given choice as to what questions to answer. But there is only local evidence that choice leads to better performance. There was also the worry that if students have negative feelings about the approach they might impact on the ratings they give instructors. The recommendation in the first paper (Sangelkar et al., 2014) to improve the rubric is also repeated and the idea of building up an item bank is suggested.

The concepts of continuous assessment and mastery learning are intimately related. This is no better illustrated than by the design of a first-year course for aerospace, civil, and mechanical engineering students in the core subjects of mathematics, fluids, solids, and structures at the Queen's University, Belfast. Cole and Spence (2012) describe how on the basis of a substantial knowledge of research in engineering and higher education they responded to the challenge of teaching a large class by using continuous assessment. They were influenced by American Research that showed that subject matter learning "depended strongly on the quality of the students effort or engagement with learning opportunities" (Pascarella and Terenzini, 2005). American research also shows that active learning enhances understanding and written guidance in how to run active learning in large groups was taken (Felder and Brent, 1999, 2003; Smith et al., 2005). Assessment was designed on the basis that it should support learning and for this reason they accepted that the assessment activities should direct students learning toward the significant aspects of the course (Gibbs and Simpson, 2004). The result: in the lectures, formal teaching was interspersed with elements of active learning. These were supported by smaller group tutorial classes of between 25 and 30 students. During these tutorials, the students were given worksheets with five or so questions of increasing difficulty. These ranged from requirements for short and long answers, the long answers having the purpose of promoting "deep" understanding of fluid behavior. Between weeks 3 and 11, the students also took a 10-minute test. The lecture material was also supplemented by 3.5-hour laboratory classes. At the end of the course, a 2-hour examination comprising 20 multiple-choice questions (40%) and three structured questions (chosen from 5) worth 20% each was conducted. This counted for 80% of the final mark. The "partial" mastery element comprised the 10-minute tests because a condition was made that a student must pass six of the nine tests. In a "pure" mastery learning approach, the student would have had to have passed each test before moving on to the next. Cole and Spence provide substantive evidence to show that as might be expected, tutorial attendance improved, and that examination performance improved significantly. Notwithstanding the fact that students thought that the structure encouraged continuous learning and helped them build confidence, some might consider the weighting of the coursework to the final examination to be somewhat harsh!

In the 1970s, it was reported that any form of test, including oral questions, could be used with the Keller model and that testing was much longer than tests with the equivalent mastery learning approach (Heywood, 1989). The relationship of mastery approaches to computer-assisted learning and assessment is self-evident. One of these is adaptive learning that is a form of personalized instruction and some papers report activities in this area. One, a Scottish work in progress is proposing to use it to help students master foreign languages (Rimbaud et al., 2014). In Europe, it is a core competency for computing students; therefore, any technique that fosters mastery is welcome. Rimbaud and his colleagues describe adaptive learning as follows "these systems diagnose what learners individually know and don't know (Intelligent) then generate learner specific content and interfaces, including learning paths and features that match the learner's preferred approach (Adaptive). Learners then explore articles, eBooks, audio files, videos, quizzes, and courses covering fresh topics instead of unnecessary repetition (while repetition is effective in some situations it can be demotivating in others)." It is claimed that learners find it saves them time and that it can either supplement or replace individual tutors—a much-discussed topic in respect of e-learning. Similar technologies are being used in the development of "serious" games for educational environments. Moreno-Ger et al. (2014) consider that assessment is the "key issue for converting serious games into mainstream educational content." With these developments comes the concept of "stealth assessment." It is the gathering of data during the playing of the game so that "inferences can be made about the level of relevant competences." It is intended to blur the limits between assessment and learning that would seem to integrate assessment into the design of the curriculum.

6.5 Discussion

The literature discussed in this chapter draws attention to several features that should be a cause for concern. The first relates to the issue of validity. It seems that there is a widespread assumption that if an outcome statement is agreed and written that it is correct. This is a face validity judgment. As was shown in Chapter 3, students did not perceive all the items in a criterion-referenced scale in the same way as the assessors. Similarly, the study reported by Squires and Cloutier (2011) and other studies give cause for similar concern. The agreement between instructors and students was poor. How many courses are being offered where there are differences that are unknown to the assessors?

The value of using other measures such as the MBTI to assign students to teams is highlighted (Barr, Krueger, and Aanstoos, 2006). Similarly, the comparative study of entrepreneurially minded engineers and students focused on personal and professional competencies and as such the affective domain. It is one of the few studies that linked the professional with the personal. It suggested that much more could be done in the curriculum to enhance the attributes of leadership, employee development, conflict management, flexibility, goal orientation, persuasion, and futuristic thinking. While this applies to entrepreneurship studies, employers complain that graduates generally do not have the "soft" (professional) skills of the kind listed. The central question is whether they can be expected to gain them without real-world experience in the unreal world of academia? Are competencies a function of the situation or innate to the person? This issue will be discussed in Chapter 11.

In 1962, Furneaux could not establish what engineering examinations tested and suggested that engineering educators should determine what objectives they wanted to assess. Few papers have been published in subsequent years about what examinations and tests actually measure. In a different but complementary vein, Jones et al. (2009) analyzed examination questions to distinguish between low-, intermediate-, and higher-order cognitive questions in the light of Bloom's *Taxonomy*. It was not simply a face validity exercise but took into account the examinee's scores per question. As Chapter 3 shows, there were other techniques that could have been used (e.g., factor analysis). The important point is not so much criticism of the analysis but the fact that they recognized that the design of examination questions is not a trivial activity. This writer is led to believe that many teachers have not moved beyond the situation described in the opening paragraphs of Chapter 1 or Section 7.1 to the effect that setting examinations and tests is the afterthought of the educational process. For them to be able to take this matter seriously, they have to have an understanding of how students develop and how they learn. This is the subject of the next chapter.

References

Almarshoud, A. F. (2011). Developing a rubric based framework for measuring the ABET outcomes achieved by students of electrical machinery courses. *International Journal of Engineering Education*, 27(4), 859–866.

Anderson, L. W., Krathwohl, D. R., Airasian, P. W., Cruikshank, K. A., Mayer, R. E., Pintrich, P. R., Raths, J., and Wittrock, M. C. (eds.) (2001). *A Taxonomy for Learning Teaching and Assessing. A Revision of Bloom's Taxonomy of Educational Objectives.* Addison-Wesley/Longman.

Anderson, L. W., and Sosniak, L. A. (eds.) (1994). *Bloom's Taxonomy A Forty-Year Retrospective.* Chicago: National Society for the Study of Education. University Press of Chicago.

Ashour, O. M., Sangelkar, S., Russell, L., and Onipede, O. (2014). Redesign the engineering teaching methods to provide more information to improve students education. In: ASEE/IEEE Proceedings frontiers in Education Conference, 1815–1820.

Barr, R. E., Krueger, T. J., and Aanstoos, T. A. (2006). Continuous outcomes assessment in an introduction to mechanical engineering course. In: ASEE/IEEE Proceedings Frontiers in Education Conference, S1E-9 to 14.

Besterfield-Sacre, M., Shuman, L. J., Wolfe, H., Atman, C. J., McGourty, J., Miller, R. L., Olds, B. M., and Rogers, G. M. (2000). Defining outcomes. A framework for EC-2000. *IEEE Transactions in Education*, 43, 100–110.

Blair, P. (1977). Cost-efficiency of self-paced education. An experiment with Keller PSI. *Engineering Education*, 68(7), 763–764.

Cameron, L. A. C. (1984). Standardisation techniques in the aggregation of marks. *Studies in Education (Dublin)*, 2(2), 56–54.

Cameron, L. A., and Heywood, J. (1985). Better testing: give them the questions first. *College Teaching*, 33(2), 76/77.

Carter, G., and Lee, L. S. (1974). University 1st year electrical engineering first year examinations. *International Journal of Electrical Engineering Education*, 11, 149.

Castles, R., Lohani, V. K., and Kachroo, P. (2009). Utilizing hands-on learning to facilitate progression through Bloom's taxonomy within the first semester. In: ASEE/IEEE Proceedings Frontiers in Education Conference, W1F-1 to 5.

Cheville, A., Yadav, A., Subedi, D., and Lundeberg, M. (2008). Work in progress—Assessing the engineering curriculum through Bloom's taxonomy. In: ASEE/IEEE Proceedings Frontiers in Education Conference, S3E-15/16.

Cole, J. S., and Spence, S. W. T. (2012). Using continuous assessment to promotr student engagement in a large class. *European Journal of Engineering Education*, 37(5), 508–525.

Cowan, J. (2006). *On Becoming an Innovative University Teacher*, 2nd edition. Buckingham: Open University Press.

Creed, C. J., Suuberg, E. M., and Crawford, G. P. (2002). Engineering entrepreneurship. An example of a paradigm shift in engineering education. *Journal of Engineering Education*, 91(2), 185–195.

Duval-Couetil, N., Reed-Rhoads, T., and Haghighi, S. (2010). Development of an assessment instrument to examine outcomes of entrepreneurship education on engineering students. In: ASEE/IEEE Proceedings Frontiers in Education Conference, T4D-1 to 6.

Felder, R. M., and Brent, R. (2003). Designing and teaching courses to satisfy ABET engineering criteria. *Journal of Engineering Education*, 92(1), 7–25.

Ferro, P. (2011). Use of Fink's taxonomy in establishing course objectives for a resigned materials engineering course. In: Proceedings of Annual Conference of the American Society for Engineering Education, Paper 309.

Fink, L. D. (2003). Creating Significant Learning experience. An Integrated Approach to Designing College Courses. San Fransisco, Jossey Bass.

Flinders, D. J., and Thornton, S. J. (2009). *The Curriculum Studies Reader*, 3rd edition. New York: Routledge.

Furneaux, W. D. (1962). The psychologist and the university. *Universities Quarterly*, 17, 33–47.

Gessner, P. (1974). Mastery. An experiment in self-paced learning. *Engineering Education*, 64(4), 368–369.

Gibbs, G., and Simpson, C. (2004). Conditions under which assessment supports student learning. *Learning and Teaching in Higher Education*, 1, 3–31.

Hereford, S. (1979). The keller plan (PSI) within a conventional academic environment. *Engineering Education*, 70(3), 250–260.

Heywood, J. (1989). Assessment in Higher Education. Chichester, Wiley.

Heywood, J. (2000). Assessment in Higher Education: Student Learning, Teaching, Programmes and Institutions. London, Jessica Kingsley.

Heywood, J. (2005). *Engineering Education. Research and Development in Curriculum and Instruction*. Hoboken, NJ: IEEE/John Wiley & Sons, Inc.

Heywood, J. (2008). *Instructional and Curriculum Leadership. Towards Inquiry Oriented Schools*. Dublin: National Association of Principals and Deputies/Original Writing.

Highley, T., and Edlin, A. E. (2009). Discrete mathematics assessment using learning objectives based on Bloom's taxonomy. In: ASEE/IEEE Proceedings Frontiers in Education Conference, M2J-1 to 6.

Jones, K. O., Harland, J., Reid, J. M. V., and Bartlett, R. (2009). Relationship between examination questions and Bloom's taxonomy. In: ASEE/IEEE Proceedings Frontiers in Education Conference WIG-1 to 6.

Khan, H. (2009). Measuring learning outcomes from the ethic course. In: ASEE/IEEE Proceedings Frontiers in Education Conference M4G-1 to 6.

Koen, B. V. (1985). The Keller Plan. A suuccessful development in engineering education. *Engineering Education*, 75(5), 280–281.

Lang, C. R., and Gurocak, H. (2008). Assessment methods for upcoming ABET accreditation criteria for computer science programs. In: ASEE/IEEE Proceedings Frontiers in Education Conference, S4G-6 to 11.

Mansor, W., Hashim, H., Abdullah, S. A. C., Kamaluddin, M. U., Latip, M., Mohd Yassin, A. I., Rahman, T., Zakaria, Z., and Kamal, M. M. (2008). Preliminary results on the implementation of outcome-based education on the non-examinable computer engineering modules. In: ASEE/IEEE Proceedings Frontiers in Education Conference S4B-20 to 25.

MacFarlane-Smith, I. (1964). *Spatial Ability*. London: University of London Press.

McCahan, S., and Romkey, L. (2014). Beyond Bloom's. A taxonomy for teaching engineering practice. *International Journal of Engineering Education*, 30(5), 1176–1189.

Miller, G. E. (1962). *Teaching and Learning in Medical School*. Cambridge, MA: Harvard University Press.

Moreno-Ger, P., Martinez-Ortiz, I., Freire, M., Manero, B., and Fernandez-Manjon, B. (2014). Serious games: a journey from research to application. In: ASEE/IEEE Proceedings Frontiers in Education Conference, 391–394.

Oldiran, M. T., Pezzota, G., Uzjak, J., and Gizejowski, M. (2014). Aligning an engineering education to the washington Accord. *International Journal of Engineering Education*, 29(5), 1591–1603.

Oliveira, C. G. de, and André, P. (2014). The impact of active learning strategies in second cycle students of an engineer course: a case study. In: ASEE/IEEE Proceedings Frontiers in Education Conference, 1725–1731.

Otter, S. (1991). *What Can Graduates Do? A Consultative Document.* Sheffield: Employment Department, Unit for Continuing Adults Education.

Otter, S. (1992). *Learning Outcomes in Higher Education.* London: HMSO for the Employment Department.

Pascarella, E. T., and Terenzeni, P. T. (2005). *How College Affects Students. Vol 2, A Third Decade of Research.* San Fransisco: Jossey Bass.

Pistrui, D., Layer, J. K., and Dietrich, S. L. (2012). Mapping the behaviors, motives and professional competencies of entrepreneurially minded engineers in theory and practice: an empirical investigation. In: Proceedings of Annual Conference of the American Society for Engineering Education. Paper 4615.

Poyner, C., Court, M., Phane, H., and Pittman, J. (2011). A 3-D prism/taxonomy for viewing knowledge when teaching language focused undergraduate simulation courses. *International Journal of Engineering Education*, 27(1), 128–137.

Reeves, P. M., Zappe, S. E., Kisenwether, E. C., Follmer, D. J., and Menold, J. (2014). Comparison of faculty and student definitions of entrepreneurship. In: Proceedings of Annual Conference of American Society for Engineering Education, June 2014. Paper 9496.

Sangelkar, S., Ashour, O. M., Warley, S. I., and Onipede, O. (2014). Mastery learning in engineering education. In: Proceedins Annual Conference of the American Society for Engineering Education. Paper 10917.

Serra-Toro, C., Travers, J., and Amengual, J.-C. (2014). Promoting student commitment and responding through self and peer based assessment. In: ASEE/IEEE Proceedings Frontiers in Education Conference, 1811–1814.

Shartrand, A., Weilerstein, P., Besterfield-sacre, M., and Olds, B. M. (2008). Assessing student learning in technology entrepreneurship. In: ASEE/IEEE Proceedings Frontiers in Education Conference F4H-12 to 17.

Slamovich, E. B., and Bowman, K. J. (2009). All, most or some: Implementation of tiered objectives for ABET assessment in an engineering program. In: ASEE/IEEE Proceedings Frontiers in Education Conference T3C-1 to 6.

Smith, K. A., Sheppard, S. D., Johnson, D. W., and Johnson, R. J. (2005). Pedagogies of engagement: classroom based practices. *Journal of Engineering Education*, 94(1), 87–101.

Spivey, G. (2007). A taxonomy for learning, teaching and assessing digital logic design. In: ASEE/IEEE Proceedings Frontiers in Education Conference, F4G-9 to 14.

Squires, A. F., and Cloutier, A. J. (2011). Comparing perceptions of competency knowledge development in systems engineering curriculum: a case study. In: Proceedings of Annual Conference of American Society for Engineering Education. Paper 1162.

Stice, J. E. (1979). PSI and Bloom's mastery model: a review and comparison. *Engineering Education*, 69, 175–177.

Thambiyah, A. (2011). On the design of learning outcomes for undergraduate engineer's final year project. *European Journal of Engineering Education*, 36(1), 35–46.

Tyler, R. W. (1949). Achievement testing and curriculum construction. In: E. G. Williamson (ed.), *Trends in Student Personnel Work.* Minneapolis: University of Minnesota.

Whitehead, A. N. (1932). *The Aims of Education.* London: Ernest Benn.

7

Student Variability

The Individual, the Organization, and Evaluation

This chapter is a continuation of Section 1.11, which shows that factors other than cognitive ability contribute to student performance and considers student variability as it relates to cognitive and affective (emotional) development and the role that the curriculum plays in development in both dimensions. It is noted that the curriculum has tended to neglect the affective dimension of behavior at the expense of the cognitive yet engineering activities often require engineers to work with others and draw on both of these dimensions of behavior. This chapter challenges some of the major assumptions that are made about education. The first is that students all learn in the same way. Research on learning styles shows that they do not. This finding has consequences for assessment and teaching as does the finding that a variety of study strategies will be found among students.

Whereas it was once thought that the last stage of Piaget's model of cognitive development was complete, this is no longer the case as Perry's model of post-Piagetian intellectual development shows. Efforts by engineering educators to design curricula that would respond to Perry's model are discussed. A curriculum designed to follow the stages of this model should lead students from dependence to independence where they take responsibility for their own learning and are able to solve ambiguous problems. In today's parlance, they should be critical thinkers.

There has long been a critical thinking movement in the United States and tests have been developed for its assessment and used by engineering education. This movement

The Assessment of Learning in Engineering Education: Practice and Policy, First Edition. John Heywood.
© 2016 The Institute of Electrical and Electronics Engineers, Inc. Published 2016 by John Wiley & Sons, Inc.

is briefly described and is followed by a section on the assessment of development. King and Kitchener's distinction between critical thinking and reflective judgment is given and their theory of intellectual development described. The merits of their Reflective Judgment Interview are briefly considered.

Students have to learn the art of reflective practice if they are to be able to make the choices that the solution of engineering problems demands, that is, to have adaptive expertise. Reflection requires the ability to be self-aware and its key skill is to be able to self-assess (see Section 8.9). How engineering subjects are taught matters for the development of adaptive expertise depends on judgment and judgment needs to be reflective. Work is both cognitively and emotionally construed for which reason it is incumbent on employers and employees to understand how organizations and individuals interact at this level.

Intellectual, emotional, and professional development cannot be completed within college alone because a person continues to develop and will do so in response to any situation. For this reason, employers and managers have as much responsibility for the development of their engineers as do the colleges from which they come, and in these days of rapid turnover have an obligation to help them prepare for their next work assignment.

Assessment remains problematic and there is a need for considerable investment in research and development in these areas. The theme of this chapter continues in Chapter 8.

7.1 Introduction

The term *evaluation* is used in the sense that it was used in the 1960s and 1970s and as it is used in the model of the curriculum process in Exhibit 3.9. It is that process over and above the assessment of achievement that uses other measures to determine if the goals of the curriculum are being achieved. It takes into account all those factors that contribute to achievement (performance). The starting point is with the individual and the factors that influence a person's achievement. It is fair to say that many engineering educators have naïve views about learning. This should not surprise anyone, since in most countries there is no requirement that educators in higher education *per se* should be trained for the job. The result is that many teachers, probably unconsciously, hold the view that the mind is a slate on which knowledge has to be impressed. This would account for the continuing widespread use of the lecture method supported by an equally unconsciously held epistemology that there is a fixed body of knowledge to be imparted that can be changed only within limits.

A major effect of such attitudes that are often reinforced by the department and/or school in which the academic works is that students are thought to be individuals who can be manipulated and when that manipulation does not work that is, they do not achieve, the blame is laid at the student's door. It is a theory X model of student learning.[1] A theory Y model of student learning views the student as an active being that has something to offer the institution. Given this to be the case, the educational problem is to match what the student brings to his or her courses with what the department and/or

Students/Employees Bring to Their Work	Educational or Other Organization Helps People to Be Effective by

1. Knowledge
 	
  (a) General
  (b) About his or her specialization or subjects of study
  (c) About business, school, or college
 	
2. Physical skills
 	
  (a) Health
  (b) Related psychomotor skills
 	
3. Cognitive and affective skills
 	
  (a) Abilities to recognize and solve a problem
  (b) Ability to make judgments
  (c) Ability to communicate
  (d) Ability in the development of satisfactory interpersonal relationships
 	
4. Personality and drive
 	
  (a) A certain activity level norm
  (b) A certain level of risk taking
  (c) Aspirations and expectations
  (d) Acceptability
 	
5. Values
 	
  (a) Interests
  (b) A moral disposition
	Job or lesson analysis
	
	  (a) Providing a definition of the key results the job is required to produce, or the aims and objectives of a curriculum program
	  (b) A definition of the knowledge and skills required for performance of the task
	  (c) Details of information and resources necessary for the completion of the job or, homework, a project or practical class
	
	School, college, or organization
	
	  (a) Provides a structure in which people can work or learn
	  (b) A management or teaching style that will motivate whether in groups or individually
	
	Recruitment, education, and training
	
	  (a) Matching what an individual brings to the job needs of the job, or matching student abilities to the requirements of the program
	  (b) Background knowledge and experience of similar work or instructional situations likely to be of use
	  (c) Training or instruction in specific knowledge and skills for a defined function

EXHIBIT 7.1. What a person can bring to an organization and what an organization can give to that person. Adapted from W. Humble and cited in Heywood, J. (1989). *Learning Adaptability and Change*. London: Paul Chapman/Sage.

school has to offer the student as outlined in Exhibit 7.1 (Heywood, 1989). The idea of the organization as a learning system (Section 4.2) is also to be found in the works of Argyris and Senge (Argyris and Schön, 1978; Senge, 1990). It is all the more surprising that we do not talk about universities as learning organizations.

The educational system is governed by a number of questionable assumptions. It is assumed everyone learns at the same pace. So it is expected that everyone will reach 12th grade with same quantity of knowledge. Therefore, the system of schooling is staged by the year and it is the fault of those who do not reach that stage, not the school. It is assumed that primary purpose of schooling is to imbibe knowledge. Therefore, in some schools the rooms are arranged for the teacher to talk and the students to listen. In higher education, they go to lectures for the same purpose. These assumptions condition the design of buildings and the structure of variables that have to be taken into account when considering human learning. Section 7.2, which is a development of some of the points made in Section 1.11, challenges that assumption and shows that learning preferences may influence performance, and that teachers need to take this into account when they design examinations and tests. The power of expectations to influence the ways in which we approach assessment is illustrated by the considerable amount of research that has been conducted on study habits and strategies and the influence of curriculum and assessment on those strategies.

Among the major variables that influence learning is intellectual development. In the 1950s and 60s, teachers were accustomed to think that intellectual development ended with Piaget's stage of formal reasoning and some investigations in the United States suggested that some university students had yet to reach the stage of formal reasoning. However, by the beginning of the 1970s, it was clear that intellectual development continued post-Piaget.

Since then, there have been several attempts to design developmental curricula and to test students for their level of development, and it has become important in discussions about engineering education (Felder and Brent, 2003). Moreover, there has been at least one international comparison (Culver et al., 1994). Baxter-Magdola (1987) reported that students operating at high levels of the Perry positions preferred a working relationship with faculty akin to that of being a colleague, where as those at the lowest levels of the Perry model preferred a more distant but positive relationship.

As might be expected, Perry has caused others to suggest alternative models. The Models of Baxter-Magolda (1994)[2] and King and Kitchener (1994) have been of interest to engineering educators. At the same time the higher stages of these models require the use of reflective thinking. King and Kitchener's model is of particular interest because they developed a Reflective Judgement Interview to assess a person's level of development. Their approach may be considered as a development of Perry's theory. It is clearly in the same tradition.

For many years in the United Kingdom, more recently in the United States, research workers have been persuaded that the development of reflective thinking is an absolute necessity, and as Warhurst (2008) opines, it has become an article of faith in professional development. He cites Rigg, Steward, and Trehan (2007), who say that the fervor for "reflection borders on moral evangelism." The position taken here is that the writings of engineer and educator John Cowan (1998, 2006) make the case not only for the part it can play in the pedagogy of higher education but in student learning. That said, it is not the be all and end all of learning.

Given that a prime purpose of learning is to learn a method that is a "normative pattern of recurrent and related operations yielding cumulative and progressive results where there are distinct operations, where each operation is related to the others, where the set of relations forms a pattern, where the pattern is described as the right way of doing the job, where operations in accord with the pattern may be repeated indefinitely, and where the fruits of such repetition are, not repetitious, but cumulative and progressive" (Lonergan, 1973, p. 4) there is a method. It follows that the educational process is responsible for all the operations in the pattern. Lonergan (p. 6), in the same chapter, considers these operations to be "seeing, hearing, touching, smelling, tasting, inquiring, imagining, understanding, conceiving, formulating, reflecting, marshalling and weighing the evidence, judging, deliberating, evaluating, deciding, speaking, writing." In the language of assessment, these operations are competences and reflection is one of many competences that require to be developed for life and work. Currently, education is better at developing some competences than it is others: it is evidently very good at conveying knowledge, less good at ensuring that it is understood and often criticized because it does not develop skill in critical thinking. In respect of the latter, the prevailing view is that reflection is key to critical thought; however, in terms of the responsibility for learning, it is not the only factor that has to be considered in student development, for it is apparent that the affective domain is of considerable importance and in this respect Culver and Yokomoto (1999) were quick to highlight the importance of emotional intelligence. Thus, it is that this chapter considers intellectual development and reflective practice together with critical thinking and how they might influence achievement (performance). Hicks, Bumbaco, and Douglas (2014) invite us to consider them to be important in the acquisition of adaptability for adaptive expertise is an end goal of engineering education.

It is argued that the social milieu in which people learn is a powerful influence on their intellectual developments and emotional response. An individual lives in a plurality of social systems each with its own goals. The dispositions that make a person and what the person brings (gives) to life matter. It is about how they are challenged by the organization in which a person works or studies, and what an organization can bring to him or her. It is primarily about the evaluation studies that lead to these conclusions and the view that attention to these factors indicates the importance of self and formative assessment that are also considered in Chapter 8.

7.2 Learning and Teaching Styles

It is well documented that we have predispositions to learn. Perhaps the best known are the contrasting styles of convergent and divergent thinking. Many such styles are to be found in the literature. Some of them relate to abilities, as for example, spatial ability that is important for learning engineering and mathematics. Personality is thought to influence our learning behavior. For example, students who are extraverts may approach learning quite differently to introverts and this may influence their examination performance (see Sections 1.11 and 2.5). Understanding student differences is an obligation of every teacher (Felder and Brent, 2005).

The idea of learning styles seems to have originated with David Kolb (1984), who proposed a theory of experiential learning in which learning is described as a process moving through four modes known as concrete experience, reflective observation, abstract conceptualization, and active experimentation. In practice, some authors propose that faculty should design their lectures to follow the Kolb cycle (Todd, Sorensen, and Magelby, 1991); it has also been suggested that classroom assessments should be designed to assess the different components of the cycle (Heywood, 2008). Experience continually changes or develops our ideas and should not therefore be thought of in terms of fixed outcomes. It is a holistic learning process. Kolb's theory continues to be attractive to engineering educators (e.g., Hirsch and McKenna, 2008; see Section 9.8).

But there are other learning styles and these are well documented in the engineering literature (Grasha, 2002; Heywood, 2005).

No other engineering educator has done more to promote the use of learning style inventories than Rich Felder. His work dates back to the 1980s when with Linda Silverman they described a model of learning styles (Felder and Solomon Index of Learning Styles) that combined some components of the Kolb Model and the Myers–Briggs Type Indicator derived from Jungian personality theory (Felder and Silverman, 1988; Felder and Spurlin, 2005).

The parameters of the Felder–Solomon Index are:

Visual/verbal. Contrasts those who receive information visually with those who prefer verbal explanation.

Sequential/global. Contrasts those who like a step-by-step presentation of knowledge with those who like knowledge to be "presented in a broad potentially complex manner that allows them to fill in blanks through 'ah-ha' moments."

Active/reflective. Contrasts those who receive knowledge through hands-on activities while internal reflection drives the reflective learner. In the Kolb model, the hands-on learner is one who learns through concrete experiences.

Sensing/intuitive. Contrasts those who like factual knowledge and experimentation with those who like theories and principles. This comes from the Jungian model developed by Myers–Briggs. Within the MBTI are the dimensions of extraversion and introversion popularized in the Eysenk tests, and used by Furneaux in his study of engineering students (Sections 1.11 and 2.5).

Learning styles models have been used to design classroom activities. For example, at Loughborough University (United Kingdom), it was applied to the development of process control laboratory education. Abdulwahed and Nagy (2009) argued that often poor laboratory outcomes were due to poor activation of the prehension dimension of the Kolb cycle before the students came to the laboratory. The Kolb cycle follows from concrete experience through reflective observation and abstract conceptualization through active experimentation to concrete experience. The prehension dimension relates to concrete experience and abstract conceptualization. They found that the introduction of a virtual laboratory into the prelaboratory preparation improved the conceptual understanding of the students during the hands-on laboratory session.

A Chinese study describes how academics and students went to help to rebuild a primary school that had been destroyed in an earthquake. The activities of four groups were observed and members of the student teams interviewed. The analysis of the data was related to graduate attributes and the stages of Kolb's learning cycle. No guarantee could be given that each student went through each phase. But the value of concrete experience was highlighted. The students were able to link academic theory to the real world and make sense of their experiences (Chan, 2012). In Hong Kong, experiential learning is to become a major feature of the university curriculum, which means that new ways of assessment will have to be considered including direct observation and oral and interview assessment.

There are difficulties with the Learning Styles Inventory and one Finnish study with engineering students found no correlation between predicted styles among engineering students and what the test recorded. Holvikivi (2007) concluded that in spite of modifications to the test present understandings of the functioning of the mind lead to the view that people do not have fixed sets of behavior. While this might discredit the test, it does not discredit the theory. It would be very difficult to argue that individuals do not have preferences for learning in one way or another or for teaching in one way or another.

Of interest in this context is the role learning preferences might play in student performance. There is considerable evidence to suggest that students' responses to teaching are influenced by their different learning preferences or orientations. A primary conclusion is that effective teaching requires a teacher to understand the learning styles that are present in his classes and respond to them appropriately (see Heywood, 2005 and 2008, for summaries), and at a minimum this will require variety in their teaching as suggested by Svinicki and Dixon (1987). Related to this work is the view of Gulek (2003, cited by Parsons, 2008) that test anxiety can be alleviated by using a variety of assessment formats to cope with a variety of learning styles but this requires high-level skill in test design and a willingness to pilot new instruments.

Grasha (2002), over a period of 30 years, has researched both teaching and learning styles and the Grasha–Reichmann Teaching Style Survey enables teachers to establish their teaching style. A comparative study of engineering teachers at Khulna University in Bangladesh ($N = 23$) and the University of Michigan Flint ($N = 23$) obtained the teaching styles from self-assessments using the Grasha–Reichmann survey. Notwithstanding that no independent evaluation, as for example by students was made, or longitudinal data obtained, it was found that there were no significant differences in teaching styles based on the age, gender, degree earned, the number of years teaching, or academic rank. One could read into this or hypothesize that teaching style is a consistent feature of a teacher's performance. The authors recognize the need for much more research (Mazumder and Ahmed, 2014).

7.3 Study Habits/Strategies

As with the previous section, this section develops some of the points made in Section 1.11. Just as there are a variety of predispositions to learn, so there are a variety of study habits/strategies. The strategies we adopt are influenced by our expectations of the curriculum and particularly the system of assessment from which our grades are derived.

In the 1970s in Scandinavia (Sweden), Marton and Säljö (1976, 1984) studied how respondents reacted to a reading comprehension exercise. They were asked to read a text and told that they would be questioned on its context and *how they tackled the task*. An examination was also set and there were several variations in the methodology.

Marton and Säljö concluded that "the students who did not get 'the point' failed to do so simply because they were not looking for it. The main difference which was found was whether the students focused on the text itself or on what the text was about, the author's intention, the main point, [and] the conclusion to be drawn." Marton suggested that the strategies they found students using are indicative of different perceptions of what is wanted in learning in higher education. For some, "learning is through the discourse and for others learning is the discourse. Those who adopt the former strategy get involved in the activity while those who take the latter view allow learning to 'happen to them.'" It is the second group who are surface learners who do not seem to appreciate that understanding involves effort. This work was taken up by research workers led by Noel Entwistle in the United Kingdom (Marton, Hounsell, and Entwistle, 1984) and Australia (Biggs, 1987; Biggs and Tang, 2011; Ramsden, 1988 [and UK]). Research in this field continues and the concepts of deep and surface learning have become embedded in the literature and the training of university teachers. Given that the perceptions that students have of what is expected of them the design of examination questions to secure deep learning is paramount. A study in Buenos Aires that made considerable use of Classroom Assessment Techniques changed a class from teacher-centered teaching to student-centered teaching. It was found that the complaints of deep knowers were reduced once they became the leaders of their design and made decisions but surface learners got lost in nonthreatening environments (Feldgen and Clua, 2009). An Australian study showed that senior year students who demonstrated deep learning were more likely to perceive their course as encouraging independence in both attitudes and approach to learning, and as not overburdening (Watkins, 1981). A key question for engineering educators is "do they think their quizzes, tests and examinations encourage deep or surface learning and what strategies do they think their students will use?"

Fitkov-Norris and Yeghiazarian (2014), in summarizing work in this field, pointed out that there had been no formal studies of the relationship between preferred mode of learning (by which they meant visual, auditory, read/write, and kinesthetic) and the study strategies and habits students adopt to manage their learning. They propose that in order to bridge this gap, research should be undertaken that is based on recent findings in neuroscience using brain scanning. Their first finding relates to how students react when placed in a familiar or unfamiliar context, which shows student behavior to be governed by a reflective C-System and a reflexive X-system. Fitkov-Norris and Yeghiazarian report that "the C-System is used in situations which are unfamiliar and provides sequential and effortful assessment for a particular choice of course of action." In contrast, the "X-system is used in familiar situations where actions are automatic and relatively effortless and uses parallel processing." They cite as an example students attending a lecture course for the first time who have to work out what to do whereas those who have attended a number of times automatically know what to do. (It will be apparent that this finding may be related to Sternberg's theory of intelligence as discussed in Section 3.2.) Fitkov-Norris and Yeghiazarian use this finding to explain why some students could be trapped in using inefficient study habits. They suggest that students who come to use the efficient and less effortful reflexive system may through repeated actions trigger the same habitual response and become trapped into using "inefficient study habits and strategies even when they make a conscious effort to study hard or work efficiently." This parallels the idea of set mechanization put forward by Luchins (1942), who found that pupils solving problems tended to use a solution they had been shown and assimilated to solve other problems for which a more simple approach would have sufficed. In that case, the "set" had become a routine that prevented them from seeing other ways of solving the problem. But Wittrock (1963) pointed out that "set" can be used to assist learning. The point is that we easily accept sets in our learning.

The second finding from neuroscience shows that "visual learners convert words into images in their brains before processing information further. Read/write learners convert images into text before processing information further." Fitkov-Norris and Yeghiazarian note that when a learner receives information other than in their preferred style they have to translate the information into their preferred style that involves them in an additional step. This suggests that their optimal study strategy is dependent not only on their learning preference but in the way the subject matter is presented. Thus, the answer to Grasha's question, should learning and teaching styles be matched, is in the affirmative. Related to study habits, Fitkov-Norris and Yeghiazarian also conclude that because of the complexity of learning situations, the same study strategy may not work in every learning situation for the learner. Turned into assessment, this means that those learners with a strong preference are more at risk of failure than those learners who tend to a multimodal approach. While it has been found that learners taught in their preferred mode are more satisfied and achieve better that will not always be possible for which reason teachers should respond with a variety of styles in the presentation of information in their classes, and the learning strategies they wish their students to adopt. The implications for the design of test questions and the structuring of examinations are profound.

It seems self-evident that the study strategies adopted will be related to intellectual development. Overall Parsons (2008) suggested in a paper on test anxiety that students would be helped if they were given courses in study skills. Contrary to expectations, I found that when group discussion meetings were arranged on the mechanics of study that "able" students took more interest in the meetings than "poor" students (Heywood, 1977).

7.4 Intellectual Development

In the late 1970s, the theory X model was challenged by some engineering educators and in particular by Richard Culver then working at the Colorado School of Mines (Culver and Hackos, 1982). Culver was persuaded by William Perry's (1970) study of the intellectual development of students that the engineering curriculum should heed the results of that study, namely that there are post-Piagetian stages of development and that if students do not negotiate these they will not be able to handle the complex (now often called "wicked") scenarios and problems that engineering poses. To that end, he and his colleagues, notably Mike Pavelich, Barbara Olds, and Ron Miller, set about designing an engineering curriculum that would meet the criteria outlined by Perry. Over the years the curriculum has been subjected to evaluation using an inventory for evaluating the Perry Stages at which students are at that had been developed by W. J. Moore (1989; Pavelich and Moore, 1996).

Perry, following in the footsteps of Piaget, argued that Piaget's stage of formal reasoning was not the end of the matter. He had found that freshmen students brought with them attitudes from school that demanded from their tutors black and white right answers, which is exactly what university study is not designed to do. (Indeed earlier in the 1960s, this writer had found at a new university in the United Kingdom that some first-year students in the social sciences had great difficulty in coming to grips with the kind of arguments and theories with which they were presented. Some were faced with conflicting value systems that sometimes caused cognitive dissonance; (Marshall, 1980; Heywood, 1989; see also Purser, Hilpert, and Wertz, 2011). Perry showed that students went through several stages of development to reach a position where they perceived that good choices are possible and commitments have to be made. The student moves from a position of dualism when students begin by thinking all knowledge is absolute to a stage of relativism so that by stage 5 some answers are found to be better than others and everything has to be considered in context. It is held that many students never get beyond stage 5 of the nine-stage model. In particular, Perry argues that much curriculum and teaching practice reinforce the dualistic mode.

Perry's model has proved attractive to some engineering educators. As indicated by the early 1980s, it was being used by Culver at the Colorado School of Mines for course design (Culver and Hackos, 1982). He and his colleagues found that many students, including some seniors, had great difficulty in solving open-ended problems. They did not understand them as a professional does. Many freshmen, for example, do not understand why evidence has to be used to justify a decision. Sophomores and seniors see no need to devise alternative solutions. They found that professional understanding could be acquired if they designed their courses to be responsive to the levels in the model (Pavelich and Moore, 1996).

Marra, Palmer, and Litzinger (2000) reported a longitudinal study at Penn State University. They found that it was not necessarily the brightest students who were at the highest level of the Perry scale. Unfortunately, their sample was rather small and there was need for verification studies. However, in another study they accounted for the lack of growth in cognitive ability between first and third years as being due to the particular cognitive load that traditional courses placed on students, that is, the demand for memorization and the application of formulae. They suggested that the reason why there was a jump in the level between years 3 and 4 was that the students became engaged in projects and team activities (Wise et al., 2001). The acquisition of skill takes time and the provision of experience is essential to any learning, be it of design or of moral purpose (Kallenberg, 2013).

A review conducted by this writer of research in this area pointed to the high value that students attached to "real" engineering, and therefore, the importance of design work in the freshman year, but it also noted that if students are not motivated in subsequent years, that which was gained in the first year will be lost in subsequent years. It needs to be remembered that growth is not linear (Heywood, 2005).

Perry found that growth occurred in spurts followed by periods of stabilization and Culver and Sackman (1988) called these spurts "marker events." They are events that bring about what Lonergan (1957) calls "insight" revealing understandings that have been hidden (Frezza, 2014, 2015). The problem is that such events that come as a result of mental activity may not come when the instructor desires, so students have to be provided with activities that have the potential to cause marker events, and that is more likely to come from activity-based learning (Brown et al., 2013). This is consistent with the views of such distinguished scholars as Bruner and James in the United States and

> Is a significant event that influences an individual's development.
> Results in a change or expansion of the personal belief system.
> Provides new insight and, frequently a change in priorities.
> Serves as an anchor for new learning and long-term memory recall.
> Usually involves a concrete experience and reflective observation.
> Can be positive or negative.
> Can't be forced but can be programmed.

EXHIBIT 7.2. Marker events as described by Culver, R. S., and Sackman, N. (1988). Learning with meaning through marker events. In: Proceedings Frontiers in Education Conference. They have also been called "Aha" events by C. F. Yokomoto, D. R. Voltmer, and R. Ware in Incorporating the "Aha" experience into the Classroom laboratory (1993). In: Proceedings Frontiers in Education Conference, pp. 200–203.

Mascall in the United Kingdom who believe the learner (novice) has to try to experiment with what it is like to become an expert. If one wants to be an engineer, then one has to be provided with opportunities to behave as an engineer. This, of course, is one of the justifications for cooperative (sandwich) courses, internships, and the provision of work as an engineer early in an academic program and during vacations. It is also a justification for appropriately designed laboratory work, project, and problem-based learning. The characteristics of a marker event are set out in Exhibit 7.2.

The Perry model applied to design that was the purpose of the Culver, Woods, and Fitch (1990) investigation meant that the student has to arrive at a stage where he or she is prepared to accept responsibility for the engineering they propose to do. It also meant that they could cope with ambiguous situations. They argued that "most college programs while successfully teaching facts and procedures do not help students grow to intellectual maturity." That is, to take on these responsibilities, Culver also argued (personal communication) that, students at lower levels of the scheme, while able to do problems that require highly structured analytical techniques cannot cope with synthesis. This means that students at lower levels of intellectual maturity can pass examinations that emphasize analysis but not questions that seek the higher skill of synthesis and/or multiple solutions (wicked problems) for which they have to argue a case. For this reason, courses should be designed to cater for intellectual development that is appropriately assessed, that is by "wicked problems."

Finally, Hill and Brown (2014) studied the relationship between learning styles and intellectual development. As they put it, "the types of knowledge one concentrates on and how they like to perceive knew knowledge." Hill and Brown derived Perry levels for students in each of the sophomore, junior and senior years from a Cognitive Complexity Index (CCI) calculated for each student and related to his or her learning preference as measured by the Felder and Solomon Learning Styles Index (ILS). They found that with the exception of the visual/verbal dimension there was a link between learning preference and intellectual development. Most gains occurred between the sophomore and junior years and those who benefited most were the active learners, but they did not progress as far as the reflective learners. They noted that the "dimensions of the ILS where participants have higher CCI scores (reflexive, intuitive) correlate to the teaching styles of many engineering courses where theories are taught in lecture format and students are expected to process information." While in the future they intend to correlate their data with GPA which gives them a measure of intelligence, it is to be hoped they or someone else develops a test of "wicked" engineering items. In the meantime, notice should be taken of the attempts to measure intellectual development that have been made.

7.5 Critical Thinking

King and Kitchener (1994), who described a model of intellectual development that in many respects is similar to Perry's, pointed out that Dewey, who thought that *reflective thinking* was a major goal of education, also used the term *critical thinking* interchangeably with *reflective thinking*. They disagreed. Put at its simplest, Dewey thought that reflective thinking could not be initiated without an awareness of uncertainty. That seems not to be the case with critical thinking as it has come to be understood in the United States for which reason critical thinking will be considered first.

Nowhere are differences between educational cultures more exposed than in considerations of critical thinking. In the United States, critical thinking can be assessed by using standardized tests. In the British Isles, many academics would say, "well our tests do that anyway" (see Chapter 3). Critical thinking is acquired by osmosis! Cajander (2014) and his colleagues of the University of Uppsala write, "It is not uncommon to view competencies such as critical thinking and communication as something that develops as a side effect while learning is the knowledge associated with a subject e.g. computer science." For which reason, much of the literature that relates critical thinking to assessment is from the United States.

As with assessment, the concept suffers from the fact that not all those who study critical thinking agree about what it means (Hicks, Bumbaco, and Elliot, 2014). The focus of some participants is on inductive and deductive logic, whereas others simply consider it to be a process of problem solving. As might be expected, this is the perspective that has been adopted by some engineering educators. Pascarella and Terenzini (2005) write, "it would appear that most attempts to define and measure critical thinking operationally focus on an individual's capability to do some or all of the following: identify central issues and assumptions in an argument, recognize important relationships, make correct references from the data, deduce conclusions from the information or data provided, interpret whether conclusions are warranted based on given data, evaluate evidence or authority, make self-corrections, and solve problems" (p. 156). The developers of the Engineering Science examination (Section 3.1) would hold that they sought all those attributes, and the colleagues who work with this writer certainly believe with him that their examinations test critical thinking. Whether they do or not is a different matter!

Recently, Rocke et al. (2014) of the University of the West Indies have promoted the case for *argument maps* to develop critical thinking. In addition to a group-based research paper, they propose a final written examination for assessment "in which individual students would be required to consider a given engineering problem and a proposed solution to the problem. Students would be required to investigate the validity of and strengths of the justification of design choices and to propose improvements to the arguments presented if necessary." The engineering science examiners would claim that was what they achieved by their project design paper.

Pascarela and Terenzini note that more recently some scholars have understood that in order to think critically, there has to be a disposition to think critically. That is, there is a motivational element in critical thinking. This would be consistent with research on motivation and study habits (Entwistle, Thompson, and Wilson, 1974).

In the United States, among the three most commonly used tests of critical thinking are the California Critical Thinking Skills test, The Watson-Glaser Critical Thinking Appraisal, and the Cornell Critical Thinking Test. All three are composed of multiple-choice items, but there are issues of validity associated with each test (Stein, Haynes, and Unterstein, 2003; Stein and Haynes, 2011). In the engineering literature, detailed descriptions of the Cornell test, The Paul–Elder model, The CLA Model, and the Collegiate Learning Assessment, together with critiques, will be found in Kaupp and Frank (2014). Stein developed his own test, which is referred to later in this section.

The Queens University Canada, in a striking change to its engineering curriculum, integrated Model Eliciting Activities (MEA) into the first-year engineering course (Kaupp, Frank, and Chen, 2013).

Interest in the development of such curricular in US engineering schools was fostered by the National Science Foundation during the middle of the last decade. ASEE (The American Society for Engineering Education) held a symposium on MEAs at the 2010 Annual Conference and the introductory paper explained their origins in mathematics, the theory on which they are founded, and the technique (Hamilton et al., 2010). An MEA involves a case study of 30–50 minutes, which is solved by students in groups of between three and five individuals. MEAs are intended to simulate real-world problems for which testable models or solutions may be found. The curriculum principle underlying the approach is to begin with the student's own conceptual system as a basis for modelling that is "creating representations of problematic phenomena or scenarios as a means to solve those situations." This would seem to be what a student has to do if he or she is asked to undertake a miniproject or investigation of their own thinking and to pursue it to its practical conclusion without formal instruction. (Design and make projects as they are known in some school systems; see Section 3.3.) Indeed, this is one of the merits of individual project work. Be that as it may, Kaupp, Frank, and Chen evaluated the impact of the MEA course on the development of critical thinking skills. In their paper, the focus is on pre- and posttesting using verbal protocols. The paper gives some excellent examples of verbal protocols and details the coding method of analysis used following Leyden, Moskal, and Pavelich (2004). This enabled them to identify the elements of critical thinking in Paul's theoretical framework for the elements of critical thought that are shown in full (Paul and Elder, 2006).

The students completed three MEAs during the semester—Cable ferry failure; wind turbine design; and building heat loss. "Each MEA required students to develop a model of a physical system that was used to solve a problem presented by a fictitious client, write MATLAB code to implement the model, and self evaluate their report." They were asked to self-assess their work against 3–5 of the nine elements of the Paul and Elder (2006) critical thinking model (clarity, accuracy, relevance, logicalness, breadth, precision, significance, completeness, fairness, and depth). The principles of critical thinking were discussed in class in each of the MEAs so that in one class the skill of questioning was developed by creating lists of questions that should be asked by accident investigators. This is, of course, a contributory component in the skill of diagnosis, a skill that is seldom discussed in the literature.

The investigators were able to arrange for a control group of physics students. One of their findings seems to be generalizable. It is that when students are faced with project or investigational work they find it difficult to question the validity and accuracy of their own assumptions and give the briefest consideration to the potential impact of their solutions. This was equally true of the students who undertook the coursework in engineering science 30 or more years ago (see Sections 3.2 and 3.3). By the end of the semester, that had changed and the students were questioning their own assumptions. Evidently training is required to achieve this goal and students have to be given the opportunities to develop this skill. The same thing happened with the control group who approached the problem in reverse forming recommendations prior to analysis. Their recommendations were "primarily formed on assumptions based on their own personal experience." When it came to the posttest while they approached the problem as the engineering students did, they "used personal and anecdotal experience to formulate conclusions rather than support conclusions with supplemental information."

This particular point is taken from a much more detailed analysis because it would seem to be a general experience of entering students. As indicated in section 7.4, some students in the social sciences seem to be unable to stand outside their own experience and sometimes cognitive dissonance occurs when their value systems are challenged, and a few are unable to cope (Marshall, 1980). It is particularly true of graduates who are entering the teaching profession: they often require what Hesseling (1966) calls a *chocs des opinions*. It is a finding of which every educator should be aware and it involves an understanding of perceptual learning (Heywood, 1989). It does mean that students are likely to have difficulty in questioning assumptions without the provision of learning experiences that help them do just that. Kaupp, Frank, and Chen come to the conclusion that taking into account all the differences between the two groups at the posttest stage that the relative improvement in critical thinking of the engineering group could be attributed to the explicit instruction in critical thinking given that group. So much for osmosis!

This activity raises some questions about future behavior and grading? First, other studies lead to the questions "Are the skills acquired transferable?" "What happens to their critical thinking skill in subsequent years?" "How does it contribute to their study approaches?" "Should courses in subsequent years be designed so as to reinforce these skills?" Finally, how is the final grade to be constructed? Should it be a measure of a student's improvement or an assessment of what the student has achieved? Kaupp and Frank (2014) continue to report on this development.

Arising from criticism of existing assessment instruments and the desire to create an assessment that was more directly related to faculty objectives, and hence could engage the faculty in quality enhancement Tennessee Technological University embarked on a 10-year program to develop a test of critical thinking. Stein, Haynes, and Unterstein report on the experience of 3 years of that program. The work was done in collaboration with faculty who established an interdisciplinary committee that identified 10 important skills that are required for effective problem solving, life-long learning, and critically evaluating information. Similar items will be found in other lists but not always together. Among them are "identify inappropriate conclusions and understand the limitations of correlational data," "Separate relevant from irrelevant information when solving a problem." And "analyze and integrate information from separate sources to solve a complex problem." The assessors of engineering science would argue that the rubrics for the project would have allowed them to check each of these items. Note how they apply to any subject in the curriculum which was one of the purposes of the test's development although it was not expected that the test would assess all aspects of critical thinking. Notwithstanding American objections to essay tests on grounds of their reliability, it was decided that the tests would mostly involve essay answers that would enable the assessment of communication and "leave opportunities for creative answers to questions that don't always have a single correct response." And, this is one of the reasons given by university teachers, in particular in the humanities in the British Isles for essay tests. However, whereas in the British Isles, question setting and grading skills are often acquired by osmosis in this study faculty members from across the disciplines were given a 1-day workshop on scoring the examinations and evaluating the test. Similarly, faculty who were paid a stipend also participated in the scoring workshop. A minimum of two faculty members read each question and where there was a difference between them, a third grader also read the answer. "The final score is either based on the two raters that agree with each other or on the average of all three if there is no agreement. During the scoring workshops, individual tests are rotated across faculty graders after every two or three questions to increase the generalisability of the scoring. We continually analyze inter-rater agreement patterns so as to identify questions that might be problematic to score, and explore ways of rewording the question or scoring criteria to improve inter-rater agreement," all of which is to take essay examining seriously.

It was expected that the engineers would perform best on the test and they did among the sample of seniors. The ACT score accounted for about 40% of the variability. The data also show a significant increase in critical thinking scores between the freshmen and senior test scores. They also conducted pre- and posttests with smaller samples. The results led them to believe that the test could be used to identify courses and pedagogies that promote critical thinking and problem-solving skills.

Related to critical thinking is the ability to argue. A Portuguese study found among first-year engineering students a preference for abductive and deductive reasoning: moreover, the argumentative characteristics are dependent on the methods of assessment that may need to be changed. They considered their results to reflect the lower level of maturity of the students, that the ability to argue is developed progressively, and that course structures have to be designed to meet this need (Leite et al., 2011).

An Irish study in which academics from a variety of disciplines set out to establish how these disciplines understand critical thinking concluded that while critical thinking is important to most disciplines the understanding of what it is varies. While those in the humanities had the clearest definition of what it is, those in the engineering and technical disciplines were not so sure of their definitions (Ahern et al., 2012). This points up the advantages of a collegial organization in which academics from different disciplines can converse.

Pascarrela and Terenzini reviewed mostly American research on critical thinking and postformal reasoning together and concluded "that there is evidence that a curriculum experience that requires the integration of ideas and themes across courses and disciplines enhances critical thinking over simply taking a distribution of courses without integrative rationale" (p. 157). Culver has pointed out that this applies to traditional engineering programs where students do not see the whole. An assessment would try to provide for the integration that is needed by designing ill-structured problems that cut across the different subdisciplines. This is what the engineering science course and examination could do which is difficult or not impossible in a credit system where every course is assessed separately. Pascarrela and Terenzini also found that structured course interventions may enhance the level of development of postformal reasoning that is in keeping with this writer's findings in respect of the training of student teachers (Heywood, 2008). That study showed that reflective thinking—called evaluation in his study—is a skill that has to be developed, a point that has been made in respect of software engineering by Upchurch and Sims-Knight (1999). King and Kitchener thought that "teaching students to think reflectively is an institutional goal that is best met when it is built into the whole curriculum and co-curriculum of the college." It does require change from traditional teaching and assessment practices although there is unlikely to be one best way to help students develop intellectually. But as indicated earlier, King and Kitchener hold that reflective thinking (judgment) is different from critical thinking because the latter does not take into account certain epistemic assumptions. They write, "Those who see critical thinking as problem solving fail to acknowledge that epistemic assumptions about knowledge play a central role in recognizing a problematic situation. They often see a close relationship between such thinking and the scientific method. Typically, they specify a set of steps for approaching a problem, such as formulating and then testing hypotheses. What is missing from this approach is the understanding that such steps cannot be applied if the individual fails to recognize the problem exists and that this recognition itself is predicated on other assumptions about knowledge (e.g., that it is gained through inquiry). By contrast, we argue that epistemic assumptions constitute a fundamental difference between children's and adult's problem solving and that it is only in adulthood that individuals hold epistemic assumptions that allow true reflective thinking" (King and Kitchener, 1994, p. 9). That is an assumption that some of those engaged in the philosophy for young children movement might find untenable. Notwithstanding, that potential for disagreement King and Kitchener developed an assessment procedure for reflective judgment that is considered in the next section.

7.6   The Assessment of Development

As has already been indicated, a distinguishing feature of the UK and US systems of education is the willingness of teachers in the United States not only to use standardized tests but to use instruments designed for a specific purpose to evaluate their courses. At the Colorado School of Mines, Pavelich used Moore's inventory that had been established to explore the construct validity of the Perry Model (Moore, 1988, 1989; Pavelich and Moore, 1996). Here, a test that is independent of the subject is used to determine a

Stage	Description
Stage 1	Knowing is limited to single concrete observations. What a person observes is true.
Stage 2	Two categories for knowing: right answers and wrong answers. Good authorities have knowledge; bad authorities lack knowledge…
Stage 3	In some areas, knowledge is certain and authorities have that knowledge. In other areas, knowledge is temporarily uncertain. Only personal beliefs can be known.
Stage 4	Concept that knowledge is unknown in several specific cases leads to the abstract generalization that knowledge is uncertain.
Stage 5	Knowledge is uncertain and must be understood within a context; thus, justification is context specific.
Stage 6	Knowledge is uncertain but constructed by comparing evidence and opinion of different sides of an issue or across contexts.
Stage 7	Knowledge is the outcome of a process of reasonable inquiry. This view is equivalent to a general principle that is consistent across domains.

EXHIBIT 7.3. Stages of the King and Kitchener (1994) Reflective Judgment Model (adapted) (King, P. M., and Kitchener, K. S. (1994). *Developing Reflective Judgment.* San Francisco, CA: Jossey-Bass).

person's level intellectual development. If such tests have reliability and validity should an employer take these as a referent that is as valuable as tests in a subject-knowledge domain? Given the finding in the study by Wise et al. (2001) that sometimes students with not the best grade point averages do better on such measures as Moore's than those with higher GPAs, should their measures of intellectual development be preferred to their scores in subject tests? At the very least, they might be recorded in a profile of the student. It is for this reason as much as any other that some educators advocate the use of profiles of performance and portfolios that can provide evidence of that competence. Some large employers in the United Kingdom now consider that information conveyed by a degree class is too limited and are setting their own selection tests (see Chapter 12).

Another independent measure of intellectual development is the Reflective Judgment Interview (RJI) designed by King and Kitchener (1994). This instrument measures reasoning along the seven stages shown in Exhibit 7.3. The RJI focuses on the ability of students to solve "wicked" problems, hence the relevance of the model to professional engineering. The model has many similarities with Perry's in the first three or four stages and may be describing the same things. The stages or levels of the King and Kitchener model are representative of pre-reflective thinking, quasi-reflective thinking (Stages 4 and 5), and reflective thinking (Stages 6 and 7). King and Kitchener say that the RJI typically consists of four ill-structured questions that focus on the concepts of the Reflective Judgment model. "The four standard problems concern a range of issues: how the Egyptian pyramids were built, the objectivity of news reporting, how human beings were created, and the safety of chemical additives in food" (King and Kitchener, 1994, p. 100). They have also used the problem of nuclear waste. They describe a problem from two contradictory points of view with the purpose of studying how persons reason about intellectual issues. The interview is semistructured and after the question has been read out, a series of probe questions are asked which might be followed up by other questions to clarify or refocus the response (p. 102). It has been criticized for being gender biased.

It should not be assumed that the stages in the model are fixed and somehow related to the structure of the curriculum. Moore and Hjalmarson (2010) reported that given an appropriate learning environment using an MEA first-year engineering students showed that they were capable of working on complex problems. This has its parallels with the work of the philosophy for young children movement, which has shown that young children are quite capable of thinking philosophically but in their own language (e.g., Matthews, 1980).

In their second synthesis of research on the effects of college on students postformal development, Pascarella and Terenzini (2005) considered that the differences found between freshmen and senior scores on the RJI could not be explained by differences in academic ability although it was possible that the net effects were confounded by age. The fascination for the teacher of King and Kitchener's (1994) study lies in their examples of the characteristics of the assumptions that accompany each stage, the instructional goals of each stage, difficult tasks seen from the perspective of each stage, sample development assignments, and developmental support for the instructional goals of each stage (pp. 252–256; see Exhibit 7.4).

7.7   The Reflective Practitioner

There are several traditions of reflective practice and like most things one can be traced back to the Greeks. Another derives from the philosophy of John Dewey and currently prevails in higher education (Kinsella, 2009), while another is deeply embedded in monasticism and the religious congregations (e.g., Lonergan, 1957).

In 1983, Donald Schön, in the tradition of Dewey, put forward the view that reflection is the essence of professional activity (Schon, 1983). Professionals engage in a process of design: problems are framed and reframed; designs are proposed and evaluated in a process of reflection. As might be expected, this model has been applied to engineering design and numerous studies have related to individual designers and group design (Borgford-Parnell, Diebel, and Atman, 2013). Schön (1987) has also applied it to teaching, and others have sought to develop the skill of reflection among student teachers (LaBoskey, 1994; Heywood, 2008). LaBoskey writes that following Dewey's definition of reflection "one reflects in order to know whatever one wants to know, wherever a state of perplexity arises. The method of reflection is a three step process including problem definition, means-ends analysis, and generalization that are carried out with attitudes of open mindedness, responsibility and whole heartedness" (p. 5).

Schön relates reflection to action and distinguishes between reflection-in-action and reflection-on-action. Reflection-on-action seems to be what the examiners of engineering science called evaluation. They wanted the candidates when they had completed their

---

Instructional Goals for Students
Promoting Reflective Thinking—Stage 3 Reasoning
Characteristic Assumptions of stage 3. Reasoning
Knowledge is absolutely certain in some areas and temporarily uncertain in other areas.
Beliefs are justified according to the word of authority in area of certainly and according to
what "fees right" in areas of uncertainty.
Evidence can be neither evaluated nor used to reason to conclusions.
Opinions and beliefs cannot be distinguished from factual evidence.

Instructional Goals for Students
Learn to use evidence in reasoning to a point of view.
Learn to view their own experiences as one potential source of information but not as the only
valid source.

Promoting Reflective Thinking–stage 6 Reasoning
Characteristic Assumptions of Stage 6, Reasoning
Knowledge is uncertain and must be understood in relation to context and evidence.
Some points of view may be tentatively judged as better than others.
Evidence on different points of view can be compared and evaluated as a basis for
justification.

Instructional Goals for Students
Learn to construct one's own point of view and to see that point of view as open to
reevaluation and revision in the light of new evidence.
Learn that though knowledge must be constructed, strong conclusions are epistemologically
justifiable.

EXHIBIT 7.4. Promoting reflective thinking in the King and Kitchener model—Stages 3 and 6. Adapted from King, P. M., and Kitchener, K. S. (1994). *Developing Reflective Judgment*. San Francisco, CA: Jossey-Bass, pp. 251 and 254, respectively. Each tabulation also included sections for difficult tasks from the perspective of the stage, sample developmental assignments, and developmental support for instructional goals.

projects to look back at what they had done. While they knew the candidate would not be able to change what they had done, they believed that he or she would learn from comparing the final product with the original specification. But they also wanted the candidate to continually look backward and forward during the project which seems to be reflection-in-action.

Cowan (2006) has suggested that Schön's model needs to include reflection-for-action. When a person reflects on a problem they hope to solve in the future, the activity is anticipatory: it is reflecting-for-action. Lonergan (1957) makes the same point "Reflection occurs because rational self-consciousness demands knowledge of what one proposes to do and of the reasons one has for doing it. Its normal duration is the length of time needed to learn the nature of the object of the proposed act and to persuade oneself to willingness to perform the act" (p. 612). In preparing this text, the writer reflected for action but was continually caused to reflect by nonfocused activities (i.e., specifically on reflection) during the writing of the text with the result that it had to be reordered several times. Such reflection has been called by Currano and Steinert (2012) "reflection-out-of action" in relation to design activity (see Section 9.7).

A report of an assessment-driven cooperative learning exercise to enhance critical thinking at Trinity College, Dublin, is a reminder that groups can and should reflect on their work. The authors note that the activity at the beginning of a project when the skill sets of the members of the group are identified, and task responsibilities and reporting roles are distributed, and the evolution of the project planned the members are engaged in a reflection-for-action (Huggard, Boland, and McGoldrick, 2014). Using a comprehensive scheme of assessment, they concluded from six presentations of a module that students became aware of the traits common to highly successful groups, and that skill in reflection had been developed. A major problem for research is how the skills acquired transfer to context-specific situations in industry, and how industry can assist with the further development of these skills.

Cowan also uses a model of reflection due to Kolb (1984). It has found favor among engineers because of his experiential theory of learning and its expression in a model of learning styles. As taught, learning follows a cycle that begins with a concrete experience and is followed by reflective observation, abstract conceptualization, and active documentation respectively (alternatively divergent thinking is followed by assimilation, by convergent thinking, and by accommodation). In order to know more about that experience, the learner has to reflect on that experience from as many different points of view as possible. From this reflection, the learner draws conclusions and uses them to influence decision making or take action. Cowan noted that the cycle draws the learner into a form of reflective practice. The stage of reflection is a purposive bridge between the experience and the generalization, whereas Schön's reflection is an open-ended activity.

At the time of writing, Cowan was engaged in conversations with a Rosalind Gould who had attended a lecture that he gave where he recommended journaling as an aid to reflective practice. She decided to keep a journal and sent Cowan some of her entries one of which began: Today I read "our senses take in experience, but they need the richness of sifting for a while through our whole bodies"—"composting". Out of this fertile soil bloom our poems and stories. But this does not come all at once. It takes time. You need to explore and compost the material. "We must continue to work the compost pile, enriching it and making it fertile so that something beautiful may bloom" (Goldberg, 1986). This analogy seems in tune with the current BBC Gardening Page (BBC), which advises gardeners to "turn the heap with a fork, mixing up the contents thoroughly and adding water, if it is drying out. If turned regularly and in warm conditions, your compost will be ready in about 2–4 months" (BBC, 2015).

In their subsequent conversation, Cowan and Gould came to the conclusion that the notion of composting reflection can enrich our affordances for reflection to good avail. Gould said that "the composting metaphor made her think about the 'raw materials' that constitute reflection, and also about the process by which reflection enriches and enables future growth." An outsider shown in this text responded with the view that that

Levels of Reflective Thinking	Examples of Thinking That Characterize The Level.
A. Summarizing	What is worrying me most, and why?
B. Analyzing	What conclusions can I draw from what I have thought already?
C. Closing the review	What points emerge from these reflections that are likely to be of use to me?
D. Distinguishing	What questions here might open up reflections that could be valuable to me?
E. Reasoned selection	Which is the best option for me at this moment?
F. Ongoing evaluation	Have I considered all that I should have thought about?
G. Self-awareness	Have I let myself be unduly influenced by personal preferences when choosing what to do?
H. Creative thinking	What will be the best way to do this?
I. Free thought	Wait a minute, wait a minute; I think I feel another blue flash coming in....

EXHIBIT 7.5. John Cowan's nine levels of reflection each accompanied by one of the several questions he uses to illustrate each level.

is what we do when we think. Which brings us back to the hypothesis of metacognition that if individuals understand how they think (learn) their approach to learning will be enhanced. It is surely very difficult to separate reflection from metacognition. If individuals are trained in reflective thinking, they are, provided it is pointed out to them, being trained in metacognition. It is all about learning-how-to-learn.

A couple of years earlier, Cowan (private communication) suggested that there are nine levels of reflective thinking. They are shown in Exhibit 7.5 and in Exhibit 7.6 his differences between analysis and reflection are shown. If the elements of action are taken into account, then one way of looking at reflective thought is as a powerful form of self-assessment for which Cowan's tabulations provide a rubric.

Cowan is of the view, one with which this writer partially concurs, that most learners do not know how to reflect, yet up until now that has been what is asked of them by the educational elite who believe that students require a reflective capability if they are to develop high-level abilities and competencies. I say partially aware because I believe that everyone "cogitates" in their daily lives or should I say composts. But because we do not explore the possibilities of such activities, we do not develop a capacity for critical thought. We do not turn it into a competence. If the view is taken that this competence can be developed, then necessarily individuals can be helped to develop that competence, and in recent years engineering educators have begun to promote training for this purpose both among students and among themselves (see, e.g., Shull, 2014).

Barroso and Morgan (2014) incorporated "reflections" as the topic has become known into a range of analytical engineering courses. The courses in civil engineering for which they were responsible involved the provision of active interaction and cooperative

	Analysis	Reflection
1.	Conceived and written in the first person	Conceived and written in the first person
2.	Cognitive; a higher level of thinking activity	Metacognitive; thinking about thinking
3.	Seeks findings of general relevance	Seeks findings of personal significance
4.	Process standard and well established	Process personal, and likely to be handled flexibly
5.	Findings emerge from the process	Search should address a specific predetermined question
6.	Works on data assembled before the analysis begins	Often accumulates or draws in ideas and facts about context during reflective analysis
7.	Yields impersonal findings	Yields findings of personal value and use

EXHIBIT 7.6. John Cowan's distinction between analysis and reflection written in response to a question put in an online discussion that sought to have clarification of the distinction between analytical review of a semester of learning and development, and reflection on that development.

learning the purpose of which was to engage the students in critical thinking through-out the problem-solving process. A major course project enhanced these activities. Incorporated in each course were written reflections that targeted specific activities as well as the whole course experience. To assess the reflections, they used a *Reflection Depth Rubric* that had been described by Dalal et al. (2012). They also analyzed the responses and essays following a technique of qualitative analysis. Barroso and Morgan found that their students, which included two classes of graduates, could "write thoughtful reflections regarding their learning experiences." But this depended upon a thorough introduction to the exercise coupled with class discussion. One of the things that bothered them was the length of the prompts that should be given. In one of the graduate classes, the written instructions and prompt were kept simple, whereas in the other graduate class and senior elective the written instructions were more detailed and more specific prompts were given that related first to the completion of the whole project (e.g., "How did these experiences tie into your learning process and development? What impacts/experiences did working with a group vs. working individually have?" Second, the students were asked to assess a number of items related to their own submittal. For example, they were asked—"what would you work on more if you had additional time?" I chose this particular item to illustrate the confusion in terminology discussed in Chapter 1. This was expected of students completing the engineering science project (Section 3.1) but was called "evaluation." Barroso and Morgan also asked their students to reflect on the learning gained from the project with the interesting question "How do you think this will tie into your future professional career?" The second set of prompts related to the overall course. The list was preceded by the limiting phrase "you may want to reflect on the following components." Two of the prompts are quite long. One reads, "reflect on topical knowledge gained or developed throughout the course? What do you know about structural behavior and design of structures that you did not know at the beginning of the semester? Or did you already know something but now have a deeper understanding?" Their analysis of the two approaches to prompting showed that the "actual length of the prompt is not essential to receiving a thoughtful response." The rubric that they used for assessing the essays used the scale—*exemplary, very proficient, proficient, partially proficient,* and *not reflective.* In only one course is a single response reported as not reflective. The analyses showed differences between the genders. For example, females wrote longer reflections than males (average word counts of 1200 and 762, respectively). Females also obtained higher reflective ratings than males (3.00 as compared with 2.517). The authors were unable to account for these differences but conjectured that the women were less uncomfortable reflecting and writing about their experiences.

The value of academics subjecting themselves to reflective activities has been shown by Kuriakose and Vermaak (2014), who described how academics in a South African University successfully used reflective strategies to redesign a first-year program with a view to reducing the number of dropouts. There is also much interest in the United States in reflective activity and engineering educators are engaged in a substantial project at the University of Washington, Seattle, and have devised training activities for this purpose (Turns, Sattler, and Kolmos, 2014).

As presented above, the competence that is being developed may be categorized as "inside" (Section 1.3 and Chapter 11) although the authorities to whom I have referred may object to this positioning. However, Boud and Walker (1998) consider that context may be the most important influence on reflection. A view that is clearly "outside" has been expressed by Warhurst (2008). He obtained data from 29 early career lecturers at a research intensive university who were in a mandatory Post-Graduate Certificate in Academic Practice in which Schön's concept of reflective practice was discussed along with case studies of lecturers' reflective learning. The lecturers voluntarily completed a journal of their teaching practices in which they were asked to reflect on the influences on their use of those practices. Warhurst pointed out the difficulties in defining reflection, citing Kahn et al. (2006) in this respect, and chose to concentrate on reflection-on-action because there was considerable agreement about what it was. Furthermore, in order to examine the contexts in which the reflections occurred, he used an approach based on the situated learning theory of Wenger (1998), who construes learning in terms of practice, identity, community, and meaning that are interconnected and mutually defining. "Learning occurs, therefore, through experiencing, becoming, belonging and doing" (Warhurst, 2008). Learning does not take place in isolation of its social context, and it certainly depends on relationships as MacMurray pointed out 50 years ago. "We come to be who we are as personal individuals only in personal relationships" (Costello, 2002, p. 326).

Warhurst reports that the journals showed evidence of reflection that advanced the participants' understanding and practice. But he reported that most of the lecturers were also reflecting at a deeper level in that "Theory and practice are seen as existing in a dialectical relationship with formal knowledge being reflectively interpreted into practice and in turn formal knowledge evolving practice" (Warhurst, 2008). While this evidence supports the inclusion of a pedagogy of reflection in university teacher training the journals and interviews revealed the power of the social situation in which these lecturers worked. The three extracts that follow say it all:

Just over half the lecturers were plunged into practice in ways that afforded little time for reflective learning from experience.

Heads of department appeared further to limit the scope of new lecturer's independent reflective learning through allowing experienced staff to "off-load" particularly demanding teaching.

Established colleagues also appeared to deny the new lecturers access to certain opportunities for informal pedagogic learning.

Clearly, these lecturers were not placed in an environment where their pedagogical activities were encouraged. In Section 6.5, Borrego (2008) argued the need to create a culture of assessment. Warhurst formalizes the conversations that many academics interested in pedagogy have namely—that it is against the prevailing culture. Thus, there is need to create a culture of pedagogy in which assessment is part of the practice. From the student's perspective the same applies, the learning environment has to be conducive to reflective practice and a conditioning factor in the way students are assessed. The same argument applies to industrial organizations. Very often the environments created by managers inhibit learning. One wonders if industrialists really want engineers to be reflective thinkers rather than conform to what the firm wants, for reflective activity necessarily leads to the open-ended thinking that characterizes creativity. At the same time work is an activity that is both cognitively and emotionally construed for which reason it is incumbent on employers and employees to have a clear understanding of emotional intelligence (see Section 8.1) and, how on that level organization impacts with the person and vice versa.

7.8 Adaptive Expertise

Adaptive learning and adaptive expertise refer to two different aspects of learning. Adaptive learning in computing refers to intelligent and adaptive learning systems designed to individualize the learning of each student (see Section 6.4). Interactive tutorials have been designed that adapt the learning environment to the current needs of a student as a function of their previous responses (Khawaja et al., 2013). These authors found that there was an overall improvement in student performance. But they also identified a point in the course where with the intention of being helpful they caused overload for some students.

Adaptive expertise in this context is related to the competency of adaptability. It has replaced the term "transfer of learning" (skill). It requires a high level of conceptual knowledge and is a characteristic that distinguishes novices from experts. Hicks, Bumbaco, and Elliot (2014) perform a valuable service by linking it with critical thinking and reflective practice. They argue that there is a deep connection between the three. The adaptive expert is required to think both critically and reflectively if he or she is to function at the level of deep conceptual understanding required to redefine the problem. They point out that one of the common threads between the three in the literature is "disposition." "[…] some people may be just more naturally inclined to be critical thinkers, reflective thinkers, or adaptive experts. But does this mean that some individuals are less capable? Perhaps an examination of incentives to engage in critical and reflective thinking may be necessary to answer this question, and more importantly, to find ways to practice both those skills. Another common thread—lifelong learning—clearly also requires this disposition." *Capable* needs to be defined for it may be that in certain areas of activity such as entrepreneurship reflective practice may be a hindrance to the development of the kind of competencies required to be a successful entrepreneur.

We might also ask whether or not the higher-order skills of the Bloom *Taxonomy* (and its revision) more especially evaluation require critical thinking and reflective skills, and if for assessment purposes the way forward is to design assessments and examinations that measure the higher-order skills when so many do not seem to get beyond the lower-order competencies.

Hicks, Bumbaco, and Elliot come to much the same conclusion as this writer that the continual loading of extra content into the curriculum acts as a disincentive to teachers to provide the type of action learning that help develop these abilities. But they do not go as far as this writer to suggest that engineering educators should like teachers, be trained for the job.

7.9 Discussion

The social milieu in which people learn is a powerful influence on their intellectual development and emotional response. An individual lives in a plurality of social systems each with its own goals. The dispositions that make a person and what the person brings (gives) to life matter. Higher education has by and large given little attention to these influences. It is the few rather than the many who have taken the trouble to investigate the value of and relevance to their teaching of learning styles.

There is considerable evidence to suggest that students' responses to teaching are influenced by their different learning preferences or orientations. A primary conclusion is that effective teaching requires a teacher to understand the learning styles that are present in his classes and respond to them appropriately, and at a minimum this will require variety in their teaching. Test anxiety may be alleviated by using a variety of assessment formats to cope with a variety of learning styles but this requires high-level skill in test design and willingness to pilot material. At the very least, teachers should be aware of their teaching style.

It seems self-evident that the study strategies adopted will be related to intellectual development. Just as there are a variety of learning styles so students approach study in different ways that are influenced by their expectations of assessment. Early research showed that students responded to assessment by either "depth" or "surface" learning. These concepts are now part of the dialogue of higher education. One might assume that if courses are provided on learning-how-to-learn (see Section 8.5), they would embrace study strategies.

Similarly, little attention has been given to the design of courses that meet the criteria of developmental models such as those put forward by Perry, and King and Kitchener. At the very least, note should be taken of the findings that students who are in the early stages of development respond well to lectures in which the answers are given. To put it in another way, much teaching does not encourage intellectual development just as it does not encourage critical thinking with which it is associated.

Among the goals of higher education that are being promoted are that students should be able to think critically, become reflective practitioners, be able to easily adapt, and take responsibility for their learning. It is perceived that if students are able to self and peer assess they will better achieve these goals.

However, the continual loading of extra content into the curriculum acts as a disincentive to teachers to provide the type of action learning that may help develop these abilities. The design of the curriculum is a fundamental issue that needs to be addressed, but it is not a problem for this text that continues with a different look at student variability in the next chapter.

Notes

1. Douglas McGregor proposed these to describe two different orientations to management. They equally apply in teaching as the exhibit shows. Column A is adapted from a description by Schein of Theory X and Theory Y is in its original form as it is clearly related to the potential for learning. A teacher who believes Theory X explains student learning is much more likely to be committed to a monologue form of teaching than a teacher who thinks Theory Y approximates to the truth. The key question that faculty have to answer, irrespective of their beliefs, is number 5. If the answer is "no," what are they going to do about it?

A. Theory X	B. Theory Y
1. The student is primarily motivated by academic incentives and will do whatever gets him or her the greatest gain.	1. The expenditure of physical mental effort is as natural as play or rest. The ordinary person does not inherently dislike work: according to the conditions, it may be a source of satisfaction of punishment.
2. Since academic incentives are under the control of the institution, the student is essentially a passive agent to be manipulated, motivated, and controlled by the organization.	2. External control is not the only means for obtaining effort. A person will exercise self-direction and self-control in the service of objectives to which he is committed.
3. The student's feelings are essentially irrational and must be prevented from interfering with his or her rational calculation of self-interest.	3. The average human being learns, under proper conditions, not only to accept but to seek responsibility.
4. Institutions and their organizational (curriculum) arrangements can and must be designed in such a way as to neutralize and control their feelings and therefore their unpredictable traits.	4. Many more people are able to contribute creatively to the solution of organizational problems than do so.
	5. At present, the potentialities of the average person are not fully used.

2. Baxter-Magolda's Epistemological Reflection Model relates to the way in which knowledge is perceived, which is understood to be in four dimensions. These are *Absolute knowing*: the correspondence with the early stages of the Perry model is apparent. *Transitional knowing* when it is realized that some knowledge is uncertain. Magdola believes that many learners are at this stage in their middle college years and spend much time in learning how to understand another person's perspective. The ability to understand another person's point of view was considered to be particularly important by those who designed the 1976 history syllabus in Ireland when the troubles were at their peak, and it was written as an objective for the public examination that tested the abilities of 15-year-olds in history. *Independent knowing*: most knowledge is perceived to be uncertain and decisions have to be taken on the basis of their own beliefs and understandings. Magdola considers that some students reach this level by the end of their period in college. She believes that very few reach the level of *contextual knowing* when knowledge is based on the context on which evidence supporting the knowledge is used.

References

Abdulwahed, M., and Nagy, Z. K. (2009). Applying Kolb's experiential learning cycle to laboratory education. *Journal of Engineering Education*, 98(3), 283–294.

Ahern, H., O'Connor, T., McRuairc, G., McNamara, M., and O'Donnell, D. (2012). Critical thinking in the university curriculum (2012). *European Journal of Engineering Education*, 37(2), 125–132.

Argyris, C., and Schön, D. A. (1978). *Organizational Learning*. Reading, MA: Addison-Wesley.

Barroso, L. J., and Morgan, J. (2014). Implementing reflection in technical courses. In: ASEE/IEEE Proceedings Frontiers in Education Conference, pp. 1444–1449.

Baxter-Magolda, M. B. (1987). Comparing open-ended interview and standardized measure of intellectual development. *Journal of College Student Personnel*, 28, 443–448.

Baxter-Magolda, M. B. (1994). Making their Own Way. Narratives for Transforming Higher Education to Promote Self development. Sterling, VA. Stylus.

Biggs, J. (1987). *Student Approaches to Studying and Learning*. Hawthorne, Victoria: Australian Council for Educational Research.

Biggs, J., and Tang, C. (2011). *Teaching for Quality Learning at University*, 4th edition. New York, NY: McGraw International.

Borgford-Parnell, J., Deibel, K., and Atman, C. J. (2013). Engineering design teams. Considering the forests and the trees. In: Williams, Figueiredo, and Trevelyan (eds.), *Engineering Practice in a Global Context: Understanding the Technical and the Social*. Leiden, the Netherlands: CRC Press.

Borrego, M. (2008). Creating a culture of assessment within engineering education. In: ASEE/IEEE Proceedings Frontiers in Education Conference S2B-1 to 6.

Boud, D., and Walker, D. (1998). Promoting reflection in professional courses: the challenge of context. *Studies in Higher Education*, 23(2), 191–207.

Brown, A. O., and 13 associates (2013). Assessment of active learning modules: an update of research findings. Proceedings Annual Conference of the American Society for Engineering Education. Paper 7462.

Cajander, A., Daniels, M., and Von Konsky, B. R. (2011). Development of professional competencies in engineering education. In: ASEE/IEEE Proceedings Frontiers in Education Conference, S1C-1 to 5.

Chan, C. K. Y. (2012). Exploring an experiential learning project through Kolb's learning theory using a qualitative research method. *European Journal of Engineering Education*, 37(4), 404–415.

Costello, J. E. (2002). *John MacMurray. A Biography*. Edinburgh: Flora Books.

Cowan, J. (1998). *On Becoming an Innovative University Teacher.* Buckingham, Open University Press.

Cowan, J. (2006). *On Becoming an Innovative University Teacher*, 2nd edition. Buckingham: Open University Press (1st Ed-1989).

Culver, R. S., Cox, P., Sharp, J., and FitzGibbon, A. (1994). Student learning profiles in two innovative honours degree engineering programs. *International Journal of Technology and Design Education*, 4(3), 257–288.

Culver, R. S., and Hackos, J. T. (1983). Perry's model of intellectual development. *Engineering Education*, 73(2), 221–226.

Culver, R. S., and Sackman, N. (1988). Learning with meaning through marker events. In: Proceedings Frontiers in Education Conference.

Culver, R. S., Woods, D., and Fitch, P. (1990). Gaining professional expertise through design activities. *Engineering Education*, 80(5), 533–536.

Culver, R. S., and Yokomoto, C. (1999). Optimum academic performance and its relation to emotional intelligence. In: ASEE/IEEE Proceedings Frontiers in Education Conference, 13b7-26 to 29.

Currano, R. M., and Steinert, M. (2012). A framework for reflective practice in innovation design. *International Journal of Engineering Education*, 28(2), 270–275.

Dalal, D. K., Hakel, M. D., Sisiter, M. T., and Kirkendall, S. R. (2012). Analysis of a rubric for assessing depth of classroom reflections. *International Journal for ePortfolio*, 2, 75–85.

Entwistle, N. J., Thompson, J., and Wilson, J. D. (1974). Motivation and study habits. *Higher Education*, 3(4), 379–396.

Felder, R. M., and Brent, R. (2003). Designing and teaching course to satisfy the ABET Engineering Criteria. *Journal of Engineering Education*, 92(1), 7–25.

Felder, R. M., and Brent, R. (2005). Understanding student differences. *Journal of Engineering Education*, 94(1), 57–72.

Felder, R. M., and Silverman, K. L. (1988). Learning and teaching styles in engineering education. *Engineering Education*, 78(7), 674–681.

Felder, R. M., and Spurlin, J. (2005). Applications, reliability and validity of the Index of learning styles. *International Journal of Engineering Education*, 21(1), 102–112.

Feldgen, M., and Clua, O. (2009). The use of CATs and case-based teaching for dealing with different levels of abstractions. In: ASEE/IEEE Proceedings Frontiers in Education Conference, T4F-1 to 7.

Fitkov-Norris, E., and Yeghiazarian, A. (2014). Measuring the relationship between study habits and preferred learning style in higher education. In: ASEE/IEEE Proceedings Frontiers in Education Conference, pp. 1323–1326.

Frezza, S. T. (2014). A knowledge base for engineering design. In: ASEE/IEEE Proceedings Frontiers in Education Conference, pp. 158–165.

Frezza, S. T. (2015). Engineering insight. The philosophy of Bernard Lonergan applied to engineering. *Philosophical and Educational Perspectives in Engineering and Technological Literacy, 2*: Technological and Engineering/Philosophy Division of ASEE.

Grasha, A. (2002). *Teaching with Style: A Practical Guide to Enhancing Learning by Understanding Teaching and Learning.* San Bernadino, CA: Alliance Publishing.

Hamilton, E., Besterfield-Sacre, M., Olds, B. M., and Siewiorek, N. (2010). Special session. Model eliciting activities in engineering. A focus on model building. In: Proceedings of Annual Conference of American Society for Engineering Education, June 2010. Paper 1501.

Hesseling, P. (1966). *A Strategy for Evaluation Research.* Aassen: Van Gorcum.

Heywood, J. (1977). *Assessment in Higher Education,* 1st edition. Chichester, UK: John Wiley & Sons, Ltd.

Heywood, J. (1989). *Learning, Adaptability and Change. The Challenge for Education and Industry.* London: Paul Chapman/Sage.

Heywood, J. (2005). *Engineering Education. Research and Development in Curriculum and Instruction.* Hoboken, NJ: IEEE/John Wiley & Sons.

Heywood, J. (2008). *Instructional and Curriculum Leadership. Towards Inquiry Oriented Schools.* Dublin: National Association of Principals and Deputies/Original Writing.

Hicks, N., Bubaco, A. E., and Douglas, E. P. (2014). Critical thinking, reflective practice, and adaptive expertise in engineering. In: Proceedings of Annual Conference of American Society for Engineering Education, June 2014. Paper 10737.

Hill, A., and Brown, C. M. (2013). An examination of the relationship between intellectual development and learning preferences in electrical and computer engineering (ongoing). In: Proceedings of Annual Conference of American Society for Engineering Education, June 2013. Paper 6980.

Hirsch, P. L., and McKenna, A. F. (2008). Using reflection to promote team work understanding in engineering design. *International Journal of Engineering Education,* 24920, 377–385.

Holvikivi, J. (2007). Learning styles in engineering education: the quest to improve didactic practice. *European Journal of Engineering Education,* 32(4), 401–408.

Huggard, M., Boland, F., and McGoldrick, C. (2014). Using cooperative learning to enhance critical reflection. In: ASEE/IEEE Proceedings Frontiers in Education Conference, pp. 1899–1906.

Kahn, P., Young, R., Grace, S., Pilkington, R., Rush, L., Tomkinson, B., and Willis, I. (2006). *The role and effectiveness of reflective practices in programs for new academic staff: A grounded practitioner review of the research literature.* York, UK: The Higher Education Academy.

Kallenberg, B. J. (2013). *By Design. Ethics, Theology and the Practice of Engineering.* Cambridge, UK: James Clarke.

Kaupp, J. A., and Frank, B. M. (2014a). Potential, authentic and sustainable development and assessment of critical thinking in engineering through model eliciting activities. In: Proceedings of Annual Conference of American Society for Engineering Education, June 2014. Paper 9806.

Kaupp, J. A., and Frank, B. M. (2014b). Investigating the impact of model eliciting activities on development of critical thinking. In: Proceedings of Annual Conference of American Society for Engineering Education, June 2014. Paper 6432.

Kaupp, J. A., Frank, B. M., and Chen, A. S.-Y. (2013). Investigating the impact of model eliciting activities on the development of critical thinking. In: Proceedings of Annual Conference of American Society for Engineering Education, June 2013. Paper 6432.

Khawaja, A., Gangadhara, B., Prusty, R. A. J., Ford, N., Russell, C., and Russell, M. (2013). Can more become less. Effects of an intensive assessment environment on student learning performance. *European Journal of Engineering Education*, 38(6), 631–651.

Kinsella, E. A. (2009). Professional knowledge and the epistemology of reflective practice. *Nursing Philosophy*, 11, 3–14.

King, P. M., and Kitchener, K. S. (1994). *Developing Reflective Judgment*. San Francisco, CA: Jossey-Bass.

Kolb, D. A. (1984). *Experiential Learning. Experience as the Source of Learning and Development*. Englewood Cliffs, NJ: Prentice-Hall.

Kuriakose, R. B., and Vermaak, H. J. (2014). Using reflective practices to reduce dropout rates among first year students at University of technology, a South African perspective. In: ASEE/IEEE Proceedings Frontiers in Education Conference, pp. 1708–1713.

LaBoskey, V. K. (1994). *Development of reflective Practice. A Study of Preservice Teachers*. New York: Teachers College Press.

Leite, C., Mouraz, A., Trindade, R., Ferreira, J. M. M., Faustino, A., and Villate, J. E. (2011). A place for arguing in engineering education: a study on student's assessments. *European Journal of Engineering Education*, 36(6), 607–616.

Leyden, J. A., Moskal, B. M., and Pavelich, M. J. (2004). Qualitative methods used in the assessment of engineering education. *Journal of Engineering Education*, 93(1), 65–72.

Lonergan, B. J. F. (1957). *Insight. A Study of Human Understanding*. London: Darton, Longman and Todd.

Lonergan, B. J. F. (1973). *Method in Theology*. London: Darton, Longman and Todd.

Luchins, A. S. (1968). Mechanisation in problem solving. The effect of "einstellung." *Psychological Monographs,* 248 (cited by McDonald).

Marra, R., Palmer, B., and Litzinger, T. A. (2000). The effects of first year design course on student intellectual development as measured by the Perry Scheme. *Journal of Engineering Education*, 89(1), 39–45.

Marshall, S. (1980). Cognitive– affective dissonance in the classroom. *Teaching Political Science*, 8(1), 111–117.

Marton, F., and Saljö, R. (1984). Approaches to learning. In: F. Marton, D. Hounsell, and N. J. Entwistle (eds.), *The Experience of Learning*. Edinburgh: Scottish Academic Press.

Marton, F., Hounsell, D., and Entwistle, N. J. (eds.) (1984). *The Experience of Learning*. Edinburgh: Scottish Academic Press.

Matthews, G. (1980). *Philosophy and the Young Child*. Cambridge, MA: Harvard University Press.

Mazumder, Q. H., and Ahmed, K. (2013). The effect of teaching style and experience on student success in the USA and Bangladesh. In: Proceedings of Annual Conference of American Society for Engineering Education. Paper 8407.

Moore, T. J., and Hjalmarson, M. A. (2010). Developing measures of roughness: problem solving as a method to document student thinking in engineering. *International Journal of Engineering Education*, 26(4), 820–850.

Moore, W. S. (1988). *The Learning Environment. Preferences and Instruction Manual*. Olympia, WA: CSID.

Moore, W. S. (1989). The learning environment preferences. Exploring the construct validity of an objective measure of the Perry scheme of intellectual development. *Journal of College Student Development*, 30, 504–514.

Parsons, D. (2008). Is there an alternative to exams? Examination stress in engineering. *International Journal of Engineering Education*, 24(6), 1111–1118.

Pascarella, E. T., and Terenzini, P. T. (2005). *How College Affects Students. Vol. 2. A Third Decade of Research*. San Francisco, CA: Jossey-Bass.

Paul, L. R., and Elder, L. (2006). *A Guide for Educators in Critical Thinking. Competency, Standards, Principles, Performance Indicators and Outcomes*. Vol. 8. Tomales, CA: Foundation for Critical Thinking.

Pavelich, M. J., and Moore, W. S. (1996). Measuring the effect of experiential education using the Perry model. *Journal of Engineering Education*, 85, 287–292.

Perry, W. B. (1970). *Forms of Intellectual and Ethical Development in the College Years*. New York: Holt, Reinhart and Winston.

Purser, S., Hilpert, J. C., and Wertz, R. E. H. (2011). Cognitive dissonance during engineering design. In: ASEE/IEEE Proceedings Frontiers in Education Conference, S4F-1 to 5.

Ramsden, P. (1988). Context and strategy. In: R. R. Schmeck (ed.), *Learning Strategies and Learning Styles*. New York: Plenum Press.

Rigg, C., Stewart, J., and Trehan, K. (2007). A critical take on a critical turn in HRD. In: C. Rigg, J. Stewart, and K. Trehan (eds.), *Critical Human Resource Development. Beyond Orthodoxy*. Harlow: Prentice-Hall.

Rocke, S., Radix, C. A., Persad, J., and Ringle, D. (2014). Use of argument maps to promote critical thinking in engineering education. In: ASEE/IEEE Proceedings Frontiers in Education Conference, pp. 588–591.

Schön, D. A. (1983). *The Reflective Practitioner*. New York: Basic Books.

Schön, D. A. (1987). *Educating the Reflective Practitioner. Toward a Design for Teaching and Learning in the Professions*. San Francisco, CA: Jossey-Bass.

Senge, P. M. (1990). *The Fifth Discipline. The Art and Practice of the Learning Organization*. New York: Doubleday.

Shull, P. J. (2014). Using the engineering design process as metacognitive learning strategy to improve student performance. In: ASEE/IEEE Proceedings Frontiers in Education Conference, pp. 1800–1804.

Stein, B., and Haynes, A. (2011). Engaging faculty in the assessment and improvement of students' critical thinking assessment test. *Change*, March–April 2011.

Stein, B., Haynes, A., and Unterstein, J. (2003). Assessing critical thinking skills. In: SALS/COC Annual Meeting, Nashville, Tennessee.

Svinicki, M. D., and Dixon, N. M. (1987). The Kolb Model modified for classroom activities. *College Teaching*, 35(4), 141–146.

Todd, R. H., Sorensen, C. D., and Magelby, S. P. (1993). Designing a senior capstone course to satisfy industrial customers. *Journal of Engineering Education*, 82(2), 92–100.

Turns, J., Sattler, B., and Kolmos, A. (2014). Designing and refining reflection activities for engineering education. In: ASEE/IEEE Proceedings Frontiers in Education Conference, pp. 1607–1610.

Upchurch, L., and Sims-Knight, J. E. (1999). Reflective essays in software engineering. In: ASEE/IEEE Proceedings Frontiers in Education Conference, p. 3, 13a6-15 to 18

Warhurst, R. (2008). Reflections on reflective learning in professional formation. *Studies in the Education of Adults*, 40(2), 176–186.

Watkins, D. (1981). *Factors Influencing the Study Methods of Australian Tertiary Students*. Canberra: Australian National University, Office for Research in Academic Methods.

Wenger, E. (1998). *Communities of Practice. Learning, Meaning and Identity*. Cambridge, UK: Cambridge University Press.

Wise, J., Lee, Sang Ha., Litzinger, T. A., Marra, R. M., and Palmer, B. (2001). Measuring cognitive growth in engineering undergraduates longitudinal study. In: Proceedings of Annual Conference of the American Society for Engineering Education.

Wittrock, M. C. (1963). Effects of certain sets on complex verbal learning. *Journal of Educational Psychology*, 54, 85–88.

8

Emotional Intelligence, Peer and Self-Assessment, Journals and Portfolios, and Learning-How-to-Learn

The concept of emotional intelligence is discussed. Whatever view you take about it as unitary concept or not, it is clear that we have to handle its components every day. Its development can be assisted in both education and training, but it cannot be left to education alone because education cannot simulate the everyday situations that have to be faced in industry. Therefore, industry has a critical role to play in the development of their employees particularly in the initial stages of their employment.

There is a growing use of self and peer and self-assessment. The studies reported here, while delineating potential and problems, support their continued use. For example, peer assessment has been found to be useful in providing feedback. Many students remain to be convinced about their merits but this could be overcome by training students not only in the procedures but in reflective practice (Chapter 7). Such training might be a component of a course in learning-how-to-learn. Just as every engineering educator should have a defensible theory of learning so to should students. An example of a course designed to develop Senge's five disciplines of learning is given.

The chapter ends with a brief discussion of the value of journals and portfolios that are also being increasingly used. It notes that the e-world has revolutionized how they are used in the classroom situation. Writing remains a key instrument in the development of reflective practice and critical thinking.

The Assessment of Learning in Engineering Education: Practice and Policy, First Edition. John Heywood.
© 2016 The Institute of Electrical and Electronics Engineers, Inc. Published 2016 by John Wiley & Sons, Inc.

8.1 Introduction

This chapter is a continuation of Chapter 7, the division having been made for the convenience of the reader. Emotional intelligence is a complex concept that embraces a long list of skills that suggest it is not a unitary concept. Culver (1998) argues that the promotion of emotional intelligence is necessary for any program in engineering. Nevertheless, many short courses have been given in industrial organizations that attend to these skills. They try to influence the "tacit" knowledge that we have acquired about ourselves and our relationships with other people: it is a knowledge that influences how we manage ourselves, others, and the tasks we are given. Clearly, it can help or hinder aspirations to think critically or reflect actively.

Intellectual, emotional, and professional development cannot be completed within college alone because a person continues to develop and will do so in response to any situation. For this reason, industrialists have as much responsibility for the development of their engineers as do the colleges from which they come, and in these days of rapid turnover have an obligation to help them prepare for their next work assignment (see Chapter 12).

The abilities of reflection and critical thinking are helped by self- and peer assessment. Ford and Rennie (1999) found this to be the case. They had required their students to self-assess the industrial component of a cooperative program. Although their students did not like doing it, they appreciated its value. Ford and Rennie make the point that derivation of the final grade is not the most important point of the scheme.

It is argued that involvement in grading helps develop critical faculties, but Cowan and Boyd, reflecting on their collaborative self-assessment, were led to wonder if the learner should be involved in setting goals. This is what happens in programs of independent study. At its simplest, self-assessment is evaluative; at its most complex, self-assessment is deeply reflective.

There are positive responses to peer assessment and peer learning, but it is clear that if they are going to be used, care has to be taken with course design. They can change the role of the instructor.

Learning journals are often used as a means of developing self-assessment, and George and Cowan (1999) have shown how reflective journals can be fitted into the Kolb Learning Cycle, and how it may be used to plan "active experimentation of their generalizations."

Portfolios provide a mechanism for assessment, but as so often with the introduction of new techniques of assessment, the commitment of the organization in which it is to be used is essential, it is a *sine qua non* that teachers should be adequately trained for this purpose. It is crucial to define the purpose of the portfolio, its structure, and assessment. Much confusion can arise when it is not clear whether the portfolio is being used as a process or a product.

Both the last chapter and this one have focused on various aspects of student learning yet probably the best kept secret from students in higher education is about what learning is and how it is accomplished. Not only should every teacher have a defensible theory of learning but so too should every student. It may be argued that to achieve that goal some provision should be made for the study of learning-how-to-learn early in a student's college career. The example given later of the use of Senge's theory is a reminder that firms should be learning systems. Senge's (1990) frame of reference could be used by students on internships to analyze the organization.

8.2 Emotional Intelligence

Studies of the effect of temperament on learning and performance have a long history. It is clear that the emotions influence our response to learning in the same way they influence our response to others. We have also learnt that we need to be able to govern (control) our emotions, particularly in critical incidents. The ability to do this is sometimes called emotional (or social) intelligence. Since 1995, when Goleman (1994) published a bestseller on the topic it has been the subject of much discourse (Bar-On and Parker, 2000). There are nontraditional intelligences that appear to overlap with it such as "practical intelligence" (Sternberg, 2000).

There is a long history of training in industry that has focused on the skills commonly embraced by the term *emotional intelligence*, be it a unitary intelligence, or not (Cherniss, 2000). It is clearly tested in the affective domains that Freeman and Byrne (1973) described for the assessment of general practice in medicine (Chapter 3.5). Culver (1998) argued that promoting emotional intelligence is necessary if a successful engineering program was to be achieved. He cited a list of components that contributed to emotional intelligence that were given in a "self-science" curriculum at Nueva School in California. An adaptation of this list is given in Exhibit 8.1. They could be regarded as a list of personality traits, in which case they might be better called personal transferable skills. One way of looking at emotional intelligence is to consider it to be the interplay between the cognitive and affective domains in the conduct of living, if you accept that living is problem solving, which embraces critical thinking. But, as Hedlund and Sternberg (2000) point out, the competencies required for solving a problem will be a function of the type of problem to be faced.

Within practical intelligence, Sternberg and Grigorenko (2000) include practical problem solving, pragmatic intelligence, and everyday intelligence (see Exhibit 5.9). Sternberg and Hedlund considered that what differentiates emotional from social and practical intelligence is "tacit" knowledge. That is the knowledge that is not taught but acquired as part of everyday living. Their categories of "tacit" knowledge are managing self, managing others, and managing tasks. It will be understood that management training is about trying to influence these three types of knowledge. Such knowledge is *procedural* in the sense that it is associated with particular situations, and is transferred to similar situations and this is why the content of knowledge and the problem solved differentiate the three constructs. They write, "The ability to acquire knowledge, whether it pertains to managing oneself, managing others, or managing tasks, can be characterized appropriately as an aspect of intelligence. It requires cognitive processes such as encoding essential information from the environment and recognizing associations between new information and existing knowledge." (Which is what we try to do in

Self-awareness	Observing yourself and recognizing your feelings with a view to action or trying to change action in specified circumstances. This can include mode of study, reactions to people, etc.
Personal decision making	Examining one's actions and predicting the consequences. Knowing the basis of the decision, i.e., cognition or feeling. This covers the gamut of small and large decisions that relate to everyday actions.
Managing feelings	Requires self-awareness in order to be able to handle anxieties, anger, insults, put-downs, and sadness.
Handling stress	Use of imagery and other methods of evaluation
Empathy	Understanding how people feel and appreciating that in the learning situation students can become stressed, and that such stress can be reduced by the mode of instruction
Communications	Becoming a good listener and question asker: distinguishing between what someone does or says and your own reactions about it; sending "I" messages instead of blame.
Self-disclosure	Building trust in relationships and knowing when one can be open.
Insight	This is different from cognitive insight. It is about understanding one's emotional life and being able to recognize similar patterns in others so as to better handle relationships.
Self-acceptance	Being able to acknowledge strengths and weaknesses and being able to adapt where necessary.
Personal responsibility	Being able to take responsibility for one's actions. This relates to personal decision making. Learning not to try and pass the buck when the buck really rests with one's self.
Assertiveness	The ability to be able to take a controlled stand, i.e., with neither anger or meekness. Particularly important in decisions involving moral issues in engineering on which professional ethic demands that a stand should be made.
Behavior in groups	Knowing when to participate, lead, or follow.
Conflict resolution	Using the win/win model to negotiate compromise. This is particularly important in industrial relations and it applies to both partners in industrial conflicts.

EXHIBIT 8.1. Culver's (1998) adaptation of the items in Nueva School's components of emotional intelligence in Culver, R. S. (1998). A review of emotional intelligence by Daniel Golman. Implications for technical education. In: Proceedings Frontiers in Education Conference, pp. 855–860.

critical incidents.) "The decision to call this aspect of intelligence, social, emotional or practical intelligence will depend on one's perspective and one's purpose." What is not in doubt is that engineers need to be introduced to this kind of thinking and behavior during their education and training (see, for example, Shull, 2014). It cannot be done in college alone because it is not possible to wholly simulate the everyday reality of industry although this is not to deny the value of simulations and the application of gaming to engineering problems as many publications show.

8.3   Self- and Peer Assessment

Self-assessment has come to have at least two meanings. The first is the self-assessment of an organization or unit of an organization. For example, a school of engineering might evaluate its instructional practices using self-questionnaires (Vidic, 2008; Zappe et al., 2014). Several papers at recent conferences have outlined strategies for sustainable student outcomes assessment (Sundarajan, 2014).

Second, are the practices of self-assessment adopted by students to check for themselves their progress and/or performance, to learn what they are competent and incompetent in, and are able to do. This section is concerned with the latter.

As used in engineering education, self-assessment often relates to the practice of evaluating a product and asking questions about whether or not it could have been improved. Inspection of many engineering science project reports (Section 3.1) showed that such evaluations could be either superficial or in depth, and confirmed the need for training. Or, it relates to the evaluation (scoring) of one's performance either formatively or summatively. Or, it relates more generally to the learning one undertakes during a course that may or may not involve an instructor. Or, it relates to one's self and one's performance in life and the way one relates to others. At its simplest, it is evaluative, and at its most complex, it is deeply reflective. Like everything else, electronic techniques are being increasingly used in support of its practice (e.g., Berry and Carlson, 2010).

As has been shown in Chapter 3, in the 1960s and 70s, attempts were made to design multiple-objective (strategy) approaches to assessment in engineering science. Much note was taken of the semicriterion rubrics used to assess performance in laboratory and project work. Later, the same kind of approach was used in the development of rubrics for the assessment of the industrial training period of sandwich (cooperative) courses. Exhibit 8.2 shows one such rubric (Ford and Rennie, 1999). In a similar approach to Mickelson, Hanneman, and Brumm (2002; see Section 10.2), the curriculum development from which this rubric is taken was derived from employers and a cohort of third-year students nearing the completion of their industrial placement took part in separate workshops to develop self-assessment rubrics from aims and objectives defined by themselves. The interesting finding was that the two groups produced almost identical criteria (Ford and Rennie, 1999). Two of the self-assessment rubrics are shown in Exhibit 8.3. The students were pursuing a degree in environmental health and in their professional practice they would be expected to follow health and safety legislation in respect of both themselves and their clients.

In addition to discussing the reliability and validity of the assessment scheme, Ford and Rennie report that it had had the continuing support of employers and students. They write: "Students were of the opinion that although self-assessment is not an enjoyable experience, it should be undertaken during the industrial placement period. The importance students attached to self-assessment in this project is supported by other research which asked students about the skills they expected to develop during

	X	Y	
Gives the problem no forethought			Researches background data and formulates a plan of action
Tends to be subjective and cursory, and fails to keep an overview			Endeavors to be objective throughout and keeps an overview
Is haphazard in the conduct and recording of the investigation			Is systematic and uses appropriate observational questioning and recording techniques
Ignores the fact that some pieces of data may be more valuable than others. Fails to identify relevant criteria for comparison			Weighs the value of different elements of data. Correctly selects relevant criteria for comparison
Is narrow in the identification of practicable courses of action			Identifies possible courses of action and selects appropriate one
Does not implement and monitor the selected courses of action			Carries selected courses of action through to completion, monitoring progress and taking follow-up action
Fails to evaluate own performance and the outcome			Critically appraises performance and outcomes

EXHIBIT 8.2. Schedule for the assessment of investigative skills and problem solving during an industrial placement (Ford, N. J., and Rennie, D. M. (1999) Development of an authentic assessment scheme for the professional placement period of a sandwich course. In: J. Heywood, J. M. Sharp, and M. T. Hides (eds.), *Improving Teaching in Higher Education*. Salford: University of Salford, Teaching and Learning Quality Improvement Scheme.

university education; their responses rated the ability to evaluate one's own work only second to problem solving (Boud, 1986). The value of the self-assessment scheme is that it requires the student to use and practice reflective thinking skills. These skills are fundamental if students are to learn effectively from the experience gained during professional practice." Later, they say that "despite these efforts to limit the degree and effects of subjectivity, it is the view of the authors that the derivation of a final mark for student performance (to contribute to the degree award) is not the most important outcome of the scheme" (p. 39), a remark that is food for substantial thought.

(a) *An example of a collaborative assessment.* Several years earlier, Boyd, a student of civil engineering at Herriot Watt University, and her Professor, John Cowan, had published an outstanding example of collaborative self-assessment in a course on design in civil engineering (Boyd and Cowan, 1986). Cowan had established a contract with the class that they would set their own goals week by week. He would not provide advice or direction. "In lieu of examinations, each learner would prepare self-assessment in the form of a criteria list of her desired goals, a description of her actual learning and a reconciliation of these (in relation to agreed bench marks), leading to the choice of mark.

4.2. Do you approach your work with active interest seeking and accepting a level of responsibility which you can handle?			
Not always keen on some areas of work. Prefer to await direct instruction or take on tasks that prove difficult to handle.	Respond positively to interesting topics or staff who make an effort. Do not like to push self forward.	Willing and responsive approach to work. Accept manageable responsibility.	Keep a subject indexed record that includes a personal review of samples, visits, and relevant proformas.
4.3. Do you keep a systematic personal record of your experience?			
Keep a haphazard daily diary of visits made and photocopies of all available material.	Keep a chronological record of training program and copies of proformas and other interesting material.	Keep a subject indexed record of training with selected interesting material and proformas.	Keep a subject indexed record that includes a personal review of samples, visits, and relevant proformas.

EXHIBIT 8.3. Extracts from the rubrics for self-assessment in relation to organizational skills (Ford, N. J., and Rennie, D. M. (1999) Development of an authentic assessment scheme for the professional placement period of a sandwich course. In: J. Heywood, J. M. Sharp, and M. T. Hides (eds.), *Improving Teaching in Higher Education*. Salford: University of Salford, Teaching and Learning Quality Improvement Scheme.

Each stage in this process of assessment would be open to questioning, and discussion; but the ultimate decision would remain completely within the jurisdiction of the learner."

Their short paper is a dialogue about what happened. Boyd was shown the Perry model of intellectual development in order for her to gain an understanding of the psychological roots of learning. She wrote of her third-year experience with Cowan when he had placed the responsibility for learning and assessment on the students that "in accepting responsibility for setting my own criteria I clearly made a commitment to my eventual choice. Hence my learning must be rated at one of the higher levels of Perry's scale. Surface processing which is encouraged by cue-consciousness was absent—because I had chosen my aims and criteria and directed my learning accordingly. And deep processing was positively encouraged—because I was subject to no pressure other than those which were self imposed."

They met again after Boyd has completed 4 months of her final (fourth) year when both instruction and assessment were conventional. Their comments are in stark contrast because they concluded that now: (There was no commitment to the higher level of Perry's scale, except in part of the final year project, where she had opted to work on her own criteria, rather than those of her supervisor. Hard won habits of deep-processing and, in particular, of rigorously searching for key points and issues persisted to some extent in her private reading and even in attendance at lectures, although there was a marked regression compared with the working style in the third-year design. At the same time, cue-seeking and cue-conscious activities were more frequent, more deliberate, and more purposeful.)

These points show that if one is to pursue desired learning outcomes that require a different approach to instruction then it is essential that all those involved in teaching a program should be committed to achieving those goals. It is not the conclusion they reached although it is equally challenging. They wrote: "These highly subjective and presumably biased impressions prompt us to wonder if it is possible to generate higher level commitment without involving the learners in setting goals and criteria; and also if the habit of deep processing once developed is likely to persist to some extent, even when circumstances actively encourage surface processing."

(b) *Other exercises in self-assessment.* Other exercises in self-assessment by engineering students were reported (Boud, Churches, and Smith, 1986; Polizzotto and Michalson, 2001). It was an integral part of the Problem Based learning program developed at McMaster University (Woods et al., 1988). It was also used in design that involved teamwork (Gentili et al., 1999). Exhibit 8.4 shows a scheme that was developed by Freeman, Heywood, and Humble (Heywood, 1989) for use by managers and trade union officials asked to watch a video that had been taken of their performance. Many similar rubrics have been published (Adair et al., 1978; Woodcock, 1979; see, in particular, Smith, 2000, for a 35-item schedule on "How I Act in a Conflict").

Claxton (1995) argued that involvement in grading helps develop critical faculties because the student has to internalize the functions of correction and evaluation. "This fosters the vital distinction between the 'informative' and the 'emotive' distinctions of evaluation. Being able to turn a critical eye on one's own product, while at the same time retaining equanimity towards self, is the vital factor in the development of resilience." For Claxton, intelligence is "knowing what to do when you don't know what to do" and this requires resilience, resourcefulness, reflectivity, and responsibility which is what industrialists say they want graduates to be! The literature suggests, and it is this writer's experience, that students (and indeed experienced teachers) benefit from training in peer and self-assessment particularly if they wish to grapple with higher levels of conceptual understanding (Penny and Grover, 1996; Shull, 2014).

(c) *First-year courses and self-assessment.* Notwithstanding the interest that project-based courses create among first-year students, they also bring with them their own stresses. A particular stress is the requirement in some courses that students assess their own effectiveness in a team (see Part II). This was observed among 40 different teams pursuing projects at Purdue University. Individuals rated the performance of the team with the *Team Assessment Tool*, which had categories for interdependency, learning, potency, goal setting, and validity. The *Research Observation Tool* enabled the researchers to rate the teams on three forms of interdependency, two forms of potency, and two forms of goal setting (Exhibit 8.5). Detailed notes were taken of the performance of the team. It was found that students "do only a fair job of this self-assessment and do not seem to get better at the assessment as time progresses" (Moore, Diefes-Dux, and Imbrie, 2006).

Another study by Khachikian, Guillaume, and Pham (2011) in an extensive investigation among engineering students, found a relation between ability, self-assessment, and the ability to manage their time. Students who generally perform well were found to be better at self-evaluation. They called these students "Master Students" who

Questionnaire for use in negotiating-skills development training

SELF-APPRAISAL - BEHAVIOUR IN DISCUSSION

Please study your performance from the video playback. This is a private assessment so be perfectly frank in answering the questions — otherwise you are only fooling yourself.

In this discussion I tended to:	Yes	No
Ask specific questions about the topic under discussion	☐	☐
Try to score debating points	☐	☐
Get irritated with an opponent	☐	☐
Ask for clarification/facts about a point made by the other side	☐	☐
Contribute helpful suggestions	☐	☐
Admit I was misinformed/wrong	☐	☐
Interrupt before a speaker had finished	☐	☐
Opt out of answering an opponent's question on the grounds that I would appear to give way	☐	☐
Criticize when I had not really got a real point to make	☐	☐
Close the door to further argument	☐	☐
Change my mind when my assumptions were shown to be faulty	☐	☐
Keep quiet when I had nothing constructive to say	☐	☐
Overrule the chairman/leader	☐	☐
Prepare my case before the meeting	☐	☐
Not listen to an opponent's argument because I disagreed with his or her case	☐	☐

PLACE THIS SHEET INSIDE THE FOLDED SHEET 2
AND FOLD OVER THE RIGHT-HAND EDGE SO THAT THE ANSWERS
SHOW IN CUT-OUTS

EXHIBIT 8.4. (Continued)

Now compare your results and consider whether your contribution to the meeting was:

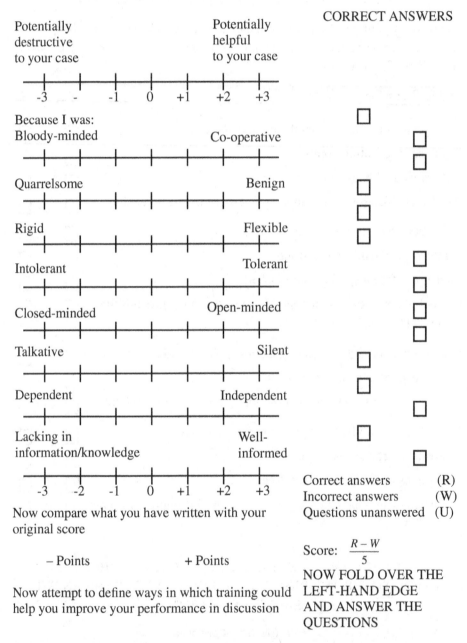

CORRECT ANSWERS

Potentially
destructive
to your case

Potentially
helpful
to your case

-3   -   -1   0   +1   +2   +3

Because I was:
Bloody-minded   Co-operative

Quarrelsome   Benign

Rigid   Flexible

Intolerant   Tolerant

Closed-minded   Open-minded

Talkative   Silent

Dependent   Independent

Lacking in
information/knowledge

Well-
informed

-3   -2   -1   0   +1   +2   +3

Now compare what you have written with your
original score

– Points   + Points

Now attempt to define ways in which training could
help you improve your performance in discussion

Correct answers   (R)
Incorrect answers   (W)
Questions unanswered (U)

Score: $\dfrac{R - W}{5}$

NOW FOLD OVER THE
LEFT-HAND EDGE
AND ANSWER THE
QUESTIONS

EXHIBIT 8.4. A questionnaire designed for use in negotiating skills development training by W. Humble, J. Freeman, and J. Heywood and cited in Heywood, J. (1989). *Learning Adaptability and Change*. London: Paul Chapman, pp. 48 and 49).

Interdependency—Cooperation among team members used to accomplish a task
This team used a process method (i.e., code of cooperation) to hold each other accountable
This team showed evidence of taking on roles and knowing the roles of other team members.
This team had an effective process to complete this MEA.   What evidence did you observe of interdependency. N
Potency—Team potency is the shared perception that the team can reach their goals.
This team thought they could actually complete the MEA.
This team believed they could produce a good response to this MEA.
What evidence did you observe of this potency? N
This team reflected on its goals during the process of solving this MEA.   What evidence did you observe of goal setting? N

EXHIBIT 8.5. Extract from the Research Observation Tool. Each statement required a response on a 5-point scale with a 6th point for *did not observe*. N indicates response in the form of a note (Moore, T., Diefes-Dux, H., and Imbrie, P. K. (2006). Assessment of team effectiveness during complex mathematical modelling tasks. In: ASEE/IEEE Proceedings Frontiers in Education Conference, T1H-1 to 6.

continually adjusted their time-on-task to match course requirements. The investigators suggested that a way of helping the less able students would be to get them to complete time logs and undertake regular self-assessments which together could be used by instructors for formative advice. In South Africa, however, Swart, Lombard, and de Jager (2010) were unable to establish a significant relation between time management indicators and ability although they felt a larger sample might change the result even though it was consistent with other studies. They identified eight time-management practices that would help students.

To reduce the pressure in project-based design courses, Carberry, Johnson, and Henderson (2014) are currently evaluating a course that is scaffolded by three miniprojects that are undertaken while simultaneously completing a semester-long project. A particular stage of engineering design is practiced in the mini-project and what is learnt is applied directly to their main project.

At Michigan Technological University, Kemppainem and Hein (2008) reported that in spite of the use of several instructional activities designed to take into account a variety of learning styles, first-year students experienced difficulty with several topics. To remedy this problem, self-assessments were introduced as a resource. They took the form of a series of nongraded questions on topics designed to relate the lecture material to real-world applications that they could study in their own time. Using Blackboard CE, the students could obtain instant feedback. The system was voluntary and the process could be repeated any number of times. They included an ethics case study. The authors presented data that showed that those who used the assessment procedure performed better in their midterm and final examinations than those who had not used the system. There was also improvement in the responses to the ethics case study.

The argument for the inclusion of self-assessment in first-year introductory engineering courses is furthered by Kaul and Adams (2014). In their course, the assessment of outcomes is obtained by direct measures of student performance in multiple assign-ments and three team projects. The investigation reported compared the results of the direct measures with student perceptions (self-assessments) obtained by surveys of the extent to which the desired outcomes were obtained. Statistical analysis showed that student self-assessment was aligned with the direct measurement of each of the four outcomes (multidisciplinary teams; formulating and solving problems; effective com-munication; and lifelong learning). Against performance indicators calculated for each outcome that for effective communication was the lowest and this was consistent with the students' assessments. In general, the self-assessments correlated with the direct measures, suggesting a role for self-assessment in course evaluation.

A key skill in all project and research work is the ability to gather information. Engineers have to be information literate. It has been reported that advanced-level students and experts are more likely to gather information than first-year students. This view was upended in a recent investigation by Douglas et al. (2014), who, using a Self-Assessment of Problem Strategies (SAPP) instrument that purports to report behaviors in gathering, locating, reflecting on, and using information, compared the performance of freshmen and junior year engineering students. They found no significant difference between the scores of the two groups; moreover, they showed relatively high levels of information literacy skill. Would there have been differences with groups of more advanced students? How would such data correlate with general intellectual ability?

(d) *Peer assessment and mentoring.* If peers could assess each other reliably—what a wonderful opportunity for reducing the work load of a tutor! As it is, while scores may be incorporated into the marks, the responsibility for grading remains with the tutor. One experiment at Imperial College challenges this marks-based culture.

In a voluntary tutorial system at Imperial College, suitably qualified undergraduate teaching assistants (UTAs) helped with first-year tutorials by marking submitted work, giving written and verbal feedback, and leading problem-solving discussions as part of the foundation to computer science program. A triangulation was made of the perceptions of the students, academic tutors, and teaching assistants (TAs). Alpay et al. (2010) reported a favorable outcome to the experiment but noted that the effectiveness of peer tutoring depends on the good course design and assessment procedures that they believe to have been achieved in this case. The advantage of the scheme was "its close integration with the course, with structured assessment exercises which supplement course outcomes and ultimately the final exam." However, their more important conclusion was that the scheme demonstrated that a marks-based culture does not motivate students whereas this zero-weighted scheme fostered intrinsic motivation, and in these days when there is much discussion about role models for students—these "students saw the UTAs as role models since they were students like them, but who had already successfully achieved mastery in the subject."

The potential of experience-led monitoring has also been shown by two small case studies at Coventry University in which students studying engineering part-time, while working in industry, worked with full-time students. Research at the university had suggested that the performance of part-time engineering students was better than that of full-time students, which suggested that there was merit in mixing the two groups because the part-time students had already gained professional skills. One case study evaluates the effects of mixing students in a third-year project, and the other of using part-time students in civil engineering in the age range 22–31 years to mentor five first-year students. It was found that in project work the part-time students would take the lead but there is the possibility of role-conflict since they will want to maximize their mark and this may make them reluctant to help others. The evidence suggested that full-time students were actively interested in the work of the part-time students and benefited from the access that was provided them to "actual physical artefacts, drawings, photos, example documents and templates" (Davies and Rutherford, 2012).

The use of peers can change the role of the lecturer substantially as Borglund of the Stockholm Royal Institute of Technology found when he replaced his lectures with more student-centered sessions based on peer learning in a course on aerolasticity. "I found the new approach to be much more stimulating than traditional teaching" (Borglund, 2007). Part of this new approach required a project report. A first draft was submitted, which he reviewed for feedback purposes. This was followed by a second report with additional requirements that was peer-reviewed. This was achieved by asking each of the teams to review five or six reports written by other students but unnamed. Each student reviewed each of the reports and wrote a comment on them relating to content, structure, clarity, and conciseness. The team then met and discussed the reports and each student prepared a review of a particular report. Borglund reviewed the final versions. He found that his workload was comparable to that of marking the 2-hour written examination that had been abandoned when the course was changed.

Peer appraisal is seen as a means for engaging students in active learning, particularly in team projects, for the development of skill in critical thinking and metacognition, and for learning about student misconceptions (O'Moore and Baldock, 2007). There are many other reports of its use with engineering students.

Wigal (2007) described an Introduction to Engineering Design Course where peer evaluations of oral presentations contributed one tenth of a student's oral presentation grades. It was found from seven semesters of evaluation that while students could distinguish between good, average, and poor performance, they could not do this when it came to marking on a standards scale. A means of converting the students' scores to realistic scores had yet to be found.

Marin-Garcia, Miralles, and Marin (2008) wanted to involve the students in the assessment process but they were concerned that student assessments should be reliable. The activity evaluated consisted of interviewing two company managers (and) comparing their answers with the theory taught in the course (management) and presenting the results of their study to other colleagues in the class. The presentation took place on the last day of the class and was not compulsory, even though it accounted for 10% of the final mark in the subject—5% was for presentation (average of the marks given by colleagues and the lecturer) and 5% was for the degree of agreement with marks given by each student when it was compared with the marks average for all of the students). Few significant differences were found between the students' marks and those suggested by the lecturer. The authors considered that the marking by the students was adequate.[1]

Wigal (see above) took the view that if students could critically evaluate their peers they would be learning skills in critical thinking. A statistical analysis confirmed that they were able to undertake critical evaluations. In another study, Enszer and Castellanos (2014) evaluated two methods of peer evaluation. The first used a traditional point division method (Michaelson et al., 2004) and the second The Comprehensive Assessment of Team Member Effectiveness instrument (Ohland et al., 2014). In both cases, a statistically significant correlation was found between the two methods, and that students evaluate themselves more highly than their peers was found to have modest significance. In contrast, Thompson (2014) found in her study that students tended to score their own reports lower than their peers. But this study, although having considerable limitations, was set up for a different purpose, which was to enable more meaningful formative feedback to be given to students in a large first-year introduction to engineering course. In teams of 4 or 5, the students reviewed each other's work using an instructor-designed rubric. Student perceptions suggested that the goals had been achieved. A weakness recognized by Thompson with this and other similar work is that no measures were made of the reliability and validity of peer assessment or of student learning gains. However, sufficient information was gained to make decisions about the future direction of the course.

In another study, Spanish students were divided into to two equally performing groups to develop problem-solving skills through continuous formative assessment. One group utilized self-assessment and the other peer assessment. It was found that the peer assessment was more effective than self-assessment and that both were more effective than instructor formative assessment, yet a survey revealed that the students had more confidence in instructor assessment than in peer assessment. The students believed that self-assessment helped them more than peer assessment! (De Sande and Godino-Llorente, 2014).

At the Universidade Positivo in Brazil, students in a graduate course based on the seminars that the students gave (Dziedzic, Janissek, and Bender, 2008) 30% of the final mark was given for their performance in these seminars. In this case study, students were given an assessment schedule that distinguished between the form of the seminar and its content. In the first course, the students assessed two seminars at the end of the course. In the second course, the students assessed all the seminars. Given that the marks were not taken into account, it is perhaps surprising to find that the quality of seminars improved when peer assessment was employed. Unfortunately, while there was agreement between tutors and students about the marks to be awarded to the best candidates, there was no such agreement at the other end of the scale. The students were reluctant to give poor marks.

In Exhibit 8.6, two different approaches to the assessment of oral presentations are shown. Columns A and B summarize a Brazilian scheme, and columns X and Y summarize the scheme used by Wigal (2007). It is of interest to note that assessments of performance in these areas have to be done in teacher education. In general, there are areas where no two assessors might agree as for example in voice and presentation. A great deal of care has to be taken with such assessments.

A. Item	B. Criteria	Weight	X Category	Y Category Criteria
Form	**Voice**	18	**Voice**	Pleasing, not distracting? Varied in pitch, intensity, volume, rate quality? Expressive?
	Tone			
	speed		**Language**	Clear, accurate, varied, vivid? Appropriate standard of usage? Helped presentation?
	confidence			
	Image	18	**Delivery**	Poised, at ease, communicative, direct? Good eye contact? Do gestures match voice and language?
	posture			
	language			
	appearance			
	Slides	18	**Visual aids**	Contain right amount of information? Easy to read? Enhance effectiveness of presentation
	Font size			
	Colors			
	Figures			
	Content			
	Writing			
	Time	18	**Length**	Between 5 and 7 minutes
	Organization	18 (100)	**Organization**	Purpose clear? Clear arrangement of ideas? Introduction, body, conclusion? Clear pattern of development?
Content	**Summary**	20		
	Critical review	20		
	Examples	15		
	Counter examples	10		
	Proposal	20		
	Assessment	15 (100)		

EXHIBIT 8.6. Two schemes for the assessment of oral presentations. Columns A and B from Dziedzic, M., Janissek, P. R., and Bender, A. P. (2008) Assessment by peers—an effective learning technique. In: IEEE/ASEE Proceedings Frontiers in Education Conference, T2F- 1 to 5. Columns X and Y extracted from Wigal, C. M. (2007). The use of peer evaluations to measure student performance and critical thinking ability. In: IEEE/ASEE Proceedings Frontiers in Education Conference, S3B-7 to 12.

It is not always the case that students are soft on those who do less well. In a comparison of peer-reviewed scores with those of a TA for a civil engineering project in structural analysis, the students gave fewer marks than the TA even though "the comments and identified weaknesses were actually the same" (Barroso and Morgan, 2010). The rubrics had been provided by the instructors and the TA used the same rubrics as the students. Without going into the details of the project cycle or the arrangements for assessment, three points are of value. First, if the peer evaluation is taken seriously, the peers value the feedback and the evaluators witness alternative points of view about both the technical content and the organization of reports. Second, the instructions for the review can be designed to avoid harsh or easy marking, especially for friends. The instructions read as follows: "part of your team's responsibility will be to review the work submitted by another team in your class. Reviewing the work of another engineer with a critical eye is an important skill that you will use frequently in your professional career. The goal is to provide constructive feedback so that future work submitted by your team is improved. Your team will be evaluated on the quality of the feedback provided—being too easy or too hard will not help anyone improve, as well as instructions on marking projects and grading rubrics." The grading was used for formative evaluation.

Third, Barroso and Morgan illustrated the difficulty of balancing what was in the reports and what had been observed. It is always difficult with rubrics of this kind. Their description is reminiscent of the marking by and discussions between the moderators of the engineering science projects 40 years earlier. Barroso and Morgan (2010) wrote: "[...] the TA had more flexibility in utilizing the rubric and could modify it to better suit the team's submission and approach. For example one of the teams did not explicitly present the loading diagrams for the beams. However, the group had gone beyond the minimum project requirements and determined the shear and moment diagrams. The peer group assigned a zero score for the diagram component as those were missing. In contrast, the TA penalized the group for missing the diagrams but gave them extra credit for having the correct corresponding moment diagrams mitigating the loss of not having the diagrams."

One reason for long lists of outcomes is to overcome such difficulties so as not to allow for shades of grey. But as teachers of Higher National Certificate engineering in the United Kingdom pointed out to me in a workshop the assessment process demanded a positive response to each outcome to gain a pass to a particular question. An excellent answer that showed some imagination but did not cover all the outcomes would not be passed (see Section 3.6).

There is considerable interest in the use of online peer review schedules as for example with MOOCS (Carey, 2014). A recent study at the University of California San Diego contrasts a home-developed program (Delson, 2012) with an earlier instrument developed by Ohland and his colleagues (2005). Delson points out that such assessments can be administered during as well as at the end of the course and he had found that poorly performing students could be identified early in the project and corrective action taken. In Australia, Willey and Gardener (2009a) argued that any method of developing attributes associated with teamwork and the ability to give and receive feedback should include self- and peer assessment. At the same conference, they reported that when self- and peer assessments were used on multiple occasions to assess individual student assignments, team contributions, and in bench marking exercises, more than two-thirds of the students said that it improved their ability to meet the learning outcomes. Willey and Gardener (2009b) had been triggered to undertake this study because many of the students perceived self- and peer assessment as a means of ensuring fairness in group activities that is, ensuring that every member of the group contributed a fair share. Even with the changes, they found that 10% continued to believe that the activity was about fairness. Anecdotal evidence suggests that this was an excellent result, but they warn that such multiple approaches can increase the mark load of academics and may over-assess some students (see Section 1.7). The analysis of their investigation involved the use of an online management tool called the SPARK-(Self and Peer Assessment Resource Kit). This package produces three assessment packages, two of which were used in the reported investigation (Beamish et al., 2009). The first is called the Self and Peer Assessment factor. It is a weighting factor that is determined by comparison of the self and peer rating of a student's contribution with the average rating of the team. It determines the student's mark. The second factor is a Self Assessment to Peer Assessment score that is the ratio of a student's own rating of himself or herself to the average rating of their contribution by their peers. This factor was also used to identify students who were overrating and underrating the scales on the marking schedule.

(e) *Peer review.* An interesting variant of peer assessment is the "peer review" as conducted for journal articles. Hanson (2014) reports on how such a process was used in an upper-level geotechnical engineering course in which the term paper accounted for 25% of the course grade. The peer review process was regarded as central to the learning objectives and "largely controlled the timeline for submittal of project deliverables." The review process was anonymous and the students received considerable information about the process which took place over a single week. The student evaluations showed a generally positive response: the value of receiving feedback on their draft papers was noted with that from the instructor receiving total affirmation. While the administration and grading of assignments could be complicated, the scheme was thought to be beneficial.

(f) *Questions of honesty.* Although cheating has been the subject of many studies, it is not a primary topic of this text. However, Serra-Toro, Traver, and Amengual (2014) report a study at a Spanish university that introduced self- and peer assessment into their continuous assessment framework that had part of its goal to encourage honesty. The other goal was to encourage reflective thinking. In this course, the self- and peer assessments were followed by a validation laboratory test the results of which were used to moderate the self- and peer assessments. This test was "exam-like, in-lab test" that was administered individually to avoid cheating (see Section 3.1 for a similar validation activity). Peer assessment was introduced after the first year because a large proportion of students showed inconsistent behavior. While they proposed to continue with the framework they had developed, they considered that students did not understand the benefits of self- and peer assessment. They suggest several approaches to deal with this problem, one of which is to "study mechanisms to make the grades between self-assessment and peer assessment and the validation test have a higher positive correlation." A major advantage was the earlier provision of feedback to the students.

8.4 Learning Journals and Portfolios

At the same time as the developments in self- and peer assessment in the 1980s and 90s were taking place, some engineering educators experimented with journals and portfolios. They are now part of the instructor's tool kit. "Written diaries and journals have often been replaced by the web log, or blog, where writers maintain an online sequence of observations or reflections that may be kept in date order. It should also be noted journals need not be written, they may be kept as video logs, known as vlogs. Hence, in today's world of social media reflective journaling has moved beyond the keeping of a physical written document to a much broader setting involving the use of youtube, Google hangouts etc." So said Huggard, Boland, and McGoldrick at FIE 2014. That said, the principles remain the same and much is to be learnt from previous understandings. For example, Yokomoto (1993) who asked his students to keep a learning journal, found that "superior students are extracting more information than I expected. It has shown me how much better the information gathering processes of superior students are, including attention span and note taking skills." Journals can inform tutors about performance and remedial action can be taken if required (e.g., Mejia, Goodridge, and Green, 2014). Journals can also give some idea of commitment (Wellington and Collier, 2002). Blogs may be used to document engineering and technical knowledge (Balascio, 2014).

In the 1980s, Cowan (2006) introduced a learning journal in a course that had as its purpose the development of skills used in professional life in civil engineering. They were a collection of weekly reflections. Put together, they might comprise a portfolio although Cowan says they are undertaken by individuals for their personal use. George and Cowan (1999) also showed how reflective journals can be fitted into the Kolb Learning Cycle. It might be suggested that students with experience of journal writing should be invited to examine their work in terms of the Kolb cycle, and to use it to plan the "active experimentation of their generalizations."

Most recently, Piergiovanni (2014) with the aid of an online classroom management system, used Bloom's *Taxonomy* to develop 15 online journal prompts that were delivered to the students for them to write short responses. The responses given electronically were coded for the abilities of questioning and reasoning; recognizing assumptions; presenting and evaluating data; and drawing conclusions. Word counts were obtained for each of the four abilities. She found that many of the responses were vague and that students did not understand the purpose of the journal responses. But she also found evidence of critical thinking "and even some deeper thinking when students linked what they had learned in this class with experiences in another class." Evidence of that elusive skill in the transfer of learning! Could that be because the students were enrolled in an engineering program in a liberal arts college? Notwithstanding the subjective method of analysis, the fact that it often led to insightful questions in class is indicative of an improvement in student learning.

Much of what is said about journals applies to portfolios. In 1993, in the United Kingdom, Wright had suggested that portfolios should be used in ratings of continuous professional development, and in the United States, Olds (1997) reported that portfolios were one of the recommended tools for outcomes assessment, this in the period running up to the introduction of the ABET 2000 requirements. Cress and McCullough-Cress (1995) suggested that a portfolio that required reflective assessment could provide for the continuous improvement of students and teachers and reconcile student needs with coursework. They believed that courses should be constructed as learning organizations. In the United Kingdom, Payne reported an engineering degree course that in its first-year students failed to understand the concept of the portfolio. The introduction of portfolios was as much a challenge for the teachers as it was for the students. Several evaluations were made, and among the findings was that portfolios encourage self-reflection (Bramhall et al., 1991; Payne et al., 1993; Ashworth, 1996).

There are now a number of reports on the uses of e-portfolios; one reports on the development of an analytic procedure for their scoring in a technical communications program that had as its objective the assessment of the components of the course (Johnson, 2006).[1] They have also been used to explore the interest of first-year students in engineering with the finding that the number of hits on a student e-portfolio was the strongest indicator of retention, and faculty were able to determine the markers of at-risk students (Goodrich et al., 2014). At the other end of the student spectrum in Scotland, they have been used to aid the development of professional identity of final-year computer science graduates (Smith, Sobolewska, and Smith, 2014). These reports are examples of showcase portfolios now in common use among young people applying for jobs. Eliot and Turns (2011) in the United States also focused on the same problem (see also Turns and Sattler, 2012; Section 10.8). It is not surprising to find that some investigations are focusing on tools to assist with assessment, and a Spanish report showed that assessment time was reduced, and the portfolio also helped the teachers understand the learning processes of their students (Groba et al., 2014). In a technology teacher education program at the University of Limerick, a small sample of students completed e-portfolios in which they used a color coding system to identify students' metacognitive awareness during a design task. The panes were color coded so that the student could indicate if a particular pane was an instance of a student identifying an idea; of an idea being developed; and/or of the provision of a solution to a particular problem they encountered or proving an idea (Rowsome, Land, and Gordon, 2014). The student could use more than one code. The definition of metacognitive awareness used was due to Tarricone (cited by Rowsome, Lane, and Gordon, 2014). It is "as awareness of learning process, reflection on learning and memory, identification of strategies for problem solving and monitoring and control of learning processes." The overall goal of the project was to obtain an indication of extent to which the students could self-regulate their learning. They found that the students used each pane to narrate the story of their project and they argued that students who narrated their thinking were providing evidence of metacognitive behavior (Whitbread and Coltman, 2009).

A major longitudinal study of the use of portfolios in the assessment of the clinical competence of nurses was reported by Griffin (2012). Her findings are relevant to the implementation of portfolios in any system. They also confirm previous findings. As so often with the introduction of new techniques of assessment, the commitment of the organization in which it is to be used is essential: it is a *sine qua non* that teachers should be adequately trained for this purpose. She found, as in previous studies that it was "crucial" to define the purpose of the portfolio, its structure, and assessment. In this case, it was used to assess the five competences required by the Nursing Board. But portfolios have other purposes, and Griffin noted that confusion can arise from uncertainty about whether a portfolio is being used as a process or a product, and in the case of the health professions confusion in the terminology relating to competence can make the use of portfolios problematic. At the same time, the results of her investigation may be judged to be encouraging.

In Griffin's study, four raters were used. Her results suggested that there need be no more than two. Her quantitative data showed that portfolios could be rated reliably. In clinical situations where a student either passes or fails, there is no need to score the portfolio. Those who score portfolios are faced with the question of the purpose of the portfolio. If it is being used to certify competence on a pass/fail basis as holistic dynamic concept, then a score is meaningless. When numerical values are attributed to competence and are also attributing value to sufficiently quantify the issue is "what portfolio score is an appropriate indicator of competence?"—which is the issue that raters of projects, for example, the moderators of the engineering science projects were faced (Section 3.1). In engineering whether or not to score the portfolio would depend on its purpose.

In contrast to a UK study that found that over the 4-year period students found their voice (Bradbury Jones, 2011, cited by Griffin), the students in her investigation showed evidence of the need to conform. She concluded that a difference found between her interviews and portfolio content was mainly due to the hospital culture that differed from that experienced by a comparison group (nonportfolio) in another hospital. Commitment to the activity by the organization is a major factor in creating a culture in which intrinsic motivation can flourish. Griffin found that nursing students did not differ from student teachers (Darling, 2001, cited by Griffin) in that they were motivated both extrinsically and intrinsically. "Self directed learning, reflection on learning and preparing for practice was evident in the students' portfolios and manifest in their clinical practice."

Student perceptions of the portfolio changed during the 4-year program. First-year students considered it to be a "Badge of Merit." "In years three and four it was experienced as a 'necessity' and as a welcome 'departure' respectively." Griffin pointed out that at the end of the 4 years the portfolio became a record of achievement that contains corroborated evidence from two sources, the student and his or her clinical assessors.

Research on portfolios continues to be necessary and experiments in their use need to be piloted. If the model of higher education proposed in Chapter 12 is anywhere near correct, portfolios will have a major role to play in its process. It should also be evident that both journals and portfolios can be used to evaluate the effectiveness with which a course is achieving the outcomes required by accrediting agencies and, therefore, for continuous curriculum improvement (Heywood, 2005; see also Dempsey et al., 2003). They may also be used as an aid to understanding learning by teachers and their students. Not only should every teacher have a defensible theory of learning but so too should every student. It may be argued that to achieve that goal some provision should be made for the study of Learning-How-to-Learn.

8.5 Learning-How-to-Learn

Some institutions provide students with Learning-how-to-Learn courses independently of the school in which they are taught. Occasionally, teachers report efforts to help students understand learning and the reflective processes they engage. One such approach was used by Mistree, Ifenthaler, and Siddique (2013) in the United States in a graduate course on *Designing for Open Innovation*.

It is only possible to give the flavor of the quite a complex curriculum design that is described in detail in their paper. The first section outlines the principles on which the course is based. The second contrasts the differences between this course and traditional course design, and how the students who come from different universities choose the competencies they wish to develop. The developmental nature of the course is outlined in the third paragraph. The fourth paragraph indicates how Senge's theory of the learning organization pervades the whole course, and the fifth paragraph shows how the lectures provide a scaffold that supports the objectives of the program.

Their starting point was the finding that effective teams share mental models (Cannon-Bower, Salas, and Converse, 1993). Individuals construct internal models that integrate the relevant semantic knowledge and the perceived requirements of the situation. In complex situations, these complex structures will be integrated at declarative, procedural, and metacognitive levels. Given that all members of a team have such internal mental models and given that these models are a source of action, a team should be more effective if these models are shared, and the more similar they are, the more effective the team will be. There will be both team and task models that are shared. The authors cite evidences that this is so, among them studies of flight simulators in laboratory settings (Mathieu et al., 2005).

In contrast to traditional course design where students are told what to learn, in this course students decide what they want to learn, they evaluate themselves instead of being evaluated. The course is organized as a collective rather than a hierarchy. The students were asked, therefore, to say what competencies they wished to develop so that they will be competitive in the world of 2030. This they do having responded to the question "What competencies do you need to develop to be successful at addressing dilemmas associated with the realization of a complex, sustainable socio-techno-eco system in a distributed engineering world?" Once the student had defined the competencies she or he wished to develop, he or she defined the skills he or she would need to become competent in performing the task. In this case, the skills in the original Bloom *Taxonomy* were used.

The course was structured at three levels: individual learning, team learning, and learning from each other in a community. This was achieved by the design of the assignments that were focused first on the individual so as to help each student decide on his or her learning objectives. A "Question for the Semester" that was made known at the beginning of the course had to be answered by the teams. Otherwise, each student had to deliver an end of semester assignment and semester learning essay.

These assignments were framed within the five disciplines (characteristics) of Senge's (1990) learning organization, the purpose of his framework being to help create a learning community of individuals, teams, and cross-teams in the class. The second characteristic of Senge's model is personal mastery that is achieved in this case by achieving the personal learning objectives that are tied to the competencies they have chosen. Senge's framework pervaded the whole of the course. Understanding of Senge's model and its limitations was supported by the second assignment that had to be completed in a collaborative virtual environment.

Class content was focused on dilemmas that resulted from economical, sociological, and environmental aspects of energy policy and bridging fuels. Each lecture began with a question ("Question for the Day") that had the same purposes as a declaration of objectives, and showed the purpose of the lecture. For example, the third lecture asked, "How do I create knowledge?" and how do I keep track of my progress in attaining my competencies?" The first question of lecture 4 asked "what is a learning community?", and the first question of lecture 4 asked "what is a dilemma?", and the third "How do you identify a dilemma?" these questions are repeated in lectures 18 and 19, respectively.

It is evident that a large amount of work had to be done to develop and implement this course, which is certainly significant in preparing students for working in learning organizations. It is certainly what many industrialists say they would want, but how many develop their organizations along these lines? There are, of course, other approaches to helping students learn-how-to-learn including short courses of a few days duration. The important thing is that students should be put on the road to developing a defensible theory of learning.

8.6 Discussion

Every student and every engineer bring not only their cognitive abilities to the solution of problems they have to solve but also the many dimensions of emotional intelligence that contribute to their motivations and values. This is a matter that is largely forgotten in the planning of curriculum programs. While aims like that of the ability to reflect in practice or to think critically are widely promoted provision for their development in the curriculum is largely ignored. As was shown in Chapter 7, much teaching reinforces low levels of development. In consequence, reflective practitioners are not produced. Fear of allowing students to take responsibility for parts of the curriculum or self- and peer assessment undoubtedly plays apart. But many teachers may be afraid because they have never been trained in alternative approaches to instruction and assessment that would enable them to form their own conceptualizations. The fear of not being able to cover the curriculum also plays apart. Many of the techniques that enhance learning and development require more time than current approaches would allow. Again teachers have not been trained in the theory and practice of curriculum design; hence, there is no understanding of the role of knowledge in learning. There are sufficient examples within engineering education to show that all these issues are resolvable.

It may be argued that to achieve that goal some provision should be made for the study of learning-how-to-learn early in a student's college career. Linked to the idea of an organization as a learning system, it would provide frames of reference with which the student could observe a firm and understand better how it functions.

But intellectual, emotional, and professional development cannot be completed within college alone: a person continues to develop and will do so in response to any situation. For this reason, industrialists have as much responsibility for the development of their engineers as do the colleges from which they come, and in these days of rapid turnover have an obligation to help them prepare for their next work assignment.

Finally assessment in these areas remains problematic and there is scope for considerable research in this domain. That will require agreement about what it is we want to measure. The portfolio is used as a vehicle for making judgments about performance and reflective capability but it requires that teachers should be trained in its use, and that its purposes and structure are made clear. The most challenging remark in these studies is to be found in Boyd and Cowan's (1986) comment on their experience of collaborative self-assessment—"These highly subjective and presumably biased impressions prompt us to wonder if it is possible to generate higher level commitment without involving the learners in setting goals and criteria." This is what happens in programs of independent study. One wonders if there is a place for independent study within engineering programs.

Note

1. This short paper contains an excellent review of self and peer assessment up to the time of its publication. The criteria and mark scales are shown, and for those unfamiliar with educational measurement—reliability.

References

Adair, J., Ayres, R., Debenham, I., and Despres, D. (1978). *A Handbook of Management Training Exercises*. London: British Association for Commercial and Industrial Education.

Alpay, A., Cutler, P. S., Eisenbach, S., and Field, A. J. (2010). Changing the marks-based culture of learning through peer-assisted tutorials. *European Journal of Engineering Education*, 35(1), 17–32.

Ashworth, P. (1996). Unpublished conference report cited in Heywood (2005).

Balascio, C. C. (2014). Engineering technology workplace competencies provide framework for evaluation of student internships and assessment of ETAC and ABET program outcomes. In: Proceedings of Annual Conference of American Society for Engineering Education, June 2014. Paper 9236.

Bar-On, R., and Parker, J. D. A. (eds.) (2000). *The Handbook of Emotional Intelligence*. San Francisco, CA: Jossey-Bass.

Barroso, L. R., and Morgan, J. R. (2010). Incorporating peer review of course term project in structural analysis course. In: Proceedings Frontiers in Education Conference, S2G-1 to 6.

Beamish, B., Kizil, M., Willey, K., and Gardener, A. (2009). Monitoring mining engineering undergraduate perceptions of contribution to group project work. In: Australian Conference for Engineering Education, pp. 318–325. Adelaide: University of Adelaide.

Benlloch-Dualde, J. V., and Blanc-Cavero, S. (2007). Adapting teaching and assessment strategies to enhance competence-based learning in the framework of the European convergence process. In: Proceedings Frontiers in Education Conference, S3B-1 to 5.

Berry, F. C., and Carlson, P. A. (2010). Assessing engineering design experiences using calibrated peer review. *International Journal of Engineering Education*, 26(6), 1503–1507.

Borglund, D. (2007). A case study of peer learning in higher aeronautical education. *European Journal of Engineering Education*, 32(1), 35–42.

Boud, D. (1986). *Implementing Self-Assessment*. Australian Higher Education Research and Development Society.

Boud, D., Churches, A. E., and Smith, E. M. (1986). Student self-assessment in an engineering design course: an evaluation. *Journal of Applied Engineering Education*, 3(2), 83–90.

Boyd, H., and Cowan, J. (1986). A case for self-assessment based on recent studies in student learning. *Assessment and Evaluation in Higher Education*, 10(3), 225–235.

Bradbury Jones, C., Sambrook, S., and Irvine, F. (2011). Nursing students and the issue of "voice." A qualitative study. *Nurse Education Today*, 31(6), 628–632.

Bramhall, M. D., Eaton, D. E., Lawson, J. S., and Robinson, I. M. (1991). An integrated engineering degree programme: student centred learning. In: R. A. Smith (ed.), *Innovative Teaching in Engineering*. Ellis Horwood.

Cannon-Bower, J. A., Salas, E., and Converse, S. (1993). Shared mental models in expert team decision making. In: J. Castella (ed.), *Individual and Group Decision Making. Current Issues*. Mahwah, NJ: Lawrence Erlbaum Associates.

Carberry, A., Johnson, N., and Henderson, M. (2014). A practice-then-apply scaffold approach to engineering design education. In: ASEE/IEEE Proceedings Frontiers in Education Conference, pp. 1845–1848.

Carey, K. (2015). The End of College: Creating Future Learning and the University Everywhere. New York. Rivershead Books.

Cherniss, C. (2000). Social and emotional competence in the work place. In: R. Bar-On and J. D. A Parker (eds.), *The Handbook of Emotional Intelligence*. San Francisco, CA: Jossey-Bass.

Claxton, G. (1995). What kind of learning does self-assessment drive? Developing a "nose" for quality. Comments on Klenkowski. *Assessment in Education. Principles, Policy and Practice*, 2(3), 339–344.

Cowan, J. (2006). *On Becoming an Innovative University Teacher*, 2nd edition. Buckingham: Open University Press.

Cress, D., and McCullough-Cress, B. J. (1995). Reflective assessment portfolios in engineering education. In: ASEE/IEEE Proceedings Frontiers in Education Conference, 4C1-7 to 9.

Culver, R. S. (1998). A review of emotional intelligence by Daniel Golman. Implications for technical education. In: Proceedings Frontiers in Education Conference, pp. 855–860.

Darling, L. F. (2001). Portfolio as practice: the narrative of emerging teachers. *Teaching and Teacher Education*, 17(1), 107–121.

Davies, J. W., and Rutherford, U. (2012). Learning from fellow engineering students who have current professional experience. *European Journal of Engineering Education*, 37(4), 354–365.

Delson, N. (2012). RATEMyTeammate.org A proposal for an on-line tool for teambuilding and assessment. In: Proceedings of Annual Conference of the American Society for Engineering Education, June 2012. Paper 5024.

Dempsey, G. I., Anakwa, W. K. N., Huggins, B. D., and Irwin, J. H., Jr., (2003). Electrical and computer engineering assessment via senior miniproject. *IEEE Transactions on Education*, 46(3), 350–358.

De Sande, J. C. G., and Godino-Llorente, J. I. (2014). Peer assessment and self assessment: effective learning tools in higher education. *International Journal of Engineering Education*, 30(3), 711–721.

Douglas, K. A., Wertz, R., Purzer, S., and Fosmire, M. (2014). First year and junior engineering students' self-assessment of information literacy skills. In: Proceedings of Annual Conference of American Society for Engineering Education, June 2014. Paper 9567.

Dziedzic, M., Janissek, P. R., and Bender, A. P. (2008). Assessment by peers—An effective learning technique. In: Proceedings Frontiers in Education Conference, T2F-1 to 5.

Eliot, M., and Turns, J. (2011). Constructing professional portfolios, sense making and professional identities. *Journal of Engineering Education*, 100(4), 630–634.

Enszer, J. R., and Castellanos, M. (2014). A comparison of peer evaluation methods in capstone design. In: Proceedings of Annual Conference of American Society for Engineering Education, June 2014. Paper 6320.

Ford, N. J., and Rennie, D. M. (1999). Development of an authentic assessment scheme for the professional placement period of a sandwich course. In: J. Heywood, J. M. Sharp, and M. T. Hides (eds.), *Improving Teaching in Higher Education*. Salford: University of Salford, Teaching and Learning Quality Improvement Scheme.

Freeman, J., and Byrne, P. S. (1973). *The Assessment of Post-Graduate Training in General Practice*. Guildford: Society for Research into Higher Education.

Gentili, K. J., McCauley, J. F., Christianson, R. K., Davis, D. C., Trevisan, M. S., Calkins, D., and Cook, M. D. (1999). Assessing student's design capabilities in an introductory design class. In: ASEE/IEEE Proceedings Frontiers in Education Conference, 3, 13b-8 to 10.

George, J., and Cowan, J. (1999). *A Handbook of Techniques for Formative Evaluation: Mapping the Students Learning Experience*. London: Kogan Page.

Goleman, D. (1994). *Emotional Intelligence. Why It Matters More Than IQ*. New York: Bantam Books.

Goodrich, V. E., Aguiar, E. M. D., Ambrose, G. A., McWilliams, L. H., Brockman, J. B., and Chawla, N. (2014). Integration of a portfolio in a first year engineering course for measuring engagement. In: Proceedings of Annual Conference of American Society for Engineering Education, June 2014. Paper 9712.

Griffin, C. (2012). A longitudinal study of portfolio assessment to assess competence of under-graduate student nurses. Doctoral dissertation, University of Dublin, Dublin.

Groba, A. R., Barreiros, B. V., Lama, M., Gewerc, A., and Mucientes, M. (2014). Using a learning analytics tool for evaluation of self-regulated learning. In: ASEE/IEEE Proceedings Frontiers in Education Conference, pp. 2484–2491.

Hanson, J. L. (2014). Use of peer review of projects to enhance upper level geotechnical engineering course. In: Proceedings of Annual Conference of American Society for Engineering Education, June 2014. Paper 10880.

Hedlund, J., and Sternberg, R. J. (2000). Too many intelligences. Integrating social, emotional and practical intelligence. In: R. Bar-On and J. D. A. Parker (eds.), *The Handbook of Emotional Intelligence*. San Francisco, CA: Jossey-Bass.

Heywood, J. (1989). *Learning, Adaptability and Change. The Challenge for Education and Industry*. London: Paul Chapman/Sage.

Heywood, J. (2005). *Engineering Education. Research and Development in Curriculum and Instruction*. Hoboken, NJ. IEEE/John Wiley & Sons, Inc.

Huggard, M., Boland, F., and McGoldrick, C. (2014). Using cooperative learning to enhance critical reflection. In: ASEE/IEEE Proceedings Frontiers in Education Conference, pp. 1899–1906.

Johnson, C. S. (2006). The analytic assessment of on-line portfolios in undergraduate technical communication. A model. *Journal of Engineering Education*, 95(4), 279–288.

Kaul, S., and Adams, R. D. (2014). Learning outcomes of introductory engineering courses: student profiles. In: Proceedings of Annual Conference of American Society for Engineering Education, June 2014. Paper 8872.

Kemppainem, A., and Hein, G. (2008). Enhancing student learning through self-assessment. In: Proceedings Frontiers in Education Conference, T2D-14 to 19.

Khachikian, C. S., Guillaume, D. W., and Pham, T. K. (2011). Changes in student effort and grade expectation in the course of a term. *European Journal of Engineering Education*, 36(6), 595–605.

Marin-Garcia, J. A., Mirralles, C., and Marin, M. P. (2008). Oral presentation and assessment skills in engineering education. *International Journal of Engineering Education*, 24(2), 926–935.

Mathieu, J. E., Heffner, T.S., Goodwin, G. F., Cannon-Bower, J. A., and Salas, E. (2005). Scaling the quality of team mates metal models: equifinality and normative comparisons. *Journal of Organizational Psychology*, 26(91), 37–56.

Mejia, J. A., Goodridge, W., and Green, G. (2014). Using web-base learning logs to analyze students' conceptual understanding of truss analysis in and engineering Statics course. In: ASEE/IEEE Proceedings Frontiers in Education Conference, pp. 2789–2792.

Michaelson, L. K., Knight, A. B., and Fink, L. D. (eds.) (2004). *Team Based Learning. A Transformative Use of Small Groups in College Teaching*. Sterling, VA: Stylus.

Mickelson, S. K., Hanneman, L. F., and Brumm, T. (2002). Validation of work-place competencies sufficient to measure ABET outcomes. In: Proceedings of Annual Conference of American Society for Engineering Education. Paper 1607.

Mistree, F., Ifenthaler, D., and Siddique, Z. (2013). Empowering engineering students to learn how to learn: A competency based approach. In: Proceedings of Annual Conference of the American Society for Education. Paper 7234.

Ohland, M. W., Layton, R. A, Loghry, M. I., and Yuhasz, A. G. (2005). Effects of behavioral anchors on peer evaluation reliability. *Journal of Engineering Education*, 94, 319–326.

Ohland, M. W., Layton, R. A., Loughry, M. L., Woehr, D. J., Ferguson, D. M., Salas, E., and Heyne, K. (2014). SMARTER teamwork: system for management, assessment, research, training, education and remediation for teamwork. In: Proceedings of Annual Conference of American Society for Engineering Education, June 2014. Paper 9102.

Olds, B. (1997). The use of portfolios in outcomes assessment. In: ASEE/IEEE Proceedings Frontiers in Education Conference, pp. 262–265.

O'Moore, L. M., and Baldock, T. E. (2007). Peer assessment learning sessions (PALS): an innovative feedback technique for large engineering classes. *European Journal of Engineering Education*, 32(1), 43–55.

Payne, R. N., Bramhall, M. D., Lawson, J. S., Robinson, I., and Short, C. (1993). Portfolio assessment in practice in engineering. *International Journal of Technology and Design Education*, 3(3), 37–42.

Penny, A. J., and Gover, G. (1996). An analysis of student grade expectations and marker consistency. *Assessment and evaluation in Higher Education*, 24(2), 173–184.

Piergiovanni, P. R. (2014). Reflecting on engineering concepts. Effects on critical thinking. In: ASEE/IEEE Proceedings Frontiers in Education Conference, pp. 1895–1898.

Pistrui, D., Layer, J. K., and Dietrich, S. L. (2012). Mapping the behaviors, motives and professional competencies of entrepreneurially minded engineers in theory and practice: an empirical investigation. In: Proceedings of Annual Conference of the American Society for Engineering Education. Paper 4615.

Polizzotto, J., and Michalson, W. R. (2001). The technical, process and business considerations for engineering design. In: ASEE/IEEE Proceedings Frontiers in Education Conference 2, F1G-19 to 24.

Rowsome, P., Lane, D., and Gordon, S. (2014). Capturing evidence of meta cognitive awareness of pre-service STEM educators' using "codifying" of thinking through portfolios. In: Proceedings of Annual Conference of American Society for Engineering Education, June 2014. Paper 10910.

Senge, P. M. (1990). *The Fifth Discipline*. New York: Doubleday.

Serra-Toro, C., Traver, J., and Amengual, J.-C. (2014). Promoting student commitment and responding through self and peer based assessment. In: ASEE/IEEE Proceedings Frontiers in Education Conference, pp. 1811–1814.

Shull, P. J. (2014). Using the engineering design process as metacognitive learning strategy to improve student performance. In: ASEE/IEEE Proceedings Frontiers in Education Conference, pp. 1800–1804.

Smith, K. A. (2000). *Teamwork and Project Management*, 2nd edition. New York: McGraw Hill.

Smith, S., Sobolewska, E., and Smith, I. (2014). From employability attributes to professional identity: students transitioning to the workplace. In: ASEE/IEEE Proceedings Frontiers in Education Conference, pp. 1197–1200.

Sternberg, R. J., Forsythe, G. B., Hedlund, J., Horwath, J. A., Wagner, R. K., and Williams, W. M. Practical Intelligence in Everyday Life. Cambridge, Cambridge University Press.

Sternberg, R. J., and Grigorenko, E. L. (2000). Practical intelligence and its development. In: R. Bar-On and J. D. A. Parker (eds.), *The Handbook of Emotional Intelligence*. San Francisco, CA: Jossey-Bass.

Sundarajan, S. (2014). A strategy for sustainable student outcomes assessment for mechanical engineering program that maximizes faculty engagement. In: Proceedings of Annual Conference of American Society for Engineering Education, June 2014. Paper 9251.

Swart, A. J., Lombard, K., and de Jager, H. (2010). Exploring the relationship between time management skills and the academic achievement of African engineering students—a case study. *European Journal of Engineering Education*, 35(1), 79–89.

Thompson, P. (2014). Peer assessment of design reports in a first-year introduction to engineering course. In: Proceedings of Annual Conference of American Society for Engineering Education, June 2014. Paper 9537.

Turns, J., and Sattler, B. (2012). When students choose competencies: insights from the competence-specific engineering portfolio studio. In: Proceedings Frontiers in Education Conference, pp. 646–651.

Vidic, A. D. (2008). Development of transferable skills within an engineering science context using problem-based-learning. *International Journal of Engineering Education*, 24(6), 1671–1677.

Wellington, S. J., and Collier, E. E. (2002). Experience using student workbooks for formative and summative assessment. *International Journal of Electrical Engineering Education*, 39(3), 263–268.

Whitbread, D., and Coltman, P. (2009). The development of two observational tools for assessing metacognition and self-regulated learning in young children. *Metacognition Learning*, 4, 63–85.

Wigal, C. M. (2007). The use of peer evaluations to measure student performance and critical thinking ability. In: Proceedings Frontiers in Education Conference, S3B-7 to 12.

Willey, K., and Gardener, A. (2009a). Self and peer assessment: a necessary ingredient in developing and tracking students' graduate attributes. In: Proceedings of Research in Engineering Education Symposium, pp. 1–9, Palm Grove, Queensland.

Willey, K., and Gardener, A. (2009b). Changing students' self perceptions of self and peer assessment. In: Proceedings of Research in Engineering Education Symposium, Palm Grove, Queensland.

Woodcock, M. (1979). *The Team Development Manual*. Aldershot: Gower Press.

Woods, D, R., Haymark, A. N., and Marshall, R. R. (1988). Self assessment in the conduct of McMaster problem solving skills. *Assessment and Evaluation in Higher Education*, 13(2), 107–127.

Yokomoto, C. F. (1993). Using journals in the engineering classroom. In: ASEE/IEEE Proceedings Frontiers in Education Conference, pp. 120–124.

Zappe, S. E., Merson, D., Hockstedt, K. S., Schott, L., and Litzinger, T. A. (2014). A self assessment of the use of evidence-based instructional practices in engineering. In: ASEE/IEEE Proceedings Frontiers in Education Conference, October 2014, pp. 2197–2220.

9

Experiential Learning, Interdisciplinarity, Projects, and Teamwork

Project work and teamwork are considered to be effective ways of developing the personal transferable skills that industry requires. In addition, carefully selected projects can help students work across the disciplines. It has been found that high levels of interdisciplinarity and integration may contribute to positive learning experiences. However, it is suggested that many students are challenged by collaboration skills. Such skills have a large affective component and might be influenced by context and an individual's personality. Communication skills are particularly challenged when groups have different perceptions of the problem. Several attempts to assess collaboration, teamwork, and individual contributions within team work have been reported.

Transdisciplinary projects are able to integrate the tools, techniques, and methods from a variety of disciplines. Impediments to collaboration include disciplinary prejudice, unwillingness to listen and ask questions, and lack of shared ideas. A key problem that is not fully understood is the level of knowledge required by each partner of the other disciplines involved. For this reason, it is suggested that there is a need for formal training in group work.

"Constructive Controversy" has been recommended as means of creating mutual understanding about a problem area. An experimental course based on constructive controversy led to the reminder that the pedagogic reasoning for the use of a nontraditional method of instruction needs to be explained to students.

The Assessment of Learning in Engineering Education: Practice and Policy, First Edition. John Heywood.
© 2016 The Institute of Electrical and Electronics Engineers, Inc. Published 2016 by John Wiley & Sons, Inc.

One study found that students entered college with ideas about work practices that were at variance with those expected by industry. Moreover, they became more committed to these attitudes as the program progressed. This highlights the value of recruiting students who have prior experience of industry in one form or another.

It has been argued that teamwork can contribute to the development of innovation skills and creativity. The research reported on innovation suggests that heterogeneous teams were not innovative teams. One study that used design notebooks found that contrary to experience the most efficient use of the creative process was in the production phase and not in the conceptual design phase. In both areas of innovation and creativity, there needs to be more research.

Another study had as its starting point the notion of design as reflective practice and introduced the concept of "reflection-out-of-action."

A brief analysis of what can be achieved in college begins a section that asks if teamwork can be taught. Teamwork has important pedagogical functions that may be undesirable or irrelevant in industry. While a project may not be completed, student's learning may progress considerably with time-on-project. Projects are a form of experiential learning and one scheme of project instruction designed to meet the requirements of Kolb's theory of learning is reported.

9.1 Introduction

The common theme of this chapter is the project and the uses to which it is put with a particular focus on teamwork. Clearly, what is said applies to problem-based learning (PBL; e.g., Jollands, Jolly, and Molyneaux, 2012). Project work began to become popular in the 1950s in the United Kingdom when final-year individual projects were required from every student on Diploma in Technology Courses (see Section 2.2). In nonsandwich (cooperative) courses, they are thought to contribute to a graduate's readiness to work. Final-year projects are now a normal: part of the education of engineers (Ku and Goh, 2010). Rubrics for their assessment notably, that of projects completed by individuals, came into use and have stood the test of time and the same debates about their structure in relation to reliability and validity continue (Thompson, Clemmensen, and Ahn, 2013). Although team projects were undertaken, there was a fear of assessment. For some educators, this fear has not gone completely away. By 2000, there was a substantial body of knowledge on the conduct of projects and problem-based learning and their assessment (Heywood, 2005; Sheppard et al., 2009). Karl Smith (2000) had produced the first of several editions of his standard work of reference on project management in which there is a mass of information for those who have to design rubrics for assessment. This chapter continues the examination of current practice begun in Chapters 3 and 6.

The first section of this chapter focuses on the use of projects in interdisciplinary studies. Since the solution of many engineering problems requires the contributions and collaboration of persons from a number of disciplines, the second section describes some interdisciplinary investigations and concludes with a study that shows that cross-disciplinary practice is experienced in different ways by the participants (Adams and Forin, 2013). Another study brought students from art and design together with students from engineering. The two groups had entirely different conceptions of design. It is thought that working in teams on projects may enhance a person's creativity (Dart, McGrann, and Stark, 2009). Approaches such as these are used in some service courses. Their analysis yields insight into the qualities required for effective collaboration that Trevelyan (2010) identifies as a key ability that expert engineers require. Some courses have had as their goal the development of team membership skills, which implies that team skills can be taught. The techniques for assessment and evaluation of such activities necessarily embrace the affective domain and it is these that are of interest.

One of the things that the engineering science assessors tried to measure was originality (see Section 3.1), which would seem to be the same thing as innovativeness that is also discussed. All of these dimensions have attached to them the question "can they be taught?" While there are specific sections that ask if innovation and teamwork can be taught (Sections 9.6 and 9.8) respectively, all of the sections touch on these issues.

9.2 Project Work as a Vehicle for Integrated Learning and Interdisciplinarity

Project and team work are now seen as mechanisms for helping students develop the "professional" ("soft") skills required by industry, and to appreciate that engineering problems are often complex and require knowledge from several disciplines for their solution. For this as well as for learning about and managing projects, project work is to be found at all levels of the curriculum, in all disciplines, and across disciplines. In Europe, after the Second World War in recognition that some subjects such as "work" should be studied by many disciplines, special institutes were created for this purpose (Heywood, 1974). In the United States, at the other end of the spectrum, freshman design projects that had an interdisciplinary base were introduced to illustrate the problems of engineering, and in the hope that student retention would be increased. The need for students to stop perceiving the subjects of the curriculum as separate and independent boxes that Culver and Hackos (1983) perceived in 1982 continues (Section 7.4). According to Laughlin, Zastavker, and Ong (2007) engineering schools were complaining about this in 1902! They reported on how integration affects first-year students' perceptions of and attitudes to introductory technical courses at Olin College. Their paper contains a good summary of issues surrounding definitions of interdisciplinarity and integration (see also Froyd and Ohland, 2005). At the college, students experienced both approaches through project focused and cooperative learning. They found among a sample of 75 students that high levels of interdisciplinarity and integration contribute to a positive learning experience. In addition to the level integration achieved between the courses, they also identified the degree of collaboration between professors in an integrated course block, and the relationships between individual professors and students as significant factors in the students' experience. The examples in the paper show how the students experienced the course in relationship to knowledge and skills and in this respect the paper is particularly valuable, but there is a need to go one step further and put the experiences to test in a transfer situation. Laughlin and his colleagues suggest that the approach may be an effective way to influence retention, and that the development of new curricula, along these lines, might cause smoother career trajectories for students. Preparation for industry begins at day 1 of a program. These remarks would apply equally to problem-based learning.

9.3 Learning to Collaborate

Collaboration is not a thing that young graduates like to do according to Trevelyan (2010). This finding is supported by Lingard and Barkataki (2011). They used peer evaluation by students who participated in team projects and found that students completed their individual tasks well. But "students are challenged when it comes to collaboration skills. They tend not to help one another or to ask for help if they needed it, and they often fail to communicate with each other when required." According to Trevelyan, these attitudes are transferred to industry. Lingard and Barkataki suggest that one reason for this is that university studies are required to be done on one's own without cheating which includes asking another student for help. As indicated in Chapters 4 and 5, communication and collaboration are skills that have a large affective dimension and to a greater or lesser degree are context dependent in which personality plays a role (e.g., Rhee, Parent, and Oyamot, 2012). One of the greatest challenges of engineering education is to prepare students to work with others from noncognate (or perceived noncognate) disciplines.

Dart, McGrann, and Stark (2009) described a project that involved engineering students working with students of fine art to "build, fabricate, test and install an interactive sculpture for Binghamton University campus." This project involved two groups of students who had radically different conceptions of design working together. This put a stress on their powers of communication in a way that would not normally be possible in a university course. For example, both groups had to understand the different meanings they gave to "function." But as the authors report when the functions are de-composed overlaps that were not immediately apparent began to be seen. But all students learnt from those in the division of building and grounds with whom they had to work. The authors argued that most of the a-k ABET outcomes were met by the project, but this would be true of much project work. They considered the project to be transdisciplinary rather than interdisciplinary because it integrated the tools, techniques, and methods from a variety of disciplines (Ertas et al., 2003).

In one of the rare papers with "performance" in its title Gorbet, Schoner, and Spencer (2008) of the University of Waterloo in Canada describe a Technology Art Studio in which fine arts and engineering students collaborated. Previously the authors had shown that interdisciplinary collaboration could act as a catalyst for transformative learning. They cite Cranton (2002) who describes transformative learning as taking place when "through some event, traumatic or ordinary, individuals become aware of holding a limiting or distorted view. If they critically examine this view and are open to alternatives, they may consequently change the way they see things. They have then transformed some part of how they make meaning out of experience," which seems to be similar to Hesseling's (1966) point about administering *Chocs des Opinions*. Gorbet, Schoner, and Spencer had found previously that there was no intrinsic relationship between transformation-level achieved and grade performance. In the study reported,

Mark Range	Criteria
80–100	An exceptional performance that demonstrates through visual form, a superior understanding of the issues pertinent to the class: the work solves the problem and reflects deep involvement on the part of the student. Work that could stand on its own outside the context of the problem. Excellent work.
70–79	A notable performance demonstrating through visual form, a thorough grasp of the issues pertinent to the class. Work that solves the problem and that reflects the expected degree of involvement and competency on the part of the student. Good work.
60–69	A satisfactory performance that demonstrates through visual form, a suitable grasp of issues pertinent to the class. Work that may be okay in one aspect but fails in one or more others. There may be a good deal of effort, but it does not solve the problem, or it may solve the problem in a minimal way, but the student has not invested the time required. Acceptable work.
50–59	A barely adequate performance that demonstrates through visual form, some familiarity with the issues pertinent to the class. Work in which there has not been sufficient investment of time and effort. Minimal effort. Poor work.
0–49	Incomplete work or totally insufficient work. Failure.

EXHIBIT 9.1. Assessment rubric used by Gorbet, R., Schoner, V., and Spencer, G. (2008). Impact of learning transformation on performance in a cross disciplinary project-based course. In: ASEE/IEEE Proceedings Frontiers in Education Conference, TC2-18 to 22.

they investigated the collaboration levels achieved by students working in pairs, levels of transformation, and grade achievement. "The students were asked to choose a major art movement and create a sculptural work made from at least 80% corrugated cardboard that would launch an egg the furthest distance." The assessment rubric is shown in Exhibit 9.1. In addition, each team was rated for collaboration by a Fine Arts instructor and an Engineering instructor (C Score). Finally, the students were asked the two questions shown in Exhibit 9.2 in order to cause the students to reflect, and the responses were rated as shown also in Exhibit 9.2 (T score). They found that at the extremes the T scores and C scores gave a consistent relationship with grades; this was not true of the midrange C and T scores.

They identified three impediments that prevented strong collaboration. These were:

1. Disciplinary prejudices and a lack of appreciation for the partners' knowledge base.
2. Unwillingness to listen, asks questions, and engage in the collaborative process.
3. Lack of sharing ideas and workload.

Reflective Questions	Levels of Transformation
"You are just finishing your first collaboration. Please describe what you thought it meant to 'collaborate' before you started the first project and how that that may have changed through the process."	1. "Statements recognizing disciplinary differences (*becoming aware of holding a limiting or distorted view*)."
"Reflecting on the collaboration you've engaged in thus far in the first project, what are the key lessons you will take from that process into your next collaboration?"	2. "Statements referring to interdisciplinary discussions, or the cross-over of information from one disciplinary partner to the other (*critically examining this view and are open to alternatives*)."
	3. "Statements suggestive of the integration of other disciplinary concepts or views into the respondents practice or thinking (*changing the way they see things*)."

EXHIBIT 9.2. Reflective questions in left-hand column and characteristics of the different levels of transformation in the right-hand column in Gorbet, R., Schoner, V., and Spencer, G. (2008). Impact of learning transformation on performance in a cross disciplinary project-based course. In: ASEE/IEEE Proceedings Frontiers in Education Conference, TC2-18 to 22.

They noted that sometimes a high collaboration score was achieved even though a project was unfinished and, therefore, obtained a lower grade. Because they thought that students would implicitly recognize the need to collaborate, no provision was made for collaboration in the assessment rubric. This led them to take the view that there should be some formal training in small group work and collaboration, the beneficial effects of transformational learning should be explained, and examples of how a perspective from another discipline can help restart a project in difficulty or solve a confounding problem should be given.

Northrup and Northrup (2006) make a distinction between team-based project outcomes and peer contributions to teamwork. They describe a semester long project to design, construct, and test a prototype solar vehicle that was completed by teams of mechanical, electrical, and computer engineering students. Learning to work together in industry is of considerable importance and requires an ability to learn each other's languages particularly in a large plant, as this writer can testify. Thus, the assessment of multidisciplinary action is important. The authors used two instruments for this purpose. The first was a Multidisciplinary Interaction Survey: answers to five questions of the 25-item inventory are shown in Exhibit 9.3. The second was Layton's survey for measuring individual contributions in the teams (Ohland, Layton, Loughry, and Yuhasz, 2005). An important aspect of the investigation was the attempt to relate their dysfunctional scale with that proposed by Kaufman, Felder, and Fuller (2000). They did find that one team was nearly dysfunctional but no team reached the definition of dysfunctionality proposed by Kaufman and Felder, so they were unable to say that dysfunctional groups are related to lower peer ratings. They did find support for the proposition that multidisciplinary

Q3	All disciplines worked together in initial brainstorming meetings.
Q5	The EEs and the MEs discussed design trade-offs during the project.
Q11	The implementation of the trailer release required all MEs and CPEs to work together.
Q17	CPEs and EEs worked together to test the vehicle or parts of the vehicle.
Q24	The vehicle would have performed better if we had had better teamwork.
This team established interim goals to complete this MEA.	

EXHIBIT 9.3. Answers to five questions from a 25-item Multidisciplinary Interaction Survey that show that the team established interim goals to complete the MEA in Northrup, S. G., and Northrup, D. A. (2006). Multidisciplinary teamwork assessment: individual contributions and interdisciplinary interaction. In: IEEE/ASEE Proceedings Frontiers in Education Conference, S2E-15 to 20.

teams that choose integrated design solutions have higher levels of interdisciplinary interaction than groups that choose nonintegrated design solutions. (See Shooter and McNeil, 2002, for a cognate study that assessed teamwork and individual contributions.)

From a phenomenographic study of 22 engineers and nonengineers working in engineering contexts, Adams and Forin (2013) found that there were four ways of experiencing cross-disciplinary practice that were hierarchically related.

At the first category, people work together to learn about what each contributes and what "I" can contribute. In so doing, they become aware of different ways of thinking. They learn the complexity of the situation and learn to ask questions about things they do not understand. They take responsibility for effective collaboration. The second category is "intentional." It is at another (higher) level of engagement because each person in the team contributes to the learning process so as to bring about a "systems perspective." They learn "how to negotiate meaning across perspectives and formulate or investigate problems through multiple lenses" (p. 112). Much of this learning is accomplished through experience and failure. The next level is reached when that which has been learnt is applied to the cross-disciplinary problem and activity, which Adams and Forin describe as strategic leadership. They say that it is experienced by the making of or enabling conceptual connections, "building allegiances, shared ownership and trust, and managing differences to create new paradigms or frameworks that leverage diverse perspectives" (p. 116). Their final category represents "transformative reflective practice that challenges prior training and ways of thinking" (p. 119). Support for this thesis is to be found in the same volume from Itabshi-Campbell and Gluesing (2013), who obtained from engineers' narratives of product development that went particularly well coupled with examples that were less effective. They highlight the social dimension of engineering knowledge and comment that "only after understanding the differences in what people know and how they communicate, can one start to challenge others' assumptions. These skills are particularly important for engineers who are in a position to facilitate or lead solution finding for a problem" (p. 153). Then they say "it is the managers' duty to enable cultivation of such skills" [...] which would seem to be part of the job of "technical coordination" that was perceived to be the central role of the engineer by Trevelyan in which a major skill is the ability to ask the right questions.

The question is: where does one learn these skills? Can these skills be learnt in the absence of experience, as for example, in a college course? Or, are they learnt by trial and error in the workplace? Or, is it possible to prepare for them? Inspection of the detailed description of King and Kitchener's (1994) stages of reasoning suggests that some preparation can be made for them through the design of courses (programs) that are framed to assist cognitive development in which assessments lead interdisciplinary or transdisciplinary thinking and interpersonal development like those at Alverno (Mentkowski and Associates, 2000). There is much to be learnt from teacher education in this respect (LaBoskey, 1994; Heywood, 2008).

Having to collaborate with others is a simple fact of everyday life. It is essential in any organization and while it is very often between two people it is often in larger teams. Clearly, Gorbet and his colleagues were seeking true collaboration and not independent activities by persons working in a group. An area for investigation is the knowledge necessary that individuals in the teams have to have of each other's disciplines before they can learn to collaborate. Or, is it through learning about each other's disciplines that competency in collaboration is developed. At the same time it is a matter of common observation that some people have impediments that hinder collaboration that we put down to that elusive concept "personality." A problem for management is that some very able persons are simply unable to collaborate. More generally, among the engineering profession many young graduates enter industry with a preference for solitary work as Trevelyan (2010) demonstrated.

Of course, it may be that true collaboration is necessary to get people working independently, as for example when a team has a choice in what each member will do in the assembly of an engine (e.g., group technology as practiced at Saab; Burbridge, 1975; Ranson, 1972). Daniels and Cajander (2010) make much of this difference in relation to learning in open-ended international student projects but their focus is on the use of "constructive controversy" in such situations. Although the idea has been around since 1981 (Smith, Johnson, and Johnson, 1981), there has been renewed interest in it in recent years (Matusovich and Smith, 2009). Its purpose in this setting was to create mutual understanding of the problem area and help reduce the confusion that can occur in open-ended projects.

9.4 Constructive Controversy

"Constructive controversy exists when one person's ideas, information, conclusions, theories and opinions are incompatible with those of another and the two seek to reach an agreement" (Johnson and Johnson, 2007). Daniels and Cajander used a staged approach as a guide to understanding how constructive controversy could be used as scaffold. They followed the six stages of constructive controversy outlined by Johnson and Johnson, which are:

1. Students are assigned problem/decision, initial conclusion.
2. Students present and listen and are confronted with an opposing position.

3. Students experience uncertainty, cognitive conflict, and disequilibrium.

4. Cooperative controversy.

5. Epistemic curiosity, information search.

6. Incorporation, adaptation to diverse perspectives, new conclusion.

Daniels and Cajander describe their experiences in two papers (see also Laxer et al., 2009). They found that things did not always work out as planned. For example, Stage 4 overlapped with Stage 2 in that controversy about how to cooperate took place in Stage 2. In Stage 5, while curiosity was sparked about what could make the project better, the students who had worked in subgroups on parts of the project felt pressed to deliver their part of the project. There was evidence of new insights and ideas, but they were not seen as something to act on but as things to be noted. "There was not enough incentive to change what they were doing." Daniels and Cajander concluded that it was possible to use the model in a less structured manner. While it seems to have potential in educational settings, what can students learn from it, that will help them in an industrial situation where there are problems of collaboration and conflict? One thing that some students might begin to learn is the art of communication. Like Gorbet et al., Daniels and Cajander emphasize that when running courses in nontraditional settings it is necessary to explain the pedagogic reasoning for the choice of instructional method.

9.5 Communication, Teamwork, and Collegial Impediments to the Development of Good Engineering Practice

Teamwork activities in the curriculum may also break down the attitude that Trevelyan (2010) found among young graduates who had entered industry from his programs who preferred solitary work. Team work was seen as splitting the assignment into parts so that each one had something to do but on his own. Anything that required collaboration was seen as an interruption. Trevelyan reported that he did not see any evidence of collaboration. He argued that the structure of the education they had received contributed to these attitudes just as the continued reception of PowerPoint lectures diminishes listening skills. But could this desire be alone, be a function of personality? For example, do those who want to work alone have a tendency toward introversion? Yet effective engineering practice requires collaboration and that requires a high level of interpersonal skill, and other studies of practicing engineers confirm this view. Moreover, the finding that communication is a significant factor in the work of the engineers is consistent over time. But the teaching of communicative skills to engineering students in his experience was not done in terms of social interaction, which is a key activity in Newman's thinking and fundamental to the Alverno ability of communication. Development of skill in social interaction requires a high level of self-awareness and therein lies the value of the peer and self-assessment of a person's performance in interpersonal activities (see Section 8.3). Neither Woollacott's (2009) taxonomy or its forbear the CDIO taxonomy (see Section 10.4) while requiring much on communication do not seem to grasp this point, at least the way it is intended. The reductive statements, many though they are, do not convey the dynamic of either the person or the social relationships in which he or she engages. The same may be said of a Spanish definition of communication (Sànchez et al., 2014), which defines communication skills as "the capacity to transmit ideas to a given audience, exchange opinion, persuade the audience, interact in dialogue, participate actively and constructively in debate, exchange ideas in teamwork and use communication tools appropriately for effective communication." These are not linked to interpersonal skills or to the fact a person's communication skills are very much a function of his or her personal dispositions: a fact that has implications for assessment. Tonso (2006) makes the point that face-to-face interactions are rarely studied in detail and that if we are to understand how teams work, we need to understand how campus culture impacts on their actions.

An American study of student work practices may throw some light on Trevelyan's findings, although it may be culture bound. It highlights the need to understand what students bring to their courses in quite profound ways (see Section 4.2). Socialization is necessary both in college and work to get rid of any inaccurate ideas that a new participant might have about the college or organization. To this end, engineering educators try to teach good occupational practices related to communication and teamwork in order that students will regard collaboration and sharing as key occupational norms. Yet investigations cited by Leonardi, Jackson, and Diwan (2009) found that "students often grow more committed to practices that reflect lay stereotypes about engineering identity than when they entered the program and have increasingly difficult time accepting practices that educators, managers and others suggest are fundamental to success in most modern workplaces."

Leonardi, Jackson, and Diwan (2009) investigated work practices among 643 software, computer, and electrical engineering students spread across the years of the program in four courses (an average of 115 students per course). They found that the interviewees engaged in eight distinct work practices that enacted two occupational work norms. The practices were as follows: (1) Delay the start of work. This was seen as a sign of technical competence and expertise. (2) Ignore instructions. This was also a sign of competence and expertise. They cited a student thus, "If you can do it by figuring it out yourself instead of following some cookie-cutter process then you're on the way to becoming an expert." Leonardi, Jackson, and Diwan point out that assessment can contribute to such attitudes if the grade is given on output and not on a combination of process and product. (3) Work without a plan. The respondents perceived a good engineer to be one who could plunge into a problem without any plan and be successful because of technical competence. (4) Monitoring task difficulty. The respondents monitored their work and that of others. The word "difficult" was used in most of the interviews but words like this were often used to describe how others perceived their own work. When they described their own difficulties, they said that the others also experienced them. (5) Completing work alone. This is the Trevelyan problem. Evidently this is learnt early "Informants generally acknowledged that to be an engineer meant that one had to know how to do certain things alone and therefore one had to learn these things by working alone and without relying on others." (6) Ensuring one's contribution stands out. It is self-evident that this can impact on the work of a group. It may also be evidence of fear of working in a group. (7) Ranking oneself against others. This is perhaps the least surprising category and is done by any group of students, or for that matter, workers. In schools, grades are the way we do this and we use words like "genious," "smart," and "stupid" to describe others. (8) Exclude those who are technically inferior. Often in group work, those who are perceived to have inferior technical skills are excluded. The best engineers are considered to be those with superior technical skills. This behavior is also well understood.

The norms that Leonardi, Jackson, and Diwan (2009) drew from this were as follows: (1) Expertise is measured by doing difficult tasks and (2) success is measured by individual (rather than team) accomplishment, and these norms are rationalized. Leonardi, Jackson, and Diwan found that while underclassmen engaged in these practices frequently they did not do so as much as the upper-class students. Therefore, behaviors were reinforced as the course developed with the result that many of these students will bring with them work practices they believe to be correct that are contrary to what are, in fact, wanted.

The finding that new entrants or nearly new entrants have already formed views about what constitutes good engineering practice is likely to be surprising to many engineering education faculty. This problem needs to be faced at the very beginning of a program. The advantage of seeking recruits who have had some experience of industry is evident as is the value of a cooperative-type education. It also seems to be the case that much more attention should be paid to understanding what is going on in the groups for which we are responsible. Survey methods are unlikely to achieve such understanding and could quite easily produce a false perception of student perceptions of their development. This is not to say that such surveys are not useful but they need to be supplemented with other investigations. Finally, the way in which student work is assessed is crucial.

9.6 The Demand for Skill in Innovation: Can It Be Taught?

The media, whether it is in Ireland, the United Kingdom, or the United States, report that there is a need for innovation. In consequence, it has come to be expected that universities will produce engineers who can innovate. So, can innovation be taught? All sorts of assumptions are made about what constitutes an innovation, and how students can learn to produce innovative solutions. The literature on design is replete with ideas. One of them is that the more diverse a team is, the more likely the solutions produced are to be innovative. This belief has been challenged by Fila, Wertz, and Purzer (2011), who conducted a substantial study among 275 four-person teams of freshmen at a large US mid-western university. They undertook two studies. The first measured the effectiveness of the team from the perspective of gender and race and found that heterogeneous teams were more effective. Race alone did not alter their effectiveness. The second study included an instrument that they had designed to measure innovativeness because no such measure was found.

The project was to "design a transportation system to replace the on-campus bus system. The system was seen as inefficient and environmentally unfriendly due to diesel engines and low gas mileage." The final report had to detail the design process and proposed solution. They found that heterogeneous teams were no more innovative than homogenous teams and a low correlation between innovativeness and effectiveness. They admit that their measure of innovativeness was subjective, with each team's solutions being judged for feasibility, viability, desirability, and creativity on a 10-point scale. The assessors were provided with guiding questions. Related to innovation is creativity—again with the question—can it be taught? (See Section 2.4 for a discussion of creativity, and Section 10.5 for a discussion of teamwork in relation to learning-how-to-learn).

Many innovations arise from what Christensen has called disruptive technologies. For example, Shoop and Ressler (2011) of the US Military Academy asked students to view disruptive technologies through the lens of military applications in order to develop skills in critical thinking, creativity, and innovation. They used the pedagogy of Socratic dialogue and a method due to Sylvanus Thayer, who was a superintendent at West Point in the early 1800s. Students are expected to use the Socratic method in their engagements with faculty and are expected to ask questions like "How do you define disruptive technology?" And "Why do you believe that a specific technology is disruptive?" Such questions produce answers that set the scene for the dialogue. Thayer required "daily grades, rigid recitation protocols, and rote, hard skill oriented standards of accomplishment rather than encouraging original thought." But Shoop and Ressler argue that small classes that Thayer advocated which provided self-study and recitation are mutually supportive of the Socratic method. In addition to *The Inventors Dilemma*, the students also had to read Kuhn's *The Structure of Scientific Revolutions*, Boorstin's *The Discoverers,* and C. P. Snow's *The Two Cultures*. The students also participated in one-to-one conversations with technology leaders brought in for the purpose. Shoop and Ressler note that the course is considerably shaped by the readings and suggest other readings that might be used. They draw attention to the fact that the Socratic method has to be learnt as students have no previous experience of it. The course includes students from the life sciences and information technology. They obtained assessments from recent graduates as well as from current students that indicated that their objectives were being met with one highly successful graduate attributing that success to the course.

9.7 Creativity, Teamwork, and Reflective Practice (See Also Section 2.4)

Ekwaro-Osire, Mendias, and Orono (2009) tried to differentiate between individual and team creativity, using as their measuring instrument the students' design notebooks. The rating scale they developed for assessing the log books was developed from a six-phase version of the creative process found in Leornard and Swap (1999). The six phases are preparation (10%), innovative opportunity (5%), divergence (25%), incubation (25%), convergence (15%), and evaluation (10%). The figures in brackets are the weights they ascribed to each activity on the basis of their review of the literature.

The students were given precise instructions on how to complete the design notebooks, including the rubrics for grading. Importantly, they were given examples of good design notebook activities. From the small number of notebooks marked, they argue "that mapping creativity in capstone design courses will establish a creative process that encourages the preparation of ideas, the recombination of innovative opportunities, thinking divergently to generate novel ideas, the incubation of ideas into maturity, the convergence of idea selection, and the reflection of original thought." They found that the most efficient use of the creative process was in the production phase and not in the conceptual design phase which contradicts some industrial thinking. Whitfield's (1975) industrial model shows continuous cycles of convergent and divergent thinking in the industrial process (see Section 2.4). Ekwaro-Osire and his colleagues recognized that when the weights are taken into account, they needed to do more to influence the divergent and incubation processes. Their study confirms the value of design notebooks suggested in other studies (Dym et al., 2005).

Currano and Steinert (2012) departed from the traditional approach to design as a rational process to consider it as reflective practice. Their theory was sprung from the fact that when we get stuck many of us do not give up but go and do something else, and sometimes we get new insights. Currano and Steinert call this "reflection-out-of-Action" but note that it occurs within Schön's "reflection-in-action." One of the objects of their investigation was to establish if it was possible for designers to learn to use reflective practice purposefully to maximize the potential for creative ideation. They interviewed two undergraduates and two graduates and found that among the reflective practices they used were "mindless activities" such as having ideas when half-asleep; "physical activities" like running of working out in a gym; "conversation" when talking with friends, or in a group; "remembering" when rethinking in a variety of situations; and "sketching" or doodling. While this exploratory study provided them with a framework for future study, it does show the importance of reflection-out-of-action in creative ideation, a concept that seems to be justified by everyday experience.

In the areas of innovation and creativity, there needs to be much more research and development. The measurement of the creativity of students remains problematic (Charyton and Merrill, 2009).

9.8 Can Teamwork Be Taught?

This question is rather like the question asked in the 1950s and 60s in the United Kingdom—can design be taught? It is also like another question that has often been asked—can management be taught? Because teamwork is very much a subject of management. The answer to both questions is "Yes and No." Clearly, you cannot teach people to react in certain situations in a certain way. What you can do is alert them to what they might expect and some of the reasons why this may be so. You can give them role-playing exercises and self-assessment schedules like the one in Exhibit 8.4 and hope that they will learn from them. It is possible, therefore, to provide a learning environment in which they can develop certain skills (see Motschnig-Pitrik and Figl, 2007). Human management study is primarily about the provision of a framework for action based on understanding people behavior. It is about the affective domain as much as it is about the cognitive. It is, therefore, as much about emotional intelligence (see Section 8.2) as it is anything else. Yes, in that sense, it can be taught. Similarly, the answer is in the affirmative if the learning environment is constructed for that purpose, as for example the provision of evaluated role plays shows. There are various ways of going about this that range from a couple of hours' preparation to a substantial course. How these courses are presented and structured will depend on how the framework is perceived. The courses that have been given by this writer to both engineering students and graduate teacher trainees have been of 30 hours' duration with the principles remaining the same while the illustrative structure varied (Heywood, 1989, 2008). For a short course, it is this writer's view that an understanding of perceptual learning and role theory is essential, and they can be easily related to design. Bucciarelli's (1994) studies of designers demonstrate the relevance of this point. He looked at the process of design among teams and points out that "engineering design is a process which engages different individuals, each with different ways of seeing the object of design but yet individuals who in collaboration, one with another, must work together to create, imagine, conjecture, propose, deduce, analyze, test and develop a new product in accord with certain requirements and goals. Participants in any design project of all but the simplest kind, working in different domains on different features of the system, will have different responsibilities and more often than not, the creations, findings, claims and proposals of one individual will be at variance with another" (Bucciarelli, 2003, p. 9). In this context, they will have different perceptions of how the design goals can be achieved. It is easy to demonstrate this by bringing a mixed group of individuals, including laypersons to solve a design problem. Understanding that we do not all perceive a problem in the same way is a competency that is fundamental to living. But Bucciarelli observed, as you would do if you brought a heterogeneous group of individuals together to find a solution to a problem, that there has to be negotiation. One of the reasons for this Bucciarelli noted was that the individuals in the group were all in competition with each other because they saw the needs of the design differently. "Each participant sees differently in accord with the standards of thought and practice within their domain of specialization. It's like as if they live in different worlds." The mechanical engineer saw things differently to the electrical engineer just as the systems engineer took a different view of the design. Even the engineers were heterogeneous. But as Belbin (1981) shows, personality plays a part because negative behavior on the part of the members can contribute to dysfunctionality (Exhibit 9.4). Belbin has also suggested that the roles shown in Exhibit 9.5 make teams effective. Schemas for peer assessment would have to take these aspects of teamwork into account.

Tzouanas and Campbell (2011) of the University of Houston–Downtown also answer this question in the affirmative. They have developed a curriculum for engineering technology students that has the goal of teaching them to be good team members, and to provide the internal leadership required to solve problems that occur in dysfunctional teams that they intend to implement. Their curriculum focuses on teamwork and is based on "The Successful Manager's Handbook" (Geberlein, 2004) and Overcoming the Five Dysfunctions of a Team (Lencioni, 2005). It is given in the five units that are shown in Exhibit 9.6.

Four strategies of assessment and evaluation were used. First, a pre- and post–multiple-choice test of 20 items has the intention of measuring gains in understanding (see Exhibit 9.7). Similarly pre- and post-attitudinal surveys are administered with the

The aggressor	A person who increases his or her status at the expense of others. He or she blames, criticizes, and is generally hostile
The blocker	A person who disagrees with everything without reason
The comedian	A person who messes about and may make negative jokes
The competitor	A person who challenges others
The Devil's advocate	A person who can be useful but may turn the group to his or her own way of thinking
The digressor	A person who cannot stick to the point whose contribution can be long winded
The dominator	A person who makes loud and lengthy interventions
The side-talker	A person who continually whispers to his or her next door neighbor
The undercontributor	A person who contributes little or nothing
The withdrawer	A person who might be expected to make a contribution but doesn't

EXHIBIT 9.4. Negative roles in meetings identified in the literature. See also Berne, E. (1964). *The Games People Play. The Psychology of Human Relationships.* Harmondsworth: Penguin.

intention of determining changes in attitudes and confidence. Third, a Peer Evaluation form is given out at the beginning of the project with the object of reinforcing expected team behaviors. The parameters assessed include "meeting engagement, participation in discussion to explore ideas and solutions to project problems, listening skills, willingness to follow ground rules regarding appropriate behavior and conflict resolution within the

Chairman	Concerned with the attainment of objectives. Characterwise, while dominant, he or she will not be assertive and use the strength of the members.
The shaper	The person who always wants to bring everything together so that a project can be initiated and completed. A bundle of nervous energy able to challenge and respond to challenge.
The plant	The person who looks for new ideas when the team is bogged down. Very intelligent.
The monitor-evaluator	Brings critical analysis to the problem. Very intelligent. Has been called a "cold fish" in the literature.
The company worker	Wants to see practical implementation. Is very disciplined which requires plenty of character.
Resource investigator	Brings in ideas from outside.
Team worker	Promotes unity and harmony.
The finisher	Wants things done properly. Worries about detail. The anxious one.

EXHIBIT 9.5. The people who make teams effective. Belbin, R. M. (1981). *Management Teams. Why They Succeed or Fail.* London: Heinemann.

Unit	Goal
1	Focuses on the pros and cons of teams and identifying situations in which teams as opposed to individual assignments are the most appropriate strategy.
2	Overview of the types of attitudes and responsibilities that individual team members need to cultivate among themselves.
3	Roles and responsibilities of the team leader and includes practical tools for communicating, guiding teams, and planning and managing work.
4	Team leadership and the resolution of common forms of team dysfunction.
5	Nuances teaming in a virtual environment in which fellow team members may be in a distant country and may come from diverse backgrounds.

EXHIBIT 9.6. The proposed five units of the training for teamwork course for engineering technology students extracted from Tzouanas, V., and Campbell, L. (2011). The teamwork conundrum: what should be taught and how can we assess team learning in engineering technology. In: Proceedings of Annual Conference of American Society for Engineering Education. Paper 2505.

team, support of team decisions and contribution to work and ability to meet deadlines serve to remind students of appropriate teaming behavior."

The fifth assessment is a postproject interview undertaken for the purpose of understanding nuances of student learning that is assessed by an analytic rubric.

Kedowicz and Blevins (2011) draw attention to the fact that in civil engineering less attention has been paid to showing the relevance of communication and teamwork in engineering to freshmen classes. For this reason, they described an approach for remediating this defect with classes of the order 100 students. In their program, they incorporate communication, teamwork, and ethics instruction at all undergraduate levels. The main purpose of the freshmen class they teach is to introduce students to civil engineering. For this purpose, they use a unique textbook *The Path between the Sea*

The difference between quality circles and self-directed work groups is that:
a. Quality circles are permanent in nature while self-directed work groups are temporary in nature.
b. Quality circles have been deemed more effective than self-directed work groups.
c. Self-directed work groups are typically more contentious than quality circles.
d. Quality circles typically focus on solving issues of quality where self-directed work groups are devoted are considered a "natural" work group in so far as members are cojoined to fulfil their daily work task.
e. Quality circles are temporary systems, while self-directed work groups are permanent.

EXHIBIT 9.7. An example of the type of multiple-choice question that might be asked extracted from Tzouanas, V., and Campbell, L. (2011). The teamwork conundrum; what should be taught and how can we assess team learning in engineering technology. In: Proceedings of Annual Conference of American Society for Engineering Education. Paper 2505.

Individual assignments	Assignment description
Contemporary issues report	Students choose 1 academic article on a civil engineering issue and write a 2- to 3-page report […].
Contemporary issues presentation	Students prepare a 5-minute presentation that informs their audience about a contemporary civil engineering issue.
Team assignments	
Team assignments report	Team writes a 4- to 6-page report that assesses the issues of river flooding in the panama Canal through a comparison of a lock canal and a sea-level canal.
Team presentation	Teams present a 10- to 12-minute presentation on the status of the Panama Canal project in 1905.

EXHIBIT 9.8. Details of individual and team assignments extracted from Kedrowicz, A. A., and Blevins, M. D. (2011). Freshmen civil engineers: a model for integrating communication and teamwork in large engineering courses. In: Proceedings of Annual Conference of American Society for Engineering Education. Paper 676.

(McCullogh, 1977), which is about building the Panama Canal, guest lectures, and team learning. Apart from the technical issues faced, the book introduces them to "the interpersonal skills and political savvy they will need to cultivate throughout their career."

Five class sessions employ guest lecturers (from the faculty) to talk about their specialisms. This gives students information about the program and connects them with the faculty who teach them. From week 1, the students are formed into teams of between five and seven persons with whom they work throughout the semester. The groups provide support for each other when they are writing the individual assignments that are required. The course is taught by a team of one civil engineering and two communications instructors. Large lectures are combined with small group (eight persons) meetings. The paper gives the structure of the course in considerable detail. The students have to complete two individual assignments and two team assignments the details of which are shown in Exhibit 9.8. As a result of feedback, Kedowicz and Blevins concluded that writing instruction could be improved by the provision of a guided peer-review component. Student criticism would add another dimension to student learning. They also suggested that visual communication could improve assignments and show the significance of illustrations for providing supporting material.

Students found that the instructions given on speaking were helpful. The average scores on the speaking assignments increased between the first assignment and the second assignment. Even though individual scores were being linked to team scores, Kedowicz and Blevins were confident that the data pointed to an improvement in individual student's skill development. Evidently the students were challenged by collaborative writing.

Davids (2012) argues that oral examinations can be more productive from the perspectives of learning and assessment than traditional examinations of 1-hour duration. She was able to construct an experiment with an experimental and control group of freshmen in an introductory course to Fluid Mechanics. Both groups took

Weight, %	Criteria	Score on a 6-Point Scale Below
10	Assumptions (listed, applicable, correct understood)	0 = no understanding or performance
20	Approach (clear understanding of problem, correct method or use of equations, and applications of assumptions)	1 = poor understanding or performance (well below average—)
15	Equations (correct governing equations listed and used)	2 = Fair understanding or performance (below average—D)
25	Variable treatment (clear understanding of variables, correct values substituted in)	3 = good understanding or performance (average—C)
5	Units (correct units in answer validity discussed with justification)	4 = high level of understanding or performance (above average—B)
10	Reasonable answer (discussion on answer validity discussed with justification)	5 = Excellent or perfect understanding or performance (well above average—A)
15	Algebra/mathematical manipulation (all math procedures carried out correctly)	

EXHIBIT 9.9. Oral examination rubric extracted from Davids, L. K. (2012). A study on the effectiveness of team-based oral examinations in an undergraduate engineering course. In: Proceedings of Annual Conference of American Society for Engineering Education. Paper 3459. Each of the criteria was rated on a 6-point scale, which is shown in ascending order in the right-hand column. Additional guidance is provided, e.g., Grade 2 describes a person who "had information written, but several mistakes or misunderstandings, hints help a little, but still unsure."

a first mid-term examination in the 6th week of the semester. The second mid-term examination took place in the 12th week. The experimental group took the examinations in a team-based oral.

The teams (four members each) were given four problems to discuss in the oral in which each member would be responsible for presenting and defending at least one problem, with the problems being randomly assigned at the beginning of the 1-hour oral. The students were given and had the rubric explained prior to the first mid-term (see Exhibit 9.9). The students performed at a much higher level than they did in the final written examination and in relation to the control group.

Davids identified four limitations. The first was that there was a year lag between the control group and the experimental group. Second, an argument was presented to suggest that there was no difference in the skills of either group. Third, until the oral examinations have been used on several occasions, it is not possible to verify their reliability. A distinct method of determining both reliability and validity is necessary. Fourth, there was a difference in difficulty between the two sets of examinations.

The questions were made much harder in the oral examination because of the prior notice questions. Therefore, the questions sought to test "transfer of knowledge to diverse contexts and the synthesis of component skills." Notwithstanding these difficulties, a promising approach to examining seems to have been developed and the student feedback was positive.

Hirsch and McKenna (2008) argue that a great deal can be learnt from opportunities that are embedded in project work for the students to reflect on their experiences. They based their course design on Kolb's model of experiential learning (see Section 7.2). The reflection and abstraction modes of thinking are emphasized through team memos, team process checks, and pre- and post-course questionnaires. They argue that by providing equal emphasis to each of the modes of learning in the Kolb model, students are provided with a holistic experience that helps them make connections "between effective teamwork and essential design activities like open-mindedness, collaboration and innovation." One of the post-responses that they cite draws attention to the importance of having a shared goal for this binds the members of the group together and causes them to work interdependently. You need that commitment to be able to work together. In this sense, commitment is "virtue" and this understanding of teamwork accounts for the recent interest in the Aristotelian virtues not only in the pursuit of ethical principles but behavior more generally. In respect of commitment, it will be interesting to see follow-up work by Rhee, Parent, and Oyamot (2012), who evaluated two approaches to project instruction (one novel and the other traditional) using a short version of the Big-Five Personality Measure (Gosling, Rentfrow, and Swann, 2003) that produced a similar result. They found that extraverted students in both groups were more likely to report that group members shared in the team's responsibilities which was in contrast to emotionally stable students. They could not explain the latter because it was contrary to other reported studies.

A major point about the assessment of projects is that when it is derived from a holistic position, it does not matter if the project is not completed so long as the students have learnt from their mistakes. Hirsch and McKenna argue "that this pedagogical function of teamwork may be undesirable or irrelevant to industry but is fundamental in the classroom."

9.9 Discussion

The value of introducing students to learning strategies or courses that require knowledge from different disciplines should be self-evident in today's world. Industry, at least large- and medium-sized firms, informs us that teamwork skills are essential and in small firms that necessarily have to function as a team one person, say in a group of 10, who is dysfunctional can bring the organization to a state of collapse. Yet there is much evidence that engineering students and young graduates do not like working in teams or, if they do, want a task that is separate and distinct from the other tasks so that they can, in effect, work on their own. A problem for instructors is that teams need to be functional from the word *go*. For this reason, selection of team members is of some importance.

Interpersonal relationships matter very much and are a reminder of the importance of the affective domain in education. It is clear that curricula can inform attitudes (Exhibit 3.7) but statements of hope are not enough and nonjudgmental methods of assessment of appropriately designed learning activities to act as triggers for reflection need to be developed and used (e.g., Exhibit 8.4).

The attitudes that students bring with them to courses are of considerable importance and those who have had some experience of industry before beginning their studies (e.g., internships) are likely to bring attitudes that can positively influence their fellow class members.

Team projects achieve many functions that are important to the individuals participating in them that may not be important for industry. A project that is not completed might achieve as much in terms of the value that it adds to a student's learning as one that is completed, and students should be assessed for their development.

There has been a tendency to undervalue individual projects but prior experience of individual projects before team projects may enable students to develop a better skill base for subsequent teamwork.

The literature is replete with rubrics for the assessment of projects, but many of them have not been tested for reliability or validity.

References

Adams, R. S., and Forin, T. (2013). Working together across disciplines. In: Williams, B., Figueiredo, J., and Trevelyan, J. P. (eds.), *Engineering Practice in a Global Context. Understanding the Technical and Social.* London: CRC Press/Taylor and Francis.

Belbin, R. M. (1981). *Managerial Teams. Why They Succeed or Fail.* London: Heinemann.

Bucciarelli, L. L. (1994). *Designing Engineers.* Cambridge, MA: MIT Press.

Bucciarelli, L. L. (2003). *Engineering Philosophy.* Delft, the Netherlands: Delft University Press.

Burbridge, J. L (1975). *The Introduction of Group Technology.* London: Heinemann.

Charyton, C., and Merrill, J. A. (2009). Assessing general creativity and creative engineering design in first year engineering students. *Journal of Engineering Education,* 98(2), 145–156.

Cranton, P. (2002). Teaching for transformation. *New Directions for Adult and Continuing Education,* 93, 63–72.

Culver, R. S., and Hackos, J. T. (1983). Perry's model of intellectual development. *Engineering Education,* 73(2), 222–226.

Currano, R. M., and Steinert, M. (2012). A framework for reflective practice in innovation design. *International Journal of Engineering Education,* 28(2), 270–275.

Daniels, M., and Cajander, A. (2010). Experiences from using constructive controversy in an open-ended group project. In: ASEE/IEEE Proceedings Frontiers in Education Conference, S3D-1 to 5.

Dart, K., McGrann, T. R., and Stark, J. T. (2009). ABET assessment of student initiated interdisciplinary senior capstone project. In: ASEE/IEEE Proceedings Frontiers in Education Conference, M3E-1 to 6.

Davids, L. K. (2012). A study of the effectiveness of team-based oral examinations in an undergraduate engineering course. In: Proceedings of Annual Conference of American Society for Engineering Education, paper 3459.

Dym, C. L., Agogino, A., Eris, O., Frey, D. D., and Leifer, L. J. (2005). Engineering design thinking, teaching and learning. *Journal of Engineering Education*, 94(1), 103–120.

Ekwaro-Oisere, S., Mendias, J. J., and Orono, P. (2009). Using design notebooks to map creativity during team activities. In: ASEE/IEEE Proceedings Frontiers in Education Conference, M1J-1 to 5.

Ertas, A., Maxwell, T., Rainey, V. P., and Tanik, M. M. (2003). Transformation of higher education: The transdisciplinary approach to engineering. *IEEE Transactions on Education*, 6(2).

Fila, N. D., Wertz, R. E., and Purzer, S. (2011). Does diversity in novice teams lead to greater innovation? In: ASEE/IEEE Proceedings Frontiers in Education Conference, S3H-1 to 5.

Froyd, J. E., and Ohland, M. W. (2005). Integrated engineering curricula. *Journal of Engineering Education*, 94(1), 147–164.

Geberlein, S. H. (ed.) (2004). *The Successful Manager's Handbook*. Personnel Decisions International 92004–edition.

Gorbet, R., Schoner, V., and Spencer, G. (2008). Impact of learning transformation on performance in a cross disciplinary project-based course. In: ASEE/IEEE Proceedings Frontiers in Education Conference, TC2-18 to 22.

Gosling, S. D., Rentfrow, P. J., and Swann, W. B. (2003). A very brief measure of the Big-Five personality domain. *Journal of Research on Personality*, 37, 504–528.

Hesseling, P. (1966). *A Strategy for Evaluation Research*. Assen, the Netherlands: Van Gorcum.

Heywood, J. (1974). New types of courses and diplomas *Education and Culture* No 10, Strasbourg, Council of Europe. (summary of a commissioned report (1973) to the Council).

Heywood, J. (1989). *Learning, Adaptability and Change. The Challenge for Education and Industry*. London: Paul Chapman/Sage.

Heywood, J. (2005). *Engineering Education. Research and Development in Curriculum and Instruction*. Hoboken, NJ: IEEE/John Wiley & Sons.

Heywood, J. (2008). *Instructional and Curriculum Leadership. Towards Inquiry Oriented Schools*. Dublin: National Association of Principals and Deputies/Original Writing.

Hirsch, P. L., and McKenna, A. F. (2008). Using reflection to promote team work understanding in engineering design. *International Journal of Engineering Education*, 24920, 377–385.

Itabshi-Campbell, R., and Gluesing, J. (2013). Engineering problem solving in social contexts. "Collective Wisdom." In: B. Williams, J. Figueiredo, and J. P. Trevelyan (eds.), *Engineering Practice in a Global Context. Understanding the Technical and Social*. London: CRC Press/Taylor and Francis.

Johnson, D., and Johnson, R. (2007). *Creative Constructive Controversy. Intellectual Challenge in the Classrooms*, 4th edition. Edina Min.

Jollands, M., Jolly, L., and Molyneaux, T. (2012). Project based learning as a contributing factor to graduates' work readiness. *European Journal of Engineering Education*, 37(2), 143–154.

Kaufman, D. B, Felder, R. M., and Fuller, H. (2000). Accounting for individual effort in cooperating learning teams. *Journal of Engineering Education*, 89(2), 133–140.

Kedowicz, A. A., and Blevins, M. D. (2011). Freshman civil engineers: a model for integrating communication and teamwork in large engineering courses. In: Proceedings of Annual Conference of American Society for Engineering Education. Paper 676.

King, P. M., and Kitchener, K. S. (1994). *Developing Reflective Judgment*. San Francisco, CA: Jossey-Bass.

Ku, H., and Goh, B. S. (2010). Final year engineering projects in Australia and Europe. *European Journal of Engineering Education*, 35(2), 161–173.

LaBoskey, V. K. (1994). *Development of reflective Practice. A Study of Preservice Teachers*. New York: Teachers College Press.

Laughlin, C. D., Zastavker, Y. V., and Ong, M. (2007). Is integration really there? Students' perceptions of integration in their project-based curriculum. In: Proceedings Frontiers in Education Conference F4C-19 to 25.

Laxer, C., Daniels, M., Cajander, A., and Wollowski, M. (2009). Evolution of an international collaborative student project. *CRPIT, Computing Education*, 95, 111–118.

Lencioni, P. (2005). *Overcoming the Five Dysfunctions of a Team. A Field Guide for Leaders, Managers and Facilitators*. San Francisco, CA: Jossey-Bass.

Leornard, D., and Swap, W. (1999). When sparks fly. Igniting creativity in groups. *Workforce*, 78(10), 87–89.

Leonardi, S. M., Jackson, M. H., and Diwan, A. (2009). The enactment–externalization dialectic rationalization and the persistence of counterproductive technology design practices in student engineering. *Academy of Management Journal*, 52(2), 400–420.

Lingard, R., and Barkataki, A. (2011). Teaching teamwork in engineering and computer science. In: ASEE/IEEE Proceedings Frontiers in Education Conference, F1C-1 to 5.

McCullogh, D. (1977). *The Path between the Seas: The Creation of the Panama Canal*. New York: Simon and Schuster.

Matusovich, H., and Smith, K. (2009). Constructive academic controversy—What is it? How to structure it? In: ASEE/IEEE Proceedings Frontiers in Education Conference, M3A-1 to 3.

Mentkowski, M., and Associates (2000). *Learning That Lasts. Integrating Learning, Development, and Performance in College and Beyond*. San Francisco, CA: Jossey-Bass.

Motschnig-Pitrik, R., and Figl, K. (2007). Developing team competence as part of a person centered learning course on communication and soft skills in project management. In: ASEE/IEEE Proceedings Frontiers in Education Conference, F2G-15 to 21.

Northrup, S. G., and Northrup, D. A. (2006). Multidisciplinary teamwork assessment: individual contributions and interdisciplinary interaction. In: ASEEE/IEEE Proceedings Frontiers in Education Conference, S2E-15 to 20.

Ohland, M. W., Layton, R. A, Loghry, M. I., and Yuhasz, A. G. (2005). Effects of behavioral anchors on peer evaluation reliability. *Journal of Engineering Education*, 319–326.

Ranson, G. N. (1972). *Group Technology*. Maidenhead: McGraw Hill.

Rhee, J., Parent, D., and Oyamot, C. (2012). Influence of personality on a senior project combining innovation and entrepreneurship. *International Journal of Engineering Education*, 28(2), 302–308.

Sánchez, F., Solar, A., Lopez, D., Martin, C., Agano, A., Balancho, L., Cabre, J., Cobo, E., Farre, R., Garcia, J., and Maros, P. (2014). Developing professional skills at tertiary level: a model to integrate competencies across the curriculum. In: ASEE/IEEE Proceedings Frontiers in Education Conference, pp. 1090–1098.

Sheppard, S. D., Macatangay, K., Colby, A., and Sullivan, W. M. (2009). *Educating Engineers. Designing for the Future of the Field*. Carnegie Foundation for the Advancement of Teaching. San Francisco, CA: Jossey-Bass.

Shoop, B. L., and Ressler, E. K. (2011). Developing the critical thinking, creativity and innovation of undergraduate engineering students. *International Journal of Engineering Education*, 27(5), 1072–1080.

Shooter, S., and McNeil, M. (2002). Interdisciplinary collaboration learning in mechatronics at Bucknell University. *Journal of Engineering Education*, 91(3), 339–344.

Smith, K. A. (2000). *Teamwork and Project Management*, 2nd edition. New York: McGraw Hill.

Smith, K. A., Johnson, D., and Johnson, R. (1981). Can conflict be constructive? Controversy versus concurrence seeking in learning groups. *Journal of Educational Psychology*, 73(5), 651–663.

Thompson, M. K., Clemmensen, C. H., and Ahn, B.-K. (2013). Effect of Rubric Rating Scale on the evaluation of engineering design projects. *International Journal of Engineering Education*, 29(6), 1490–1502.

Tonso, K. L. (2006). Teams that work: campus culture, engineer identity and social interaction. *Journal of Engineering Education*, 95, 25–38.

Trevelyan, J. (2010). Restructuring engineering from practice. *Engineering Studies*, 2(3), 175–195.

Tzouanas, V., and Campbell, L. (2011). The teamwork conundrum: what should be taught and how we can assess team learning in engineering technology. In: Proceedings of Annual Conference of American Society for Engineering Education. Paper 2505.

Whitfield, P. R. (1975). *Creativity in Industry*. Harmondsworth: Penguin.

Woollacott, L. C. (2009). Validating the CDIO syllabus for engineering education using the taxonomy of engineering competencies. *European Journal of Engineering Education*, 34(6), 545–559.

10

Competencies

The chapter begins by distinguishing between the "inside" and "outside" views of competence. It is argued that the "inside" view dominates Western thinking. Questions remain about what are and what are not competencies/competences. A summary of investigations into competencies and competency-based experiential learning (coop/internship) carried out at Iowa State University is given. Workplace competencies were defined as the application of knowledge, skills, attitudes, and values and behavior that interact with the task at hand. "Key actions" required to demonstrate a competency were described. For most of the competencies essential for the professional practice of engineering, the engineering workplace ranked highest as the best place to develop and demonstrate the competencies [...] the classroom consistently ranked last. Among other findings, the authors suggest that from a combination of supervisor and self-assessments, an e-career self-management system could be developed.

Given the value attached to experiential learning, there is no mention of taxonomies in this area. Neither the taxonomy presented nor the Iowa study makes any mention of the skill of "reflection." Two attempts to develop a universal taxonomy of competencies for engineering are discussed.

A European comparison of competencies valued by employers and faculty showed that whereas teachers valued knowledge and research skills, employers

The Assessment of Learning in Engineering Education: Practice and Policy, First Edition. John Heywood.
© 2016 The Institute of Electrical and Electronics Engineers, Inc. Published 2016 by John Wiley & Sons, Inc.

valued planning, communication, flexibility, creativity, problem solving, and interpersonal skills. This raises questions of research design because it might be thought that the acquisition of research skills would necessarily involve problem solving.

A US study applied the standards-based grading system used in schools to undergraduate STEM courses. Principles for the design of such systems are enunciated. These show great similarities with the practices developed for engineering science (Section 3.2). It was found that this scheme enhanced motivation.

European literature shows that irrespective of the term used, an outcomes approach will generate lists of competencies that are common across the globe. One Spanish study reduced 37 core competencies to 5. Three Viennese studies recognized the importance of the affective domain in the development of competencies in computer science. In another study, a technique for getting students to share their reflections with each other is described. Students in another study reported how the skills learnt had benefited their private lives as well as their professional. It was found that it was not possible to provide for team projects and assume that all that students could learn from them was learnt.

An evaluation of a competency-specific engineering portfolio studio in which students selected the competencies they wished to develop showed that pedagogies that support individual choice involve a shift in the power dynamics of the classroom and therefore in the role of the instructor.

The chapter ends with a brief account of Eisner's expressive outcomes and Miller's attempt to reconcile the views of advocates and opponents of behavioral objectives.

10.1 Introduction

As was argued in the Prologue (Section 1.3), the definition we give to competence will depend on whether we take an "outside" or "inside" view of its acquisition. How competence is conceptualized influences the associated assessment strategy. For example, those who believe that competence is a statement of what a person has achieved or has the potential to perform at a certain standard will adopt techniques of assessment similar to those undertaken by the National Council for Vocational Qualifications in the United Kingdom (see Section 5.1). Griffin (2012) writes, "when competence is viewed as a dynamic process it is defined as the ongoing acquisition and consolidation of a set of skills needed for performance where intrinsic motivation is the driver. 'Inside' (innate properties) models of competence describe individuals as 'having' varying amounts of competence whereas 'outside' models of competence focus on external contextual, social, cultural and historical factors which either hinder or contribute to the development of an individual's competence." The "inside" view dominates Western thinking.

Therefore, "competencies are the result of integrative learning experiences in which skills, abilities and knowledge interact to form bundles that have currency in relation to the task for which they are assembled" (cited by Mistree, Ifenthaler, and Siddique, 2013). They can be taught and industry can tell colleges what competencies it needs.

Thus, there are all sorts of competencies that are likely to be of value to engineers working in industry, for example, global competency and innovation competency. Nevertheless, some questions remain about whether some things called competency are, in fact, a competency, for example, "adaptability." There is a great philosophical debate to be conducted. The studies that follow speak for themselves.

Some social scientists and engineers have considered what it means to have a competence, for example, Downey et al. (2006) on what it means to be a globally competent engineer. Such definitions have a considerable bearing on the design of study abroad programs. The six most important dimensions of global competency derived from a survey by Klein-Gardner and Walker (2011) of 48 persons (27 in academia and 21 in industry) who were all experts in the field of global thinking, are:

1. The ability to communicate across cultures.
2. The ability to appreciate other cultures.
3. A proficiency for working in or directing a team of ethnic and cultural diversity.
4. The ability to effectively deal with ethical issues arising from cultural or national differences.
5. Possessing understanding of cultural differences relating to product design, manufacture, and use.
6. Possessing understanding of the implications of cultural differences of how engineering tasks might be approached.

(This writer, having moved to another culture that spoke the same language, would not like to make a sharp distinction between them relative to their importance, except to say that issues relating to them were the most important he had to face.) Klein-Gardner and Walker (2011) found no significant difference between the results for these dimensions between the academics and industrialists. At the same time there were significant differences in other areas and when the qualitative and quantitative evidence was put together they found (notwithstanding other studies, in particular, that of Parkinson, Harb, and Magleby, 2009, with whom they compared their data), they could not "provide an explicit definition for what it means to be a globally competent engineer." Some of (what might be called) subcompetencies were held in higher esteem than others.

It was also found that certain aspects of global competence might be more relevant to one branch of engineering than to another. Neither do experts easily agree on what is most important. They write "the ambiguity in our results speaks to the complexity of the issue at hand: what should universities attempt to impart on their students in order to prepare them to be competent employees? Also, whose opinion is most relevant when deciding what an engineering student should know?" Doesn't this apply to other competencies? If it does not, then their argument that universities should concentrate on the first three subcompetencies surely does, the point being that it is very easy to generate long lists of subcompetencies that can destroy any attempt to get students to think critically. Their method of survey may have application elsewhere.

10.2 The Iowa Studies (ISU)

Iowa State University (ISU) investigators began with the premise that they did not know how to directly assess "an ability," and took the view that outcomes were a collection of workplace competencies that were necessary for the practice of engineering at the professional level (Brumm, Hanneman, and Mickelson, 2006). They took a rationalist approach and defined workplace competencies as the application of knowledge, skills, attitudes, and values and behavior that interact with the task at hand. Therefore, "competencies are directly measurable through actions or demonstrations of the existence of these competencies in the individual" (Brumm, Hanneman, and Mickelson, 2006). They were informed by the fact that employers of their graduates wanted to know if a graduate demonstrated a specific set of competencies "based on the analysis of the successful practice of engineering in specific engineering positions." They wished measurements of the ABET outcomes to apply across their accredited programs and the two forms of experiential education. They wanted them to be aligned with existing employer assessment, and to be "clearly and independently defined, readily observable, immediately measurable consistent with the vision and mission of our college" [...] (Brumm, Hanneman, and Mickelson, 2006). The similarities with the aims of the British National Vocational Education Qualifications will be apparent: they are open to similar criticisms (see Section 5.1). Their approach was, however, radically different.

Prior to the implementation of ABET EC 2000 in 1999 (ABET, 1996), they invited 212 persons to engage in the creation and validation of metrics for the experiential workplace. This group comprised employers in varying roles, faculty, staff, administrators and alumni, students who participated in experiential education, parents, and international faculty from partnering institutions. The investigators collaborated with Development Dimensions International Inc., who, through focus groups and the use of a critical incident technique, obtained hundreds of examples of successful and unsuccessful demonstrations of the 11 criterion 3 ABET outcomes (ABET, 2010). From these, they derived 14 ISU Competencies that were both necessary and sufficient to demonstrate the ABET outcomes (Exhibit 10.1; Mickelson, Hanneman, and Brumm, 2002).

Engineering knowledge	General knowledge	Continuous learning
Quality orientation	Initiative	Innovation
Cultural adaptability	Communication	Teamwork
Integrity	Analysis and judgment	Planning
Professional impact	Customer focus	

EXHIBIT 10.1. The ISU Competencies. Mickelson, S. K., Hanneman, L. F., and Brumm, T. (2002). Validation of workplace competencies sufficient to measure ABET outcomes. In: Proceedings of Annual Conference of the American Society for Engineering Education. Paper 1607.

Having defined the competencies, they listed a set of observable and measurable "key actions" that a student should take to demonstrate a competency. The key actions that a student would have to take to show that they had developed and demonstrated "customer focus" are:

1. Seeks to understand customers.
2. Educates customers.
3. Builds collaborative relationships.
4. Takes action to meet customer needs.
5. Sets up customer feedback systems.

(In other studies, these statements are sometimes called outcomes.) In order to validate the competencies, they returned to their respondents and asked them to read the definition of each competency and the associated key actions, and to record using a Likert scale, how important each competency was for a student or graduate to successfully demonstrate each of the ABET outcomes. The respondents were from the disciplines within the college and industry, the response rate of 32% being typical for a survey of this kind. With one exception, no competency received a rating of less than 3 on a scale where 5 was considered to be essential. Apart from seeking opinions about the degree to which the ISU competencies collectively covered the ABET outcomes, they also asked the degree to which they covered the practice of professional engineering.

A question was asked to determine what the constituents considered the probability of a student/graduate being given the opportunity to undertake actions that would demonstrate the competencies in a number of educational settings. These were the engineering workplace, coop/internship workplace (experiential education), the traditional classroom, the classroom laboratory, the classroom capstone design, extracurricular activities (engineering profession related), and extracurricular activities (non-engineering profession related). "For most of the competencies essential to professional practice of engineering the engineering work place ranked highest as the best place to develop and demonstrate the competencies [...] the classroom consistently ranked last" (Brumm, Hanneman, and Mickelson, 2006).

In practice, each key action is rated on a Likert scale to see how often an individual performs an action. A total of 61 key actions have to be rated and, according to the writers, this takes about 10 minutes. Both the students' self-assessment and supervisor's assessment are taken into account in arriving at a ranking. The activity for one of the disciplines (agricultural engineering) is described in detail elsewhere (Mickelson et al., 2003). The authors make the point that the system could be used for the development of an e-career self-management system or record of achievement as it has been called in the United Kingdom. This approach to assessment that is conditioned by the ABET EC 2000 parameters may be contrasted with the experiential approach of Ford and Rennie (see Section 8.3), and the approach to task analysis taken by Youngman et al. (1978; Section 4.2).

While experiential learning is valued, educators and industrialists do not seem to recognize the possible significance of the affective domain in job performance yet a glance at Steinaker and Bell's (1969) *Taxonomy of Experiential Learning* which follows underlines the importance of that domain. The categories are:

1. *Exposure.*
   Seeing, hearing, reading, and recognizing.
2. *Participation.*
   Observing, discussing, listening, and ordering.
3. *Identification.*
   Classifying, explaining, experimenting, contrasting, and transferring.
4. *Internalization.*
   Generalizing, comparing, contrasting, and transferring.
5. *Dissemination.*
   Communicating, debating, presenting, motivating, and influencing.

A notable exception from the categories of this taxonomy, as it also seems to be in the Iowa study, is "reflecting." Reflecting on experience and using academic knowledge to interpret them are an essential part of the experiential learning process (Kolb and Fry, 1975; see also Section 7.2). "Only experience which is raised to the level of reflection (and therefore open to theoretical treatment can really count as experience" (Ashworth and Saxton, 1992; Ford and Rennie, 1999, p. 39; see Section 8.3).

A study in Europe compared the competencies valued by instructors with those valued by employers in respect of computer science graduates. Teachers rated general knowledge and research skills more highly than employers who valued, planning, oral communication, and the ability to deal with technology, flexibility, creativity, problem solving, and interpersonal skills as more important (Kabicher and Motschnig-Pitrik, 2009). The research design seems to be problematic since it might be supposed that if students learn research skills they would learn problem solving perhaps accompanied by skill in transferability, and recognition of the need for some creativity. These studies raise the question-how much notice should academia take of industry?: an issue that will be considered in the final parts of this text.

10.3 The Outcomes Approach in Australia, Europe, and Elsewhere

The outcomes approach also informs the European Higher Education Area (Bologna, 1999), which has led to the description of the outcomes required for the different levels of higher education (e.g., Bologna, 2005; Dublin, 2004). The influential Tuning report (2005), which summarizes competencies, is widely invoked. It argues that a competency-based system requires a student centered-approach.[1] Similar developments have taken place in Australia (Australian Council for Educational Research, 2002) and India (Kumar, Garimella, and Nalla, 2014). A competency-based approach is now prescribed for Russian Universities (Bilyadinova et al., 2014). Such statements enable the mutual recognition of qualifications between countries, as for example, the Washington

Accord (1989; Bucciarelli, Coyle and McGrath, 2009; Oladiran et al., 2013) (Chapter 6, note 1). A recent Russian study that compared Russian competencies with those of Tuning and the CDIO (Conceive-Design-Implement-Operate) concluded that there "are good prospects for convergence and collaboration between Russian universities and universities in Europe and the USA" (Lunev, Petrova, and Zaripova, 2013).

They also enable comparative research (Cajander, Daniels, and Von Konsky, 2011), and attempts have been made to develop Taxonomies of Competencies (Armarego and Roy, 2013; Woollacott, 2009). The term "competency based" is now being used more frequently in American literature (Balascio, 2014; Olague-Cabllero and Valles-Rosales, 2014). In 2000, the Organization for Economic and Cultural Development (2002) published a conceptual framework for understanding generic competencies. While these were interrelated, their significance varied from job to job. A study of engineering-specific versions in Australia by Male, Bush, and Chapman (2011) led them to question the assumption that the generic competencies will be the same in all jobs. For a world-wide comparative study, see Lucena et al. (2008) in which the authors claim that "ABET has become an energetic international organization attempting to lead by example."

Behind these initiatives is the vast increase in government expenditure on higher education during the last 20 or more years and the need for governments to give quality assurance to those who elect them.

Quite outside these remits is a study of the competencies required for managers of technical education in India (Gupta and Earnest, 2010). The questions asked were whether or not their criteria meet the present and perceived future needs of industry, whether or not programs that say they do actually do, and whether as a recent comparative study suggests that the "emphasis on competence based assessment as the primary means by which outcomes based assurance takes place means that students are placed at the centre of the process and a focus" (McDermott, Daniels, and Làrusdòttir, 2014). Are they?

10.4 The CDIO Initiative

At the end of the last century and the beginning of the millennium, engineering schools at MIT and three Swedish Universities came together for the purpose of reforming engineering education. They wanted to close the gap between engineering education and engineering practice. They took the view that the origins of engineering faculty were strongly rooted in the engineering science tradition and had not so wide backgrounds in industrial experience. To address this imbalance, they thought the focus of the engineering curriculum should be on engineering design for "every graduating engineer should be able to Conceive-Design-Implement-Operate (CDIO) complex value-added systems in a modern team-based environment" (Crawley, cited by Woollacott, 2009). "Design" became the organizing principle (key concept) around which their curriculum proposal was made. It was based on their views of what the goals of engineering education should be. Like the Iowa studies, these were formulated by senior academics in consultation with alumni and leading engineering professionals. The domains of the syllabus are (1) Technical Knowledge and Reasoning, (2) Personal and Professional Skills and Attributes, (3) Interpersonal Skills: Teamwork and Communication, and (4) Conceiving,

Designing, Implementing and Operating Systems in the Enterprise and Societal Context. These were each subdivided into five statements of increasing detail. They move from general goals to teachable and assessable skills. Armarego and Roy (2013) note that this hierarchical structure produced more than 400 learning categories or competency items. They suggest that at the most detailed level the scheme is similar to that of the Bloom *Taxonomy* but that the levels of synthesis and evaluation are amalgamated. Armarego and Roy note that a small number of competency items belong to the affective and psychomotor domains. In this respect, Humble was able to use the cognitive and affective domains of the Taxonomy to analyze the skills required by managers in a plant of the British Steel Corporation (cited in Heywood, 1970). Humble regarded the affective domain to be as important as the cognitive. (See section 4.3).

Woollacott (2007) reports that a correlation with ABET's EC 2000 requirements showed that all of the learning outcomes in those requirements were covered but that the CDIO syllabus was more detailed.

Woollacott's taxonomy of engineering competencies was developed at the University of Witwatersrand as part of a drive to reduce high attrition rates among engineering students in South Africa. The approach marks a change in operational philosophy. "At the heart of the restructuring effort was the conviction that a new way should be found to address the teaching, learning and diversity issues. The foundational premise on which the research/restructuring effort was based was that instead of thinking about language, learning, organizational and related non-technical skills as academic development issues, they should be regarded as issues of 'academic competency' and that development of these competencies should be seen as the first stage in developing critically important non-technical competencies."

In contrast to the CDIO philosophy of focusing on design, Woollacott's taxonomy focuses on engineering work and the competencies required to do it well. The former focuses on what it is "to engineer" while the latter is concerned with that it is to "work as an engineer." It also distinguishes between engineering specific and nonengineering specific activities. The first level categories are A, Engineering Specific Work; B, Non-Engineering Specific Work; C, Communication; D, Inter-personal Interactions; and E, Dispositions (E1, Personal dispositions; E2, Adaptive dispositions; E3, Advanced dispositions).

Woollacott's analysis of the two approaches show considerable overlaps between the two as well as differences that could be taken into account in the search for a universal taxonomy. A major weakness that both approaches have is that their data about what engineers do is based on opinion surveys and not on task analyses of what engineers actually do (see Chapter 4). There can be differences between what engineers think they do and what they actually do. Second, they appear to be based on a particular view of what knowledge, thus learning, is. Third, reductionist models of this kind easily lead to overassessment.

10.5   A Standards-Based Approach to the Curriculum

As a result of experimentation Carberry, Siniawaski, and Dionisio (2012) have proposed that the standards-based approach to assessment that is used in K-12 schools in the

Development toward achieving the course objectives	Homework, 1	Homework, 2	Quiz, 1	Exam 1	Quiz 2
1A Understanding the concept of stress in a body	2	2	2	3	3
2A Analyzing members subject to axial forces	2	2	3	3	4
3A Analyzing members under combined loads			2	2	3
Overall average development	2	2	2.3	2.7	3.3
Current grade	C	C	C	C	B

EXHIBIT 10.2. Snapshot example of an individual student Standard Achievement report by Carberry, A., Siniawaski, T., and Dionisio, D. N. (2012). Standards-based grading. Preliminary studies to quantify changes in affective and cognitive student behavior. In: ASEE/IEEE Proceedings Frontiers in Education Conference, pp. 947–951.

United States should be extended to undergraduate STEM courses. Authors of some of the papers already reported could claim that they were developing standards-based systems of assessments but not across the whole of the curriculum.

Carberry and his partners undertook pilot studies of the integration of such a system into five STEM courses to evaluate the impacts of the system on the affective and cognitive behavior of the students assessed. Affective behavior was obtained from measures of self-efficacy and values (Carberry, Lee, and Ohland, 2010), and cognitive behavior from epistemological beliefs (Carberry, Swan, and Ohland, 2010).

As practiced in K-12, standards-based assessment is formative, in that a student's performance in achieving specified course objectives is tracked throughout the course. Carberry, Siniawaski, and Dionisio give a snapshot of a Standards-Achievement report for an individual student. It is shown in Exhibit 10.2. They highlight the importance of feedback which this system provides. A total of 120 students in five STEM classes in two institutions were taught by four instructors. They concluded that the system of assessment had contributed to a positive change in the students' self-efficacy. Quite substantial increases in self-efficacy between pre- and posttest scores were obtained. They took as their guide for measuring values "Expectancy-Value Theory," and their inventory sought to obtain measures of interest/attainment, utility value, and cost. It seems that the students found that standards-based grading gave them high interest and attainment and utility. Feedback was valued; they quote one student as saying "the direct correlation between my courses grade and the course objectives forced me to pay attention to what I should be taking away from the course."

They also obtained before and after measures of epistemological beliefs, using a modified version of the Epistemological Beliefs Assessment for Engineering (EBAE) Instrument (Carberry, Swan, and Ohland, 2010). They found that beliefs changed from naïve beliefs about STEM at the beginning to beliefs that were more sophisticated at the conclusion of the intervention.

They will be the first to admit that many more studies of this kind are required because many assumptions are made about the impact of assessment on learning without being tested under experimental conditions. Nevertheless, they offer some guidelines for the implementation of standards-based grading. One of the limitations of many of the reported studies of practices in engineering education is their failure to draw attention to aspects of their work that support other people's findings. In this case, the evolution of the criterion referenced system for the assessment of specified coursework in engineering science described in Sections 3.1 and 3.2 is sufficiently similar to the standards-based grading system described by Carberry, Siniawaski, and Dionisio to be able to reinforce the principles suggested by them for successfully implementing standards-based grading system. These principles are highlighted by the statements in italics in the following:

1. *"Establish well-defined course objectives and list them on the course or syllabus."* In the case of engineering science, this was taken as *a sine qua non* of curriculum design.

2. *"Establish a clear course grading policy and a clear set of assessment rubrics and guidelines."* In the case of engineering science this was a requirement of the public examining authority. However, it was a new approach to assessment and several iterations were required before it became the standard. During the next 16 years, it was the subject of continuing evaluation and minor amendment (Carter, Heywood, and Kelly, 1986).

3. Students did not always perceive an item in the same way as the teachers and moderators of engineering science perceived it for this reason that students should be involved in the design of the grading instruments to ensure that teachers and students understand what is being sought.

4. *"Develop a detailed standards achievement report and share it with your students at the beginning of the course."* In engineering science, a substantial booklet of *Notes for Guidance* accompanied the official syllabus. However, given the academic style of its English, it often required teachers to offer an interpretation (Carter, Heywood, and Kelly, 1986, p. 62).

5. *"Center the course lectures, assignments and schedule on the standards achievement report."* The assessment scheme devised for engineering science undoubtedly influenced the learning and development of key skills. Students had particular difficulties with planning, the development of alternative solutions, and evaluation. Improvements occurred when students were given access to examples. But the assessment scheme differentiated among the candidates more or less as a normal distribution year in year out as it was intended to do. Moreover, both the marks of the teachers and the assessors were consistent over the years although the teachers always gave higher scores. All students gained in technical knowledge, some considerably, as they had to choose their own projects and, therefore, the area of study necessary for the completion of the project. The

correlation between a subtest in the written examination on project planning designed to test for transfer was small (Heywood and Kelly, 1973), suggesting that there should have been specific instruction for this purpose (see Section 3.2). It was also argued that the written test required different skills (Heywood, 1986). A case was made for the more formal teaching of design in the course.

The significance of Carberry, Siniawaski, and Dioniso's scheme is that it shows that rubrics of the kind developed for the assessment of engineering science coursework may be applied to the assessment of the syllabus. I have seen engineering educators using such schemes in the United States but have not seen experiments of this kind reported. In relation to coursework in engineering science, and in particular project work, high levels of motivation were witnessed among students, irrespective of levels of performance (Section 3.2). This was due to the method rather than the assessment rubric. No evaluations were made of self-efficacy or values. However, a study was made of the attitudes of teachers by Denis Hiles, who devised a 29-item Likert style inventory that was answered by 29 teachers. The nine most important items showed that teachers thought that projects gave students a sense of involvement, trained students to think for themselves, increased self-reliance through independent work, and increased readiness to accept responsibility. Cognitively, they increased student's decision-making ability, improved problem-solving abilities, and developed creativity (Hiles and Heywood, 1972).

The design of scales for rubrics is not easy. This point is illustrated by a research undertaken by Thompson, Clemmensen, and Ahn (2013) at the Korea Advanced Institute of Science and Technology. They compared 21 experienced raters' use of four rating scales to score five technical posters. In contrast to some faculty who seem to believe, without any evidence, that simple systems of measurement are unreliable, this inquiry showed that if statistical weights are assigned to individual criteria their validity is challenged if the weightings are not properly balanced. They conclude that heavily weighted rubrics should be avoided. They also found that if more responsibility was placed on the grader rather than the rubric, validity was increased but at the cost of rater satisfaction. They came down in favor of 3- or 4-point ordinal scales for individual criteria, and that numerical scores should be assigned to groups of criteria.

Finally, one finding from the observations made of engineering science project work deserves mention. When developing the objectives, the working party accepted McDonald's distinction between creativity and originality that "creative is the label we apply to products of another person's originality" for even in projects that are essentially replication (e.g., engine test beds for use in the school) it is possible to find originality (i.e., "behavior which occurs relatively infrequently, is common under given conditions, and is relevant to those conditions"; McDonald, 1968, see Section 3.1).

The combination of grading of this kind and type of coursework chosen certainly led to the measurement of qualities that written examinations (test) cannot assess, and enhanced the motivation of many students. There is one caveat: it is this writer's experience that no matter what scheme is produced, someone will always criticize it and take on the role of analytical philosopher!

10.6 Recent European Studies

Given that competence and outcome have similar meanings, it is inevitable that there will be some overlap with what has gone before, so it is of some interest to see if there are any differences in approach between recent US and European studies.

Three of the papers considered here come from Spain. They show quite clearly that when an outcomes approach is applied to a professional program such as engineering, that it is inevitable that similar lists will be generated to those being generated in the United States. The problems associated with their implementation will depend on the political and institutional cultures in which they have to be solved. The principles remain the same. Two of the Spanish papers are concerned with adapting to and embedding core competence in the curriculum so that, in particular, the courses will meet the requirements of the European Credit Transfer System.

Spanish employers like other employers seek from graduates in computer engineering skills in "communication, economics, leadership, teamwork and management" (Edwards, Tovar, and Soto, 2008).

They are now demanded by Accreditation Boards. Their references show attention to developments in the United States that might be expected, and unexpectedly to a well-known reference used in teacher education (Reigeluth, 1999). But there is recognition of European contributions to the teaching of transferable skills (Chadha and Nicholls, 2006). To determine the core competencies in computer engineering, Edwards, Tovar, and Soto (2008) conducted surveys of teacher coordinators and teachers. They found 37 core competencies that they were able to rank. They were also able to rate the competencies for the degree to which they were taught in classes (high, medium, basic, or null). They reduced the list to five core competencies:

1. Analysis and synthesis
2. Applying knowledge of mathematics, science, and engineering
3. Logical and mathematical reasoning
4. Oral and written communication in its native language
5. Writing and interpreting technical documentation

It will be noticed that analysis is considered to be of the same dimension as synthesis, and that the "soft"/"professional" skills are confined to forms of communication. One of the reviewers of this book told the author that in one course that used portfolios while they were effective in terms of student learning they were dropped after 5 years because they took up too much student time and faculty marking time. But in another situation he found that "e-portfolios in distance courses in Africa proved positive for both facilitating learning and making it visible for assessment."

Because of European requirements for convergence, the freshmen course in computer technology at the Technical University of Valencia had to be changed. Benlloch-Dualde and Clavero (2007) described a pilot study among a group of freshmen in which teaching-learning and assessment methodologies were changed. A diagnostic test was administered on the first day of the course. Coursework was assessed, a multiple-choice test was given together with a problem-solving examination, and a hands-on examination in the laboratory was set; the students were required to complete a portfolio. The pilot study led to several changes. It seems that compared with the traditional method, this approach was very demanding. The portfolio had to be abandoned.

The other paper from Spain was concerned with the prediction of performance of high school students in computer engineering on the basis of reading, analysis, comprehension, and mathematics (Tovar and Soto, 2010). They found that prior grades by themselves did not explain the data. In the future, they would include socioeconomic variables, and try and establish if motivation was a significant variable. Their own test underestimated the results of the best students and overestimated those of the poor students. The development of their model and the accumulation of such data were for the purpose of making provision to help students.

Three of the investigations reported into competence came from the University of Vienna and related to the development of soft skills and communication in project management (Motschnig-Pitrik and Figl, 2007) and competencies in computer science (Figl and Motschnig-Pitrik, 2008; Kabicher, Motschnig-Pitrik, 2009). The first paper reports the effects that an active learning course had on the development of team competence at the individual level. The course philosophy was based on Rogerian psychology (Rogers, 1961) and is "Person Centred." It is a philosophy of learner-centered-learning." Engineering educator John Cowan (2006) has paraphrased and listed Roger's principles in a question form (Exhibit 10.3) and suggests that if you answer "yes" more than "no" you are a learning-centered educator. He explains Roger's principles in some detail.

Roger's psychology of learning brings the so-called affective domain of learning into play in the understanding and development of teams although in the study reported personality traits are only touched on marginally. It also brings into play an understanding of adult literacy and life skill development (Baker, 2005). It is argued that the person-centered attitudes of congruence/genuineness, acceptance or respect, and empathetic

---

1. Have you a natural potential for learning?
2. Do you learn more significantly when what you are learning is relevant?
3. Do you find it threatening to be expected to change your view of yourself?
4. When such threats are minimized, is it easier for you to learn?
5. Do you find it easier to be perceptive when you are not under pressure?
6. Do you learn significantly by doing?
7. Do you learn better when you take responsibility for your own learning?
8. Do you learn more lastingly when your feelings are involved, as well as your intellect?
9. Are you more independent, creative, and self-reliant when evaluation by others is secondary?
10. Is learning to learn the most socially useful learning for you in the modern world?

EXHIBIT 10.3. Cowan's paraphrase of Carl Roger's principles of learning in a question form. Cowan, J. (2006). *On Becoming an Innovative University Teacher. Reflection in Action*, 2nd edition. SRHE and Open University Press.

understanding are the basis for constructive communication and mutually supported interpersonal relationships. It is this dimension, while inevitably present, that seems to be missing from or not emphasized in many of the studies examined.

The communication and soft skills course in project management reported by Motschnig-Pitrik is based on four generic scenarios that are intended to enable the expression of the core conditions. They are "Student listening," "Reaction sheets," "Shared responsibility in the choice of topics," and "Team moderates unit." In the first, an article on active listening is provided online and followed by activity learning in groups of 3. One person begins by sharing some personally meaningful experience, another person listens, and the third person observes the communication. The roles are switched on two further occasions so that each person experiences each of the roles. The experiences are then shared. We are told that the facilitator listens: in my experience of such activities, facilitation can guide the group to deeper insights. The idea of the "Reaction sheets" that are distributed several days after a course unit has been completed is to encourage the students to share their reflections with each other in the next course unit. The authors report that the students become engaged at all levels of learning. Students "openly share themselves" and, given the time available in the course, tutors can be open to students' perceptions. With reaction sheets, the real challenge lies in the interpersonal handling of the reactions and acting upon the expressed feelings, meaning, observation, and suggestions of tutors. This should help bring the course on to a track that satisfies each student. The student learns to give constructive feedback and perceive the reaction of others to that feedback. To encourage shared responsibility in the choice of topics, group dialogue and decision making are encouraged. "First we elaborate and discuss the concept of soft skills. Then the facilitator provides some expert opinions and students are asked to identify the soft competencies they consider most relevant/interesting for themselves." The facilitator then conducts a sorting procedure and the results are used to form small teams around the most wanted topics.

In the program, the first three units are moderated by the facilitator. Subsequently, the workshops are moderated by teams who are given prior knowledge and access to the facilitator so as "to align the workshop topics, activities and discuss proposed activities and interpretation."

A course such as this contributes to the specific skills that team members should bring with them to team activities. These authors follow Cannon-Bower et al. (1995), who identified three types of team competencies, namely, "knowledge," "attitudes," and "skill." They use the ALL (Adult Literacy and Life skills) framework for understanding teamwork evaluation (Baker et al., 2005). This takes into account attitudes, the skills of "group decision making/planning," "adaptability/flexibility," and "interpersonal relations" together with "communication."

The competencies that make up flexibility are:

1. Monitor/adjust performance.
2. Provide/accept feedback.
3. Reallocate tasks.
4. Provide assistance.

5. Use compensatory behavior.

6. Use information to adjust strategies.

And the competencies involved in establishing effective interpersonal relations are:

1. Manage/influence disputes.

2. Consider different ways of doing things.

3. Seek mutually agreeable solutions.

4. Share the work.

5. Resolve disputes among team members.

6. Work together instead of separately.

7. Work cooperatively with others.

Neither is a long list. Their evaluation shows that students believed that all of the skills in the ALL model had been enhanced but that the greatest effect had been on communication. Students thought they were more open and congruent as a result of the course. One student wanted to transfer what he had got from the course to other aspects of his life. Overall, the authors thought there was slight trend toward the enhancement of person-centered attitudes. "Several students reflected on how important these dimensions were in their professional and personal lives and in which way they could improve their communication." This linking of the professional and the personal and the importance of the "affective" has only made itself known once before in the previous discussion, yet it is at the heart of liberal education. Neither has much attention been paid to the role of the tutor that in this case was as a facilitator and not as a lecturer.

The course provided no specific knowledge of teamwork, yet the students perceived an enhancement in their team competencies for which reason the authors thought that knowledge of teamwork and the competencies required would improve the course and that supports the views put forward in Section 9.8. One might suggest that if attention was given to learning-how-to-learn, it might also be beneficial for students.

10.7 Impact of Subjects (Courses) on Person-Centered Interventions

Figl and Motschnig-Pitrik (2008) also compared the impact of five blended learning courses, including the soft skills course discussed in the previous section. The others were in web engineering, person-centered communication, organizational development, and project management. All of them were based on the person approach (see above). The evaluation was also based on the ALL framework. They found that many students did not believe that the development of team competencies was explicitly addressed in these courses: they rated the effect of the soft skills and person-centered communication courses highest for their overall influence on team competencies. The project management course was rated highest for the development of "sharing information" or "reallocating tasks" whereas courses like "web engineering" that are solely based on team projects without further elaboration of teamwork experience are "distinctly less potent in promoting team competencies." Evidently, it is not possible to provide for team projects and assume that all that could be learnt from participation in them will be learnt. The implications for the curriculum are evident if team competencies are to be promoted. Specific provision has to be made for addressing team competencies, team work, and reflection on the experience of working in a team.

10.8   The Potential for Comparative Studies: Choosing Competencies

Some professional education programs such as medicine are more open to comparative study than others such as teacher education. Culture and politics intervene in the latter, although international testing does force some degree of convergence. Engineering is rather like medicine and the worldwide generation of outcomes makes possible comparative studies but that does not mean that engineers perceive and think in the same way as American engineers perceive and think. A small-scale study of differences in design thinking patterns among American, Chinese, Korean, and Taiwanese students found evidence that merits further investigation to the effect that "Eastern and Western designers use different patterns of design thinking" (Okudan et al., 2008). If this were to be a general characteristic, it would have implications for faculty who teach design to mixed classes of Easterners and Westerners with corresponding implications for assessment. It would seem to be a similar problem to that of catering for different learning styles, but perhaps with more difficulty. For example, Cajander, Daniels, and Von Konsky (2011) compared the graduate attributes required by Curtin University in Australia (Exhibit 10.4) with graduates in IT from Uppsala University in Sweden.

The exercise given to students in the "IT in Society" course at Uppsala required the student to apply the Curtin Graduate Attributes too themselves. The students found this difficult to do and had difficulty in rating their own abilities in relation to their level of attribute attainment. The investigators suggested that this was partly due to the fact that the students did not know with whom (professionals, fellow students) they should make a comparison. The students had to select three of the attributes they wanted to focus on during the course. The most common one selected was "professional skills." In a final reflection, most students had difficulty in showing how they had improved their competencies. The investigators concluded that "additional learning experiences and reflection exercises with an early focus on setting goals to improve chosen competencies" was required. They suggest "that students are not likely to reflect on the development of professional competencies unless required to do so, and that students are not likely to incorporate reflection as part of their ongoing professional practice." This finding is likely to be confirmed by research worldwide. In this respect, there is something to be said for the approach used by Turns and Sattler (2012).

Turns and Sattler established a competency-specific engineering portfolio studio in which students selected the competencies they wished to develop. Students were asked "to create a professional portfolio in which they make an argument about their

Graduate Attribute	Descriptor
1. Discipline knowledge	Apply discipline knowledge; understand its theoretical understandings, and ways of thinking. Extend the boundaries of knowledge through research.
2. Thinking skills	Apply logical and rational processes to analyze the components of an issue. Think creatively to generate innovative solutions.
3. Information skills	Decide what information is needed and where it might be found using appropriate technologies. Make valid judgments and synthesize information from a range of sources.
4. Communication skills	Communicate in ways appropriate to the discipline, audience, and purpose.
5. Technology skills	Use appropriate technologies recognizing their advantages and limitations.
6. Learning how to learn	Use a range of learning strategies. Take responsibility for one's own learning and development. Sustain intellectual curiosity; know how to continue to learn as a graduate.
7. International perspective	Think globally and consider issues from a variety of perspectives. Apply international standards and practices within a discipline or professional area.
8. Cultural understanding	Respect individual human rights. Recognize the importance of cultural diversity, particularly the perspective of Indigenous Australians. Value diversity of language.
9. Professional skills	Work independently and in teams. Demonstrate leadership, professional behavior, and ethical practices.

EXHIBIT 10.4. Curtin University's Graduate Attributes that apply across the whole university extracted from Cajander, A., Daniels, M., and Von Konsky, B. R. (2011). Development of professional competencies in engineering education. In: Proceedings Frontiers in Education Conference, S1C-1 to 5.

preparedness to engage in engineering practice. The studio environment provides a supportive social environment for working on the portfolio. In five sessions, students in the studio share and provide feedback to one another on their portfolios. For example, in the second session, participants share and provide feedback on initial drafts of their portfolio statements, while in the fourth session, students share and provide feedback on the first draft of their overall portfolios." The 21 junior and senior students who engaged in one or other of the two studios that were offered had completed a studio in which they had created a "core" engineering preparedness portfolio. Some students had had additional experiences of portfolios. They came from 9 out of the 10 engineering departments in the university, and some were taking double majors.

To achieve the goal, the student chooses a particular competency. The investigators found that some students chose traditional competencies like communication (6), whereas others chose competencies like leadership (5) and adaptability (2), a variety of reasons being offered for their choices. The investigators found that some students had difficulty in selecting one competency on which to focus; this problem has some similarity with the engineering science course described in Chapter 3 because each year some of the students found it very difficult to select a single project in the situation where they were not given a list from which a choice could be made. Turns and Sattler (2012) point out that pedagogies that support individual choice involve a dramatic shift in the power dynamics. One difference between the engineering science model and this one is that the engineering science projects were graded; a poor choice of project could be the difference between pass and fail.

It is of some significance to note the finding that "in choosing a competency and then writing the portfolio content, many students articulated sub-competencies within the main competency" that helped them "balance an emphasis on a single specific competency with a desire to include content from other areas." Notwithstanding this point, the fact that they did this suggests that being aware of the subcompetencies that make up a competency is of value.

While the peer learning component was valued, the investigators did not find out as much about what the students learnt from their peers about their competencies as they would have wished. It is a subject for further investigation. At the same time, they have offered a challenge to faculty by opening up the question of the student's role in the selection of competencies they wish to emphasize.

10.9 Expressive Outcomes

It has been pointed out that in the drive for measurable outcomes, little if any notice has been taken of the criticisms that were made of *The Taxonomy*, possibly because they were made primarily by persons working in the field of school education, and possibly because of the attraction of the positivist model to engineers. Yet a better understanding of what can and cannot be achieved by an outcomes model might have been gained. The term "outcomes" in its traditional meaning is something that happens as a result of something that is done. It may be intended or unintended. Thus, when we speak of learning outcomes in the context of the objectives movement, we mean "intended learning outcomes." But often the attainment of these outcomes is accompanied by the attainment of unintended outcomes that may or may not be beneficial. Eisner (1979), while accepting the case for preformulated goals, argued that there were many activities for which we did not preformulate goals. We undertook them in the anticipation that something would happen even though we could not specify what. For example, we do not think much beyond the data, even though we could predict from the ample criteria at our disposal. What we do is to evaluate retrospectively what happened against these criteria. From this, he deduces that teachers should be able to plan activities that do not have any specific objectives. He calls these expressive activities. "Expressive activities precede rather than follow expressive outcomes. The tack to be taken with respect to the generation of expressive outcomes is to create activities which are seminal; what one is seeking is to have students engage in activities that are sufficiently rich to allow for a wide, productive range of valuable outcomes. If behavioral activities constitute the algorithms of curriculum, expressive activities and outcomes constitute their heuristics."

It seems that the studio activities of Turns and Sattler (2012) are expressive activities. The problem is that even to decide on an activity is to express a goal so that the activity has a focus, rather like wide beam radio telescope that collects much more from the sky than a narrow beam telescope, some of which turns out to be of value. Turns and Sattler had a broad aim that was nevertheless focused, and much work in education follows from "focusing objectives" (Heywood, 1989).

Eisner (1979) would probably have said to those who want to provide well-defined learning outcomes to research-based final year projects such as Thambyah (2011; see Section 6.2) that they should leave things as they are on the grounds that detailed assessment taxonomies might limit expression. There is a pressing need to examine the effect of lists of outcomes on general learning and the strategies that students adapt to meet the requirements of assessment.

Miller's (1979) attempt to reconcile the advocates and opponents of behavioral objectives, as they were called, might provide a possible way forward. He wrote: "(1) Objectives expressed in measureable, behavioral terms are appropriate for basic skills and for other areas where there is agreement about the components of an instructional program. (2) For most purposes, behavioral objectives need not be reduced to trivial detail. The degree of specificity may vary and should relate to the purpose of instruction and the understanding of students and instructors. (3) The use of behaviorally stated objectives should be contained in an instructional model which recognizes and provides for individual differences. (4) Complex and long-range objectives should be included in a set of objectives, even though they cannot be described in precise terms or measured with a high degree of accuracy. (5) Educational objectives must be appropriate to the social milieu at a given time, and students should participate with their instructors in finding objectives that make sense to them. (6) In times like the present, when technological and social changes are rapid and the future uncertain, the desired behaviors should be adaptable to situations other than the existing one. The ultimate usefulness of behavioral objectives will depend upon how effectively they may be adapted to quite different learning needs and situations." All of which seems to allow for the expressive and focusing objectives (competencies), as well as competencies that are accidental.

10.10 Discussion

The outcomes movement has come to dominate thinking and practice in education worldwide. The prevailing philosophy takes the inside view of competency development. As part of the development of programs (and courses), faculty seek the opinions of alumni and industrialists as to what is required for work, but they do not make studies of what engineers actually do at work. However, there is a steady, if small, flow of papers on engineering practice that will enable appraisals of outcomes to be made, as well as evaluate the merits of alumni surveys.

Large lists of competencies are generated, but one Spanish study suggests that they can be greatly reduced to a few domains. From the perspective of assessment, shorter lists are to be welcomed although there is a lack of research that would affirm or otherwise that shorter lists are more reliable.

Systems of standards-based grading offer the possibility of the comparison of standards between courses and programs.

Some of the studies, especially the Viennese, paid particular attention to the affective domain. It is difficult to see how "soft"/"professional" skills can be developed without attention to the affective.

While recognizing the case for preformulated outcomes (objectives), a substantial case may be made in certain circumstances for not stating such outcomes in some learning activities.

References

ABET. (1996). *Engineering Criteria 2000 (EC 2000)*. Baltimore, MD: ABET Inc.

ABET. (2010). 2010 Annual ABET report. www.abet.org.

Armarego, J., and Roy, G. E. (2013). Aligning course content, assessment and delivery. Creating context for outcomes based education in Yusef, K. M (ed). *Outcomes-Based Science and technology, Engineering and Mathematics Innovation Practice*. Hersey PA. IGI Global.

Ashworth, P., and Saxton, J. (1992). *Managing Work Experience*. London: Routledge.

Australian Council for Educational Research. (2002). *Graduate Skills Assessment*. Australian Council for Educational Research (03/02).

Baker, D., Horvath, I., Campion, M., Offerman, L., and Salas, E. (2005). The ALL teamwork framework. In: T. S. Murray, Y. Clermont, and M. Binkley (eds.), *International Adult Literacy Survey, Measuring Adult Literacy and Life Skills*. New Frameworks for Assessment. Ottawa: Ministry of Industry.

Balascio, C. (2014). Engineering technology workplace competencies provide frameworks for evaluation of student internships and assessment of ETAC and ABET program outcomes. In: Proceedings of Annual Conference of the American Society for Engineering Education, June 2014. Paper 9236.

Benlloch-Dualde, J. V., and Blanc-Clavero, S. (2007). Adapting teaching and assessment strategies to enhance competence-based learning in the framework of the European convergence process. In: ASEE/IEEE Proceedings Frontiers in Education Conference, S3b- 1 to 6.

Bilyadinova, A., Dukhanov, A., Bocherina, K., Krzhizhanovgkaya, V., Boukhanovsky, A. V., and Sloot, P. M. A. (2014). Dutch–Russian double degree master's program curricula in computational science and high performance computing. In: ASEE/IEEE Proceedings Frontiers in Education Conference, pp. 1275–1282.

Bologna. (1999). The Bologna Process. Towards the European Higher Education Area. http://cc.europa.cu/education/policies/educ/bologna/bolgna_en.html.

Bologna. (2005). Working group on qualifications frameworks. A framework for qualifications of the European Higher Education Area. http://www.bologna-bergen2005.no/Docs/00-main_doc/050218_QF_EHEA.pdf.

Brumm, T. J., Hanneman, L. F., and Mickelson, S. F. (2006). Assessing and developing program outcomes through workplace competencies. *International Journal of Engineering Education*, 22(1), 123–129.

Bucciarelli, L. L., Coyle, E., and McGrath, D. (2009). Engineering education in the US and the EU. In: Christensen, S. M., Delahouse, B., and Meganck, M. (eds.), *Engineering in Context*. Aahuis, Denmark, Academica.

Cajander, A., Daniels, M., and Von Konsky, B. R. (2011). Development of professional competencies in engineering education. In: ASEE/IEEE Proceedings Frontiers in Education Conference, S1C-1 to 5.

Cannon-Bower, J. A., Tannenbaum, S. I., Salas, E., and Volpe, C. E. (1995). Defining competencies and establishing team training requirements. In: R. Guzzo and E. Sals (eds.), *Team Effectiveness and Decision Making in Organizations.* San Francisco, CA: Jossey-Bass.

Carberry, A., Lee, H.-S., and Ohland, M. (2010). Measuring engineering students design self-efficacy. *Journal of Engineering Education,* 99(1), 71–79.

Carberry, A., Swan, C., and Ohland, M. (2010). First year engineering students' engineering epistemological beliefs. In: Proceedings of Annual Conference of the American Society for Engineering Education, June 2010.

Carberry, A. R., Siniawski, M. T., and Dionisio, D. N. (2012). Standards based grading. Preliminary studies to quantify changes in affective and cognitive student behavior. In: ASEE/IEEE Proceedings Frontiers in Education Conference, pp. 947–951.

Carter, G., Heywood, J., and Kelly, D. T. (1986). *A Case Study in Curriculum Assessment. GCE Engineering Science (Advanced).* Manchester: Roundthorn Publishing.

Chadha, D., and Nicholls, G. (2006). Teaching transferable skills to undergraduate engineering students. Recognizing the value of embedded and bolt-on approaches. *International Journal of Engineering Education,* 22(1), 116–122.

Cowan, J. (2006). *On Becoming an Innovative University Teacher,* 2nd edition. Buckingham: Open University Press.

Downey, G. L., Lucena, J. C., Moskal, B. M., Parkhurst, R., Bigley, T., Hays, C., Jesiek, B. K., Kelly, L., Miller, J., Ruff, S., Lehr, J. L., and Nicholas-Belo, A. (2006). The globally competent engineer. Working effectively with people who define problems differently. *Journal of Engineering Education,* 95(2), 107–122.

Dublin. (2004). Joint quality initiative group. Shared Dublin descriptors for the bachelor's, master's and doctoral awards. Working document 2004, www.tcd.ie/teaching-learning/academic.../assets/.../dublin_descriptors pdf 2004.

Edwards, M., Tovar, E., and Soto, O. (2008). Embedding core competence curriculum in computer in engineering. In: ASEE/IEEE Proceedings Frontiers in Education Conference, S2E-15 to 20.

Eisner, E. W. (1979). *The Educational Imagination. On the Design and Evaluation of School Programs.* New York: Macmillan.

Figl, K., and Motschnig-Pintrik, R. (2008). Researching the development of team competencies in computer science courses. In: ASEE/IEEE Proceedings Frontiers in Education Conference, S3F-1 to 6.

Ford, N. J., and Rennie, D. M. (1999). Development of an authentic assessment scheme for the professional placement period of a sandwich course. In: J. Heywood, J. M. Sharp, and M. T. Hides (eds.), *Improving Teaching in Higher Education.* Salford: University of Salford, Teaching and Learning Quality Improvement Scheme.

Griffin, C. (2012). A longitudinal study of portfolio assessment to assess competence of undergraduate student nurses. Doctoral dissertation, University of Dublin, Dublin.

Gupta, B. L., and Earnest, J. (2010). Competency profile of technical education managers of India. In: ASEE/IEEE Proceedings Frontiers in Education Conference, S3J-1 to 6.

Heywood, J. (1970). Qualities and their assessment in the education of technologists. *International Bulletin of Mechanical Engineering Education,* 9, 15–29.

Heywood, J. (1986). Toward technological literacy in Ireland: an opportunity for an inclusive approach. In: J. Heywood and P. Matthews (eds.), *Technology and Society in the School Curriculum: Practice and Theory in Europe*. Manchester: Roundthorn.

Heywood, J. (1989). Problems in the evaluation of focusing objectives and their implications for the design of systems models of the curriculum with special reference to comprehensive examinations. In: ASEE/IEEE Proceedings Frontiers in Education Conference, pp. 235–241.

Heywood, J., and Kelly, D. T. (1973). The evaluation of course work—a study of engineering science among schools in England and Wales. In: ASEE/IEEE Proceedings Frontiers in Education Conference, pp. 269–276.

Hiles, D. A., and Heywood, J. (1972). Teacher attitudes to projects in "A" level engineering science. *Nature*, 236, 61–63.

Kabicher, S., and Motschnig-Pitrik, R. (2009). What competences do employers, staff and students expect from a computer science graduate? In: ASEE/IEEE Proceedings Frontiers in Education Conference, W1E-1 to 6.

Klein-Gardner, S. S., and Walker, A. (2011). Defining global competence for engineering students. In: Proceedings of Annual Conference of the American Society for Engineering Education, June 2011. Paper 1072.

Kolb, D. A., and Fry, R. (1975). Towards an applied theory of experiential learning. In: C. L. Cooper (ed.), *Theories of Group Processes*. Chichester, UK: John Wiley & Sons.

Kumar, M. N., Garimella, U., and Nalla, D. (2014). Enabling higher order thinking and technical communication—An Indian context for OBE. In: ASEE/IEEE Proceedings Frontiers in Education Conference, pp. 195–201.

Lucena, J., Downey, G., Jesiek, B., and Elber, S. (2008). Competencies beyond countries: the reorganization of engineering education in the United States, Europe and Latin America. *Journal of Engineering Education*, 97(4), 433–448.

Lunev, A., Petrova, I., and Zarpova, V. (2013). Competency-based models of learning for engineers: a comparison. *European Journal of Engineering Education*, 38(5), 543–555.

Male, S. E., Bush, M. B., and Chapman, E. S. (2011). An Australian study of generic competencies required by engineers. *European Journal of Engineering Education*, 36(2), 151–163.

McDermott, R., Daniels, M., and Lárusdöttir, M. (2014). Subject level quality assurance in computing. In: ASEE/IEEE Proceedings Frontiers in Education Conference, pp. 1189–1196.

McDonald, F. J. (1968). *Educational Psychology*. Belmont, CA: Wadsworth.

Mickelson, S. K., Brumm, T. J., Hanneman, L. F., and Steward, R. L. (2003). The data in workplace competencies in the experiential workplace. In: Proceedings of Annual Conference of the American Society for Engineering Education. Paper 852.

Mickelson, S. K., Hanneman, L. F., and Brumm, T. (2002). Validation of work-place competencies sufficient to measure ABET outcomes. In: Proceedings of Annual Conference of American Society for Engineering Education. Paper 1607.

Miller, R. I. (1979). *Assessment of College Performance*. San Francisco, CA: Jossey-Bass.

Mistree, F., Ifenthaler, D., and Siddique, Z. (2013). Empowering engineering students to learn how to learn: a competency based approach. In: Proceedings of Annual Conference of the American Society for Education. Paper 7234.

Motschnig-Pitrik, R., and Figl, K. (2007). Developing team competence as part of a person centered learning course on communication and soft skills in project management. In: ASEE/IEEE Proceedings Frontiers in Education Conference, F2G-15 to 21.

Okudan, G. E., Thevenot, H., Zhang, Y., and Schiirman, M. (2008). Culture and systems of thought: a preliminary investigation on implications for the design process and artifacts. *International Journal of Engineering Education*, 24(2), 285–303.

Oladiran, M. T., Pezzotta, G., Uzjak, J., and Gizejowski, M. (2013). Aligning an engineering education to the Washington Accord requirements: example of the University of Botswana. *International Journal of Engineering Education*, 29(6), 1591–1603.

Olague-Cabllero, I., and Valles-Rosales, D. J. (2014). Rounding up the industrial engineering educational profiles with adaptive soft skills framed by a cultural competency approach in an industry–university partnership. In: Proceedings of Annual Conference of American Society for Engineering Education, June 2014. Paper 9643.

Organization for Economic and Cultural Development. (2002). *Definition and Selection of Competencies. Theoretical and Conceptual Foundations*. Paris, France: Organization for Economic and Cultural Development.

Parkinson, J., Harb, S., and Magelby, S. (2009). Developing global competence in engineers. What does it mean? What is most important? In: Proceedings of Annual Conference of American Society for Engineering Education, June 2009.

Reigeluth, C. M. (1999). What is instructional design theory and how is it changing. In: C. M. Reigeluth (ed.), *Instructional Design Theories and Models Vol. II*. Mahwah, NJ: Lawrence Erlbaum Associates.

Rogers, C. (1961). *On Becoming a Person*. Boston: Houghton Mifflin.

Steinaker, N., and Bell, M. R. (1969). *An Experiential Taxonomy*. New York: Academic Press.

Thambyah, A. (2011). On the design of learning outcomes for undergraduate engineer's final year project. *European Journal of Engineering Education*, 36(1), 35–46.

Thompson, M. K., Clemmensen, C. H., and Ahn, B.-K. (2013). Effect of rubric rating scale on the evaluation of engineering design projects. *International Journal of Engineering Education*, 29(6), 1490–1502.

Tovar, E., and Soto, O. (2010). The use of competences assessment to predict the performance of first year students. In: ASEE/IEEE Proceedings Frontiers in Education Conference, F3J-I to 4.

Turns, J., and Sattler, B. (2012). When students choose competencies: insights from the competence-specific engineering portfolio studio. In: ASEE/IEEE Proceedings Frontiers in Education Conference, pp. 646–651.

Tuning. (2005). Tuning educational structures in Europe project. Approaches to teaching, learning and assessment in competence based degree programmes. http:/www.uniedusto.org/tuning

Washington Accord. (1989). Washington Accord. http//www.washingtonaccord.org/Washington-accord.

Woollacott, L. C. (2009). Validating the CDIO syllabus for engineering education using the taxonomy of engineering competencies. *European Journal of Engineering Education*, 34(6), 545–559.

Youngman, M. B., Oxtoby, R., Monk, J. D., and Heywood, J. (1978). *Analysing Jobs*. Aldershot: Gower Press.

11

"Outside" Competency

The differences between "inside" and "outside" competencies are briefly discussed. The overall purpose of this chapter is to indicate the significance of "outside" competency for engineering education. It begins with a summary of the concept of "accidental" competencies that has affinities with the hidden curriculum and expressive outcomes.

The belief that students can be prepared for work immediately on graduation is challenged. A phenomenological study of engineers at work is reported by Sandberg that offers an "outside" view of competency. It is found to be context dependent and a function of the meaning that work has for the individual involved. Engineers doing the same job were found to have different perceptions of work, and competencies were found to be hierarchically ordered among them, with each level being more comprehensive than the previous level. Attributes are developed as a function of work. It follows that they are not fixed; therefore, firms should undertake training (or professional development) beginning with an understanding of the conception that the engineer has of her or his work. Professional competence should be regarded as reflection in action or understanding work or practice rather than as a body of scientific knowledge.

Little has been known about how engineers utilize the knowledge learnt in their educational programs at work. A study is reported by Kaplan and Vinck that affirms previous findings that engineers tend to use off-the-shelf solutions or start

The Assessment of Learning in Engineering Education: Practice and Policy, First Edition. John Heywood.
© 2016 The Institute of Electrical and Electronics Engineers, Inc. Published 2016 by John Wiley & Sons, Inc.

with an analogy of an existing solution for a different problem. The same authors noted that engineers switch between scientific and design modes of thinking. Yet another study reported the view that engineers who are contextually competent are better prepared for work in a diverse team.

Support for the view that a university course cannot by itself prepare a student for immediate work as an engineer is to be found in Blandin's theory of the development of competence. The co-op structured course in France from which Blandin obtained the data for his theory enables technicians to qualify as technologists. Blandin and his colleagues found that the cognitive dimension of competence had five main competency indicators. One "acting as engineer in an organization" was found to be more important than the others and was thus the core competency. The core competence develops only within the company and cannot exist without long experience within a company.

Adult learning and development are little discussed in the literature of engineering education, but there is parallel between Blandin's model and Torbert's taxonomy of developmental positions for professionals, which is described. The chapter concludes with a further description of the Alverno model and likens the Alverno subability levels to the levels of competence described by Sandberg. The Alverno curriculum takes a holistic view. It is difficult to believe that the engineering curriculum does the same and helps the "dispositional growth of a person's broadly integrated way of making meaning and commitments in moral, interpersonal, epistemological, and personal realms." In such a curriculum, assessment moves away from the traditional, to what is commonly called "authentic," and from the responsibility of the tutor to the responsibility of the student.

11.1 Introduction

So far the text has shown how the rationalist approach to the declaration of objectives has become the *sine qua non* of engineering education. Overall, little attention had been given to the affective domain and its extension into personality. Engineering education might be described as being professionally centric as opposed to person centric, vocational as opposed to liberal. Awards, so important to credentialing, are now being structured from collections of components. The trouble about this, as Sadler (2007) opines, is that putting the pieces together so that they provide a meaningful learning experience "that prepares learners to operate in intelligent and flexible ways is very difficult."

One effect of these developments has been to allow industry to believe that students can be prepared immediately for work in industry and that they have no role in the induction of a person into their organization. They too have adopted a rationalist approach in the determination of what they may require in terms of competencies. The "inside" view of competence development is deeply embedded in the Western psyche. The psychological concepts associated with this "inside" model of competency arise from deeply entrenched cultural models of intelligence (Plaut and Markus, 2005, cited by Griffin, 2012) which make it difficult for faculty to change their attitudes to teaching and assessment (Borrego et al., 2013[1]). As Griffin writes, "Machine metaphors, common in western conceptions of the mind and thinking, also define what is involved in being a competent person. In many European and American cultural contexts, the person is represented and realized as a separate bounded, autonomous entity, that is, an individual. Individual actions result from attributes of the person that are activated and, then cause behavior. Accordingly competence is 'located in the individual'. Individual actions result from the attributes of the person that are activated and then, cause behavior. Accordingly competence is located 'in' the individual, 'in' the mind, 'in' the brain. This view of competence is active, as in the machine metaphor, it cranks, works, churns, and then out comes the solution to the problem."

In contrast, the "outside" view of competence considers competence to develop through relationships with others in their social worlds. Griffin writes that in "clinical practice, student nurses will not master the clinical environment independently, they will seek the social engagement of others to obtain feedback on whether they are meeting the expectations and standards of others in order to become members of the community." She argues that inside theories of competence produce inside models of assessment with emphasis on performance rather than competence *per se*.

In this chapter, the "inside" model is challenged first by the phenomenological alternative offered by Sandberg (2000), and second by Blandin's (2011) study of a post-technician cooperative apprenticeship.

Sandberg argues that competency is context dependent that has considerable implications for the meaning of a competency and the extent to which it is generalizable. Furthermore, how an individual demonstrates a competence at work depends on how he or she perceives that work. This has considerable implications for that person's education that needs to become more person-centered. Sandberg points out that it also has implications for the task of management. Blandin found the core competency of "acting like an engineer" developed within the company and could not exist without long experience in a company.

Attention has been drawn to the fact that very little notice has been taken of relevant research in adult education. In this respect, Torbert's developmental theory, which is in five main stages (frames), holds much promise for the understanding of adult behavior at work. The third frame is called "technician." To this frame belong many professionals, including engineers. They are persons who are narrowly focused on efficient methods and the internal logic of objective standards. It is suggested that industry wants people who will function in this frame and not seek advancement to the higher frames. Finally, a theory of holistic development arising from the Alverno (1994) development is briefly discussed. It is suggested that the rationalist approach has not helped engineering educators grasp the importance of the person or to help the dispositional growth that integrates their ways of making meaning and commitments in the moral, interpersonal, epistemological, and personal realms.

This chapter begins with a reminder of the significance of the hidden curriculum for understanding industry.

11.2 Accidental Competencies

As is well understood in school education, the way schools and classrooms are organized, the way teachers teach, the organization of the timetable, the rules for discipline, and the way they are implemented, all create learning. Students learn how much they can get away with when Teacher A is in command and find it is much more than when Teacher B is in control. They learn what subjects they are good and bad at and they learn ways to study, none of which are catered for in the formal system of study. These are things learnt from the "outside" and may have a powerful influence on learning, Educators call this the "hidden curriculum" (Eggleston, 1977; Jackson, 1968). Some have gone so far as to say that it is more important than the formal curriculum and that it embraces all aspects of life. If it is grasped in college and it is understood that all organizations have their equivalent, a powerful competency is given to students with which to better understand the organizations they join on leaving college. Similarly, if they are able to observe (reflect on) the accidental competencies that occur while they are in the educational system, they should also be able to transfer that learning to the organizations in which they work.

The concept of "accidental competencies" is due to Walther and Radcliffe (2006). It has affinities with Eisner's (1979) concept of expressive outcomes and the idea of the hidden curriculum. They define "accidental competencies" as "attributes that are not achieved through targeted instruction. They are acquired through the unintentional coactions of curricular elements or aspects surrounding the educational process. In that Accidental Competencies can be included in the stated learning outcomes but can in some cases go beyond that scope." The relevance of Eisner's concept of expressive outcomes to the idea of accidental competence will be apparent. Walther and Radcliffe obtained data from two focus groups, each of three students or young professional engineers using the critical incident technique (Flanagan, 1949). They give this example of an accidental competency from a transcript:

> The chaotic system of my degree structure with parallel courses and conflicting constraints was in retrospective a blessing. Today I am able to organize myself in similar conditions, manage my time and access information through networking with others.

They suggest that this example identifies three Accidental competencies:

1. The ability to make sense of work in complex systems
2. Ability to interact socially and build relations in order to gain information or advice (the need to be able to do this was seen at all levels of work in the Youngman et al. (1978) study)
3. Time measurement

It is quite evidently a competency that derives from understanding the hidden curriculum. The example is also a reminder that university systems are complex, a point that is continually emphasized by Walther and Radcliffe. In their analysis, they argue that since there is a "competency gap between education and practice," and since there is only limited agreement about what should be catered for at university, a search for accidental competencies should lead to a closing of that gap, or as I would argue an understanding of the relative responsibilities of academia and industry in the development of students and personnel. They propose a contextual model of competence formation; how that would fit into developmental theories of competence is another matter. What is it, therefore, that is particular to industry?

11.3 Understanding Competence at Work

During the last decade, there has been an increasing interest in what it is that engineers actually do as opposed to studies of the opinions of alumni about their curriculum needs that have been made at regular intervals during the last 50 years and continue to be made (e.g., Huff et al., 2012). Sandberg (2000) looked specifically at the notion of competence and was led to a general theory of what competence is from a study of engineers of work. It makes uncomfortable reading for accrediting agencies and industry since it challenges the view that competencies are a set of attributes that can be developed in college for immediate use in industry.

It seems that many of us working in engineering education, including this writer, missed this study of engineers at Volvo. There is, for example, no mention of it in "Engineering Practice in a Global Context" (Williams, Figureido, and Trevelyan, 2013). Sandberg (2000), using a phenomenological approach to the study of engineers in a Volvo engine plant, sought to understand the nature of competence among a group of 20 engineers called "optimizers." This term derived from their task which was to develop engines "by optimizing a range of qualities such as drivability, fuel consumption, emissions and engine power." From this and other descriptive passages in the paper, it may be inferred that they were a group of high-level technicians. Sandberg's questions asked in interviews sought to establish what the task meant to the optimizers. The key question was "What is a competent optimizer for you?"

He deduced from their answers, and observations of them at work, three meanings or conceptions of what optimization meant. The first conception revealed was that of engineers who optimized the parameters separately. One parameter, for example, fuel consumption, is tested and when it is optimized, the engineer moves to optimize the next parameter. This leads to a definition of competence that is "the ability to analyze and interpret how one of several monitoring parameters have influenced engine quality" (Sandberg, 2000). It "implies an understanding of how the qualities of an engine react to changes in the parameter." With experience, the engineers acquire substantial tacit knowledge that helps them judge what parameters to adjust and by how much.

A second group of optimizers perceived the problem quite differently. They looked at the engine as a system of interacting parameters. So the question they asked: what will happen to the other parameters if I adjust parameter X? At the same time, they have to know what the right order for optimizing the parameters is. There is a skill in seeing the links between the parameters that are enhanced by tacit knowledge. Both of these groups of engineers need to be able to understand and develop monitoring systems. Sandberg draws attention to the attribute of being interested in engines and self-teaching. This group of "optimizers not only develop their own knowledge, but also include other optimizers in the learning process, so a shared understanding is built up about this ongoing work" (Sandberg, 2000).

The third concept of optimization found among the optimizers is the "customer's experience of driving." For these optimizers, the qualities of the engine and the customer's requirements are related. In order to achieve this goal, they have to know how influencing quality affects the final result, that is how the customer will experience the engine. "Although the attribute of understanding and developing monitoring systems is also essential for these optimizers, its meaning differs from that given by those who expressed the first and second conceptions. The third group wants to allow for all situations in which customers drive a car by adjusting suitable parameters and by developing new operations within the monitoring system that are better suited to meeting a particular customer requirement" (Sandberg, 2000).

Given what is known about perception and the way individuals perceive the same situation differently (Hesseling, 1966; Heywood, 1989), it is not surprising that individuals should view the same job differently. More surprising perhaps is Sandberg's interpretation that the level of competence required differs between the three types. As Sandberg points out, they represent a hierarchy of competence in which each level is more comprehensive than the previous. It is influenced by the knowledge of the engine that the groups have. Expectancy theory tells us that our dispositions to work stem in no small part from the meaning that work, and therefore the type of work, has for the individual. Thus, one of the most powerful influences on our learning behavior is our work expectations (Daniel and McIntosh, 1972; Heywood, 2009).

Sandberg's central conclusion has profound implications for the education and training of engineers. It is that "human competence is not primarily a specific set of attributes. Instead, workers knowledge, skills and other attributes used in accomplishing work are preceded by and based on their conceptions of work. More specifically the findings suggest that the basic meaning structure of workers *conceptions of work constitute human competence*" (Sandberg, 2000). The basis of training should be with the conception that the worker has of his or her job that is the reason why in elementary and secondary schools individual education plans have been introduced. This is quite contrary to the idea that attributes can be developed independently of the work organization in college because it is clear that performance of a work-based competency is a function of the perceptual match that a worker has with the task. Competency is context dependent. To understand one's place within a competency and how one could be repositioned demands reflective capability in order to be able to self-assess. Moreover, taking this view would account for some of the difficulties that young graduates find when they enter the workplace (Korte, 2009). They have not been prepared to develop *contextual competence* that is to negotiate the context in which they find themselves, hence the value of internships that provide students with a variety of tasks (see Chapter 12 and Parsons, Caylor, and Simmons, 2005).

11.4 Contextual Competence

A weakness of the Youngman et al. (1978) study is that while it provided a systematic methodology for establishing what engineers do in organizations, it did not address what knowledge was used in the solving of the problems with which they were faced.

Evidently they did not think this was part of their role. However, they did refer to a study by Langton (1961) that derived a theoretical curriculum for polymer technologists from detailed descriptions of the techniques, processes, instrumentation, and design procedures used by firms through interview and brief observation of the technologists involved. The "pictures" he obtained are in considerable detail and enabled him to derive a curriculum. Studies of this kind have considerable bearing on curriculum debates about what should and should not be included in the curriculum, and in this respect Kaplan and Vinck (2013), in one of a number of studies, brought together in *Engineering Practice in a Global Context* (Williams, Figueiredo, and Trevelyan, 2013) take us a long way forward.

Kaplan and Vinck (2013) reported two case studies in which engineers were engaged in a completely new field (digital humanities). They remind us of the need to specify the field of engineering for its engagement either with a novel idea or with traditional engineering. They contrast digital humanities with civil engineering. In their present and an earlier investigation of designers in the car industry, Kaplan and Vinck showed that engineers tend to reuse off-the-shelf solutions. Youngman et al. (1978) found likewise. Kaplan and Vinck also found that the engineers "designed new solutions, using as their starting point the analogy of an existing solution for a different problem" [...] "captivating analogies and generative metaphors seemed to frame problem formulation and problem solving when (these) practitioners are confronted with new situations" (p. 75). The brackets are mine for it may be that this approach arises from the particular technology (context). Youngman et al., in their attempt to test Hesseling's (1966) theory of autism arising from specialism, found a group of engineers who became stuck when past solutions to similar problems were unsuccessful in solving a new problem that was similar but on a very different scale. The problem was, of course, solved but it was not part of Youngman et al.'s brief to establish how. It would have been nice to posit as Kaplan and Vinck reported of their engineers that "they engaged in in-depth theoretical thinking and strategic thinking above any established knowledge or rule of work" (p. 76) to solve the problem. The problem remains that little is known about how and when engineers use the knowledge they obtained on educational programs in the solution of problems. There are clearly competencies that are psychological at work, some of which are related to their levels of adaptability and flexibility. Like the other authors in this book, Kaplan and Vinck confirm the "classical results in the social study of engineering practices: social actions between heterogeneous actors, negotiation on the purposes of the client and on technical aspects (possible solutions and their validity), distributed expertise [...], work on ill-defined problems, influence of the various perspectives of the actors involved, creativity and uncertainty, the opportunistic and contingent nature of all the moves" (p. 75). Missing from that are the underlying psychological mechanisms that relate to the individuals learning on the one hand, and on the other hand how the individual interacts with the social and organizational environment. Sandberg's study is helpful in this respect.

Behind Youngman et al.'s brief attempt to try to understand the process of innovation in terms of how engineers applied the principles learnt in their engineering science courses to the solution of a major design problem was clearly the view that an engineer would be rational and solve them in this way. That is, they would operate in a scientific mode in the fashion of Kelly's (1955) model of the individual as a scientist but as Kaplan and Vinck make clear, this model is untenable because engineers switch between scientific and design modes of thinking, behaving as a function of the problem they have to solve. As Borgford-Parnell, Diebel, and Atman (2013) point out in the same review, context is all important.

Borgford-Parnell, Diebel, and Atman (2013) report a study in which they analyzed a team's attention to context and group issues during an engineering design meeting. Everyone was found to pay attention to contextual issues and two members of the seven-person group showed the "kind of contextual competence needed in design teams" (p. 96). They argued that their findings supported the contention of Palmer et al. (2011), who concluded that engineers who are contextually competent would be better prepared to work in a diverse team. This seems to be the view of the National Academy of Engineering's report *The Engineer of 2020* (2005). It is, of course, the argument of liberal educators who follow in the tradition established by John Henry Newman, but in that tradition it applies to any graduate from any subject (Heywood, 2010; see also EHEI Section 5.6). Palmer et al. wrote, "Close working relationships with industry may also help faculty to recognize the 'expectations' for contextual competence that industry employers are coming to expect from their new industry hires." Furthermore, the curriculum should "promote interdisciplinarity and general education components that can enhance students' ability to wrestle with global and local, social, ethical, political, economic and environmental impacts of engineering practices."

These findings are rather broad and seem to take us outside of the engineering context but are nevertheless important. For example, unpublished case material available to this writer suggests that part of the learning is about grasping the perspectives that persons bring from other engineering disciplines to the solution of the problem. There is a general competency that is the ability to understand and respect another person's point of view, a competency that was written into the history curriculum in Ireland at the height of the "troubles."[2]

Borgford-Parnell and his colleagues argue that the tools developed by Atman and her group over the years for analyzing engineering design practice are valuable pedagogical tools that can be used to show students what is happening to them during the design process. This, they argue, is far better than telling them that this is what happens in a lecture. They also think that they would help project managers broaden their ideas of the skills and knowledge areas that are needed, and this fits with the Youngman et al. idea of developing labor arenas (see Section 4.4).

Support for Sandberg's theory as well as for some form of cooperative education is to be found in a study by Blandin of a post-technician course in France.

11.5 A Post-Technician Cooperative Apprenticeship

One major research that shows that there are "things" that can be learnt only in industry was conducted in the French organization CESI'*Ecole d'ingénieurs*. This college for engineers was created by five firms in the French automotive industry in 1958 in order to provide manufacturing technicians through 3 years of continuing education and training with technologist (graduate engineer) status. The program is given the title "Apprenticeship Program in Engineering." Necessarily, the program helps change the identity of the student from being a technician to being an engineer. The school had described the learning outcomes in terms of the competencies expected of its students in 1997. In 2006, the French national accreditation body *Commission du titre d'ingénieurs* asked the School "(1) How do you know that the learning outcomes achieved are those described in terms of the competencies in your competency framework? (2) How do you assess that this is so?" (Blandin, 2011). A research program established to answer these questions by the School's Research Department in Educational Sciences described by Blandin (2011) is summarized in what follows.

As indicated at the beginning of this study (Section 1.2), the terminology in use is very confused and for this reason the CESI researchers were careful to begin their study by defining competence.

Blandin comments, "[…] it can be said that 'competency' is an ingredient of 'competence'. But competence is more than the sum of its ingredients. In fact when compiling the literature, it appears that 'competence' develops and is recognized at three levels: at the level of the individual, at the level of the group in which the individual works, and at the level of the organization in which he/she works. At the level of the individual, competence has a cognitive dimension […] the result is a demonstrated level of proficiency and a feeling of self-efficacy." (The importance of self-efficacy is to be found in a number of studies; for example, it is posited that cooperative education may increase the self-esteem/self-efficacy of women and contribute to their retention in engineering; Raelin et al., 2007).

Blandin continues, "at the group level, competence has an identity level: when an action is performed according to the best practices in use in a professional group, it is recognized as such by peers […] it generates a feeling of belonging to the group." Korte (2007) has noted the neglect of "group identity" in organizational literature. Blandin continues, "At the organizational level, competence has an institutional dimension." It is part of "the legitimate field of actions of the person" and may be recognized by "title, level of salary, field of action." The model of "competency" presented takes into account these three dimensions. Hence, knowledge, skills, and procedures relate to and are affected by the context and indicators of competence. This model was confirmed by a longitudinal study across the whole 3-year curriculum. Blandin and his colleagues found that the cognitive dimension of competence had five main competency indicators. One "acting as an engineer in an organization" was found to be more important than the others and was thus the core competency. Exhibit 11.1 shows the dimensions of this core competency.

For convenience, the five competency indicators are listed in Exhibit 11.2, which also shows their relationship to Australian, Irish, and UK competencies. It also shows that CESI was able to demonstrate that the CESI competencies were the same as for an engineer studying a more traditional curriculum, with the exception that the CESI curriculum did not take into account the ethical considerations that were incorporated in the others.

As a result of the longitudinal study, Blandin concludes that there are four main steps in the development of competence that "appear" to be driven by the core

Skills	to follow the rites to apply tacit rules to apply the usages, the best practice		
Rules or procedures	Corporate regulation Standards Laws and regulation		
Context	Daily activity	Small or big project	In/out of normal field of competence
Indicators	Quality Reliability Cost Time allowance		

EXHIBIT 11.1. The core competency of an engineer "Acting as an Engineer in an Organization" extracted from Blandin, B. (2011). The competence of an engineer and how it is built through an apprenticeship program: a tentative model. *International Journal of Engineering Education*, 28(1), 58–71.

competence—"acting as an engineer in an organization." It "drives the socialization/insertion process within the company. It also triggers then fosters the development of the managerial competency (mobilizing human resources appropriate for action). It makes the development of the other competencies necessary. It also maintains students' motivation for learning, at least for learning what they feel useful at the moment to solve

CESI Competency	Australian Standard (Engineers Australia)	Irish Competency (Engineers Ireland)	UK Competency (Engineering Council)
Acting as an engineer in an organization	C2	3 and 4	C and D
Mobilizing various cognitive resources for the action	C1	1	A
Using a way of reasoning appropriate to the action	C1	2	B
Mobilizing human resources appropriate for the action	C3	3 and 4	C and D
Utilizing instruments appropriate to the action	C1.1b	1 and 2	A and B

EXHIBIT 11.2. List of CESI competencies and Blandin's comparison with Australian, Irish, and UK Standards. Extracted from Blandin, B. (2011). The competence of an engineer and how it is built through an apprenticeship programme: a tentative model. *International Journal of Engineering Education*, 28(1), 57–71.

the problems posed by their acting in professional situations." It seems reasonable to infer a parallel with the development process inherent in the first three stages of Torbett's (1987) taxonomy of developmental positions for professionals (see Section 11.6 below). The student in this system is continually faced with challenges that are not beyond his or her capabilities to overcome. This means that the period in industrial practice has to be carefully designed and on an individual basis. Blandin notes that the learning situations are closely related to an individual's identity development. Blandin draws attention to the similarities between the developmental process observed and Vygotsky's concept of the zone of proximal development. Each time the organization allows the student an enlargement of his or her responsibility, it "will de facto create a new zone of proximal development beyond the current stage of competency development."

In so far as Blandin's enquiry was concerned, the core competence "develops only within the company and cannot exist without long experience within a company." Within the company, the students developed competencies that were specific to their job. The writer takes from this finding that the interaction between periods of academic study and industrial work help students acquire professional competence in professional engineering that is not available to courses of the traditional kind that have no industrial contact.

Blandin's theory provides some indirect support for a theory of competency derived from the study of engineers at work by Sandberg. In the same way, Sandberg's theory lends support to Blandin's view. Together, they show that the development of competence is something that extends beyond college into the normal practice of work where context is all important.

11.6 Theories of Competence Development in Adult Life

While the studies reported above were being made, there were developments in the field of adult learning especially in regard to the understanding of the evolution of professional competence and its assessment (McAuliffe, 2006; McClelland, 1976; Rogers, Mentkowski, and Hart, 2006). While engineering educators are familiar with and often cite Schön's work, there is no evidence that they have knowledge of the work of either Kegan (1982; 1994) or Torbett (1987). Torbert's taxonomy of developmental positions for professionals, especially managers which comprises five main frames, is of particular interest. There is a similarity with the Perry model (Section 7.4) in that the frames move from stages of concreteness and conformity to capability in abstraction and a willingness to tolerate ambiguity. Torbert's first frame is called "opportunistic." To simplify, the current way of knowing of the opportunistic is the only way to view the world. According to McAuliffe (p. 487), opportunistic persons "experience others without empathy, as objects to be manipulated" and "tend to use force and deception to reach short-term ends." One positive aspect is that their self-interest can force them to become entrepreneurs (my interpretation). The second frame is called "Diplomatic." Professionals operating in this frame are "company men" who have loyalty to the rules of the organization, but they find it difficult to make awkward decisions. The third frame is called "technician." To cite McAuliffe, "Technician professionals are narrowly focused on efficient methods and the internal logic of objective standards. In the process techni-cians fail to see the larger systems of which they are part, for they are enamoured of the consequence of their own doctrines. To them, there is no room for alternate explanations. Their logic is the only logic. Fisher and Torbert (1995) propose that technicians embrace of 'standards' can be inspiring for co-workers" (p. 489). Technician professionals are single-minded. This frame provides an explanation of the professional behavior of many engineers. It also describes the largest single group of professionals. It is arguable and a hypothesis that requires testing that industrialists want engineers to function at this level and not to advance above it.

The other frames are "Achiever," which is held to be a wider-frame that of the "technician" and "strategist." They are guided by the goals of the field beyond their own career expectations and can provide leadership. McAuliffe does not mention McClel-land's (1971) achievement motivation in his discussion of this frame but it clearly has a bearing. He points to the negative dimension that achievers are likely to pursue their agenda to the exclusion of other goals and alternatives although being open to feed-back they can be open to learning. The next frame is the "strategist." McAuliffe uses a comment by a manager in one of Torbert's papers to describe the strategist as moving from "having very explicit goals and timetables [and] a structured organization to...the collaborative process [which] focuses on inquiry, constructing shared meanings from experience and building consensus through responsible interaction" (p. 491). Each frame is more comprehensive than the one that precedes it.

McAuliffe writes "for those who would be highly competent experts and leaders, learning should lead them toward strategist thinking with its emphasis on dialogue, expe-rience and self-reflection. Instead, however, current educational and in-service training programs teach to the technician worldview, with its ideological tunnel vision and dis-interest in stepping outside of professional standards. Thus, such professionals remain embedded in the usual practices of their own fields and are less attuned to the situational contextual dynamics that professionals must account for in good portion." He argues that expertise depends as much on the ability of "how to know" as it does on "what to know." So a key question for educators who want their students to develop expertise is, to what extent do they help their students acquire the skill of "learning -how-to-learn"?

McAuliffe concludes his review with the comment that "the minimal capacity for competence seems to lie in the professional's ability to attain relative autonomy from the rules and norms and the ability to rely on self-defined procedures to make decisions."

Reference has already been made to Sandberg and Pinnington's (2009) studies of engineers and lawyers and the hierarchies of competence they drew from their stud-ies. Sandberg and Pinnington argue that competence is only partially understood if it is attributed solely to scientific and tacit knowledge, knowing-in-action, and understanding of work or practice. They suggest that they do not explain how knowledge and under-standing are integrated into a specific professional competence in work performance. They suggest that an "existential ontological perspective" may resolve this problem. This may be difficult for some engineering educators since Sandberg and Pinnington's theory derives from the work of the German philosopher Martin Heidegger, who many philoso-phers think is extremely obscure (Collinson, 1987; Johnston, 2006). He was concerned with "being" and not "personal existence": his method was that of phenomenology.

Collinson writes, "From the phenomenological point of view the world is the condition we engage with and inhabit; it is the constitutive of our lives. We are not to see the world simply as a physical object against which we are set as individual thinking subjects: rather, we are "beings-in-the-world" and Dasein, our human reality or mode of being, is that multitude of ways in which we inhabit life; that is, by "having to do with something, attending to something, attending to something and looking after it, making use of something, giving something up and letting it go, undertaking, accomplishing, evincing interrogating, considering discussing" (Heidegger, 1962, p. 83). In this sense, it is reasonable to infer that the levels of competence described by Sandberg for engineers and Sandberg and Pinnington for lawyers are different ways of "being," and that the "whole" is greater than the sum of its parts. The same may be said of Torbert's frames. Attention to "being" necessarily links the cognitive with the affective.

The central argument presented by Sandberg and Pinnington is that there is a need to regard professional competence as knowing in action or understanding work or practice rather than as a body of scientific knowledge. One thing is certain, if a person is to move to a higher and more comprehensive level they require skill in learning-how-to-learn which is part of the developmental dimension of professional competence. Movements from one level to another require a deep structural change and such change increases the capacities to both learn and reflect (McAuliffe, 2006).

The levels of competence that Sandberg and Pinnington describe have many similarities with the levels of ability that make up the generic ability domains of the Alverno curriculum with the exception that they are job or practice oriented. But the Alverno model of the curriculum goes far beyond the description of ability domains, and the levels that lead to a comprehensive performance in those domains, because it seeks to link the pursuit of these abilities to personal growth. Mentkowski and associates (2000) model the four domains referred to above in four quadrants. The upper left is "reasoning," the upper right "performance," the lower right–"Self-reflection" and the lower left "development." At the center is the active learner. At the North of the central axis, the focus is on competence which is distinguished from the world of inner meaning at the south. "Performance and reasoning are about mastery of the external world and frameworks for understanding it" (Rogers, Mentkowski, and Hart, 2006, p. 499). To the right of the east/west dimension is the context within which the person finds themselves. To the left are the individual's dispositional thinking, feelings, and choices that shape the flow of his or her experience—the structures of the person. Taken together, they are, the authors say, a model of the whole person but they also seem to this writer to convey a notion of "being." It is the notion of "being" that seems to be absent from much engineering discourse.

In their model, "holistic" is not the same as "being" or the "whole person." Rather, it denotes "the overall direction of dispositional growth in the person's broadly integrated way of making meaning and commitments in moral, interpersonal, epistemological, and personal realms" (Rogers, Mentkowski, and Hart, 2006, p. 498). They examine the relationship between multidimensional performance and holistic development. Performance is multidimensional because it takes into account all the "individuals intentions, thoughts, feelings, and construals in a dynamic line of action and his or her entanglement in an evolving situation and its broadest context" (p. 498) such as in the family or at work. They conclude that sometimes multidimensional performance and holistic development can be mutually reinforcing domains of growth. They note that adult holistic development is sometimes helped by broadening the role of the professional. To move from one level to the next would involve a perceived change of role for the engineer in Sandberg's model even though the organization may not think of it as such. But they do not think that specialized knowledge by itself is likely to be related to holistic development.

From the perspective of education, Mentkowski and her colleagues believe their research to have shown that student learning outcomes should be made explicit. In this, they agree with many other educators but they add the caveat that they "should be open and explicit." "Students," they say, "can meta-cognitively take hold of their learning by using these performance expectations to self-assess their abilities and construct a vision of how to perform processes that they can carry with them after college" (Rogers, Mentkowski, and Hart, 2006, p. 527). Apart from the necessity of formative assessment, the achievement of such goals will demand changes in assessment away from the traditional to what is sometimes called authentic, and from the responsibility of the tutor to the responsibility of the student.

But discussion of the developments of the rationalist approach in engineering does not seem to either be aware or accept the holistic view that is presented, since it is difficult to accept that engineering education helps the "dispositional growth of a person's broadly integrated way of making meaning and commitments in moral, interpersonal, epistemological and personal realms," which is what Newman thought higher education should do, if in a different language to Rogers, Mentkowski, and Hart.

11.7 Discussion

This chapter has examined competency from the perspective of the "outside" view and shown it to have a valid claim on our considerations. The conflict between the "inside" and "outside" views of competency is reminiscent of the "nurture"/"nature" debate on intelligence. Like the outcome of that debate, we are likely to end up with a theory of both!

Sandberg's research suggests that competence is a function of the context in which a person has to utilize that competence. Professional competence may be regarded as reflection in and for action conditioned to some extent by accidental competencies. Clearly, a person begins a new stage of development when they enter industry. Treating the person as if they already possess the competencies required will be to the detriment of the industry concerned. For industry to improve its performance, it has to understand the nature of the competencies that specific jobs create. So to obtain the most from a person, industry needs to understand how that person perceives the work he or she does and devise training or professional development that starts from that base. The acquisition of expertise in any job is a developmental process that the organization can either enhance or impede. This is not to say that the educational system has no role in the development of competencies we call generic. Indeed, it has but that role is preparatory. Given that organizations are also learning systems and a major function of education is learning then learning-how-to-learn has a key role to play in that preparation as does assessment when it is designed to enhance that learning.

Much of what is learnt in college is through the hidden curriculum or the informal organization of college. Competencies are learnt accidentally and learners are helped if they know this happens and recognize them when they do.

Much of what Sandberg concluded seems to be confirmed by recent studies of the development of competence in adulthood. His hierarchical theory of competence relates to the subability levels in the Alverno abilities. Alverno's approach is holistic and in that sense counter to prevailing rationalistic competence models that believe teaching via the parts will cause the attribute required to be developed. The Alverno assessment-led liberal arts curriculum was established to help women prepare for significant roles in the world of work. Being committed to a liberal arts program means that in preparing individuals for professions such as nursing and teaching its principal focus is on "being" and the "person." The professional is developed from the person. In contrast, many engineering programs focus on the professional at the expense of the person.

These findings have profound consequences for the assessment and credentialing of individuals. Major questions have to be asked, as for example, given that formal engineering education is a component of a developmental process, what is the purpose of the final grade awarded by a university for the satisfactory completion of undergraduate education? At the very least apart from the necessity of formative assessment the achievement of such goals will demand changes in assessment away from the traditional to what is sometimes called authentic, and from the responsibility of the tutor to the responsibility of the student.

Notes

1. Borrego et al. (2013) reported on a study of engineering statics instructors' beliefs about teaching and learning. They found among a national sample of statics faculty in the United States ($N = 166$ representing 22% of the invited faculty) who responded to a survey that there was a weak to moderate relationship between beliefs about learning and classroom activities. They were able to supplement this finding with data from interviews carried out in two institutions with academics teaching the course, their departmental chairs, engineering dean, the engineering undergraduate dean, and directors of STEM teaching and learning centers. In summary, they wrote, "Instructors demonstrated deep understanding of the content (including important conceptual difficulties students frequently encounter) and the developmental needs of their students. They were also aware of some of the shortcomings of lecture-based modes of instruction and rote problem solving. However, they struggled with their role in making changes to engage students more actively in learning during formal class time." The paper contains detailed descriptions of the beliefs they found these faculty had, and includes a substantial bibliography. The authors point out that those concerned with faculty development need to understand the beliefs that faculty bring with them when they seek help to change their teaching.

   The matter goes beyond beliefs about teaching because we also have views about the aims of education and the curriculum. There are conflicting visions about the aims of education and the curricula they inspire. See Schiro, M. S. (2013). *Curriculum Theory: Conflicting Visions and Enduring Concerns*, 2nd edition. Los Angeles: Sage, for an American perspective.

2. A term given to a period in Irish History from the early 1970s to the mid-1990s that related to a war between the Irish Republican Army (catholic) against the British in general and the Ulster protestants in particular with a view to uniting Ireland. Several thousand people were killed in the process. The attitudes on both sides were very hard, hence the belief of the writer of the Intermediate Certificate in History Examination that a major objective should be that students should learn not only to understand another person's point of view but to respect it.

References

Alverno. (1994). *Student-Assessment-as-Learning at Alverno College*. Milwaukee, WI: Alverno College Institute.

Blandin, B. (2011). The competence of an engineer and how it is built through an apprenticeship program: a tentative model. *International Journal of Engineering Education*, 28(1), 57–71.

Borgford-Parnell, J., Deibel, K., and Atman, C. J. (2013). Engineering design teams. Considering the forests and the trees. In: B. Williams, J. Figueiredo, and J. P. Trevelyan (eds.), *Engineering Practice in a Global Context. Understanding the Technical and Social*. London: CRC Press/Taylor and Francis.

Borrego, M., Froyd, J. E., Henderson, C., Cutler, S., and Prince, M. (2013). Influence of engineering instructors' teaching and learning beliefs on pedagogies in engineering science courses. *International Journal of Engineering Education*, 29(6), 1456–1471.

Collinson, D. (1987). *Fifty Major Philosophers. A Reference Guide*. London: Routledge.

Daniel, W. W., and McIntosh, N. (1972). *The Right to Manage*. London: Macdonald and James.

Eggleston, J. (1977). *The Sociology of the School Curriculum*. London: Routledge.

Eisner, E. W. (1979). *The Educational Imagination. On the Design and Evaluation of School Programs*. New York: Macmillan.

Fisher, D., and Torbert, W. R. (1995). *Personal and Organizational Transformation. The True challenges of Continuous Quality Improvement*. New York: McGraw Hill.

Flanagan, J. C. (1949). A new approach to evaluating personnel. *Personnel*, 35–42.

Griffin, C. (2012). A Longitudinal Study of Portfolio Assessment to Assess Competence of Undergraduate Nurses. Doctoral Dissertation. Dublin, University of Dublin.

Heidegger, M. (1962). *Being and Time*. Oxford: Blackwell. Translated by J. Macquarrie and E. Robinson.

Hesseling, P. (1966). *A Strategy for Evaluation Research*. Assen, the Netherlands: Van Gorcum.

Heywood, J. (1989). *Learning, Adaptability and Change. The Challenge for Education and Industry*. London: Paul Chapman/Sage.

Heywood, J. (2010). Engineering literacy: a component of liberal education. In: Proceedings of Annual Conference of American Society for Engineering Education. Paper 1505.

Huff, J. L., Abraham, D. M., Zoltowski, C. B., and Oakes, W. C. (2012). Adapting curricular models for local service-learning to international communities. In: Proceedings of Annual Conference of American Society for Engineering Education. Paper 4167.

Jackson, P. W. (1968). *Life in the Classroom*. New York: Holt, Rinehart and Winston.

Johnston, D. (2006). *A Brief History of Philosophy. From Socrates to Derrida*. London, Continuum.

Kaplan, F., and Vinck, D. (2013). The practical confrontation of engineers with a new design endeavour. The case of the digital humanities. Chapter 3 of Williams, B., Figueiredo, J., and

Trevelyan, J. (2014). Engineering Practice in a Global Context. Understanding the Technical and the Social. Leiden, CRC Press. (eds.), […].

Kegan, R. (1982). *The Evolving Self*. Cambridge, MA: Harvard University Press.

Kegan, R. (1994). *In Over Our heads. The Mental Demands of Modern Life*. Cambridge, MA: Harvard University Press.

Kelly, G. A. (1955). *The Psychology of Personal Constructs*. Vols. 1 and 2. New York: Norton.

Korte, R. F. (2007). A review of social identity theory with implications for training and development. *Journal of European Industrial Training*, 31(3), 166–180.

Korte, R. F. (2009). How newcomers learn the social norms of an organization: a case study of the socialization of newly hired engineers. *Human Resource Development Quarterly*, 20(3), 285–306.

Langton, N. (1961). *The Teaching of Theoretical Subjects to Students of High Polymer Technology*. 2 Vols. A Report to the Nuffield Foundation. London: The Northern Polytechnic (now University of North London).

McAuliffe, G. (2006). The evolution of professional competence. In: C. Hoare (ed.), *Handbook of Adult learning and Development*. Oxford University Press.

McClelland, D. C. (1971). *Assessing Human Motivation*. New York: General Learning Press.

McClelland, D. C. (1976). *A Guide to Job Competency Assessment*. Boston: McBer.

Mentokowski, M., and associates (2000). *Learning that Lasts. Integrating Learning, Development, and Performance in College and Beyond*. San Fransisco: Jossey Bass.

Palmer, B., Terenzini, P. T., McKenna, A. F., Harper, B. J., and Merson, D. (2011). Design in context. Where do engineers of 2020 learn this skill? In: Proceedings of Annual Conference of American Society for Engineering Education, June 2011. Paper 2129.

Parsons, G. K., Caylor, E., and Simmons, H. S. (2005). Cooperative education work assignments. The role of organizational and individual factors in enhancing ABET competencies and coop workplace well being. *Journal of Engineering education*, 94(5), 309–318.

Plaut, V. C., and Markus, H. R. (2005). The "Inside" story. A cultural historical analysis of being smart and motivated. American style. In: A. J. Elliot and C. S. Dweck (eds.), *Handbook of Competence and Motivation*. New York: The Guilford Press.

Raelin, J., Reisberg, R., Whitman, D., and Hamann, J. (2007). Cooperative education as a means to self-efficacy among sophomores (with particular attention to women) in undergraduate engineering. In: ASEE/IEEE Proceedings Frontiers in Education Conference, F1G-20 to 24.

Rogers, G., Mentkowski, M., and Hart, J. R. (2006). Adult holistic development and multidimensional performance. In: C. Hoare (ed.), *Handbook of Adult Development and Learning*. New York: Oxford University Press.

Sadler, D. R. (2007). Perils in the meticulous specification of goals and assessment criteria. *Assessment in Education: Principles, Policy and Practice*, 14(3), 387–392.

Sandberg, J. (2000). Understanding human competence at work. An interpretive approach. *Academy of Management Journal*, 43(3), 9–25.

Sandberg, J., and Pinnington, A. H. (2009). Professional competence as ways of Being: an existential perspective. *Journal of Management Studies*, 46, 1138–1170.

Torbett, W. R. (1987). *Managing the Corporate Dream*. Homewood, IL: Dow-Jones Irwin.

Trevelyan, J. (2007). Technical coordination in engineering practice. *Journal of Engineering Education*, 96(3), 191–104.

Trevelyan, J. (2010). Restructuring engineering from practice. *Engineering Studies*, 2(3), 175–195.

Trevelyan, J. (2014). *The Making of an Expert Engineer*. London: Taylor and Francis Group.

Walther, J., and Radcliffe, D. (2006). Engineering education: targeted learning outcomes or accidental competencies. In: Proceedings of Annual Conference of American Society for Engineering Education, June 2006. Paper 1889.

Williams, B., Figueiredo, J., and Trevelyan, J. P. (2013). *Engineering Practice in a Global Context. Understanding the Technical and Social*. London: CRC Press/Taylor and Francis.

Youngman, M. B., Oxtoby, R., Monk, J. D., and Heywood, J. (1978). *Analysing Jobs*. Aldershot: Gower Press.

12

Assessment, Moral Purpose, and Social Responsibility

12.1 Introduction

A great deal of time and effort worldwide is being put into redesigning courses to meet the accreditation requirements of the appropriate agencies (Section 6.1). Because these agencies focus on assessment as the mechanism for evaluating courses and assessment for evaluating student learning, it is appropriate that they themselves should be examined. In this text, I have tried to do a number of things, but in a nonlinear way for they are all entangled with each other. First, I have added to E. J. Furst's (1958) model of what a professional teacher should possess. Namely a defensible theory of assessment (perhaps a better word is "philosophy"), in addition to defensible theories of philosophy and learning. The aim of the text has been, therefore, to provide sufficient information for an engineering educator to acquire a defensible theory or philosophy of assessment.

Examining assessment is one way of focusing on the curriculum because it cannot be divorced from learning, instruction, and content. It is integral to the curriculum process yet so, often, assessment is the afterthought of the educational process. Focus on assessment forces us to consider in detail the aims and objectives of programs and courses if our curriculum activities are to have validity. As such it is also a way of focusing on the curriculum and the process that gives it life. It is with these that this chapter is primarily concerned, internally with the problems of the curricula we have, and externally with the dictates of the sociotechnical system in which we live.

The Assessment of Learning in Engineering Education: Practice and Policy, First Edition. John Heywood.
© 2016 The Institute of Electrical and Electronics Engineers, Inc. Published 2016 by John Wiley & Sons, Inc.

12.2 Moral Purpose and the Power of Grading

In the last two sentences of the second paragraph of Chapter 1, I wrote, "When we go to the surgeon's clinic we expect to see his credentials, for that is what the accumulated certificates are, hanging on a wall. Should we not expect that from engineering educators?" One of the reviewers of this text queried "engineering educators." He thought I meant engineers. I did not, although it applies equally. I meant engineering educators because their grading decisions have the power to influence the life decisions that students and new graduates make and these are as important for young people as the life-and-death decisions that surgeons have to make. Assessment has a moral purpose that is as important as its pragmatic objective. While the literature has little to say about how teachers view the moral dimensions of assessment, there is no reason to believe that they do not take the assessment of student performance seriously. But they would seem to do so without attention to the psychometric principle of validity. While for many teachers, the setting of examinations and tests is an afterthought of the educational process, they are faced with the fact that in meritocracies assessments influence the way that students think about learning. Many students, as a matter of strategy, respond by doing what they believe the assessors want, often to the detriment of in-depth learning. There is no room for complacency about assessment. It is for this reason that "assessment" has a moral purpose as well as a "social responsibility." If engineering educators had an accepted code of conduct, this would be a major canon in that code (Cheville and Heywood, 2015; Riley et al., 2015).

12.3 From Reliability to Validity: Toward a Philosophy of Engineering Education

The major change in our understanding of assessment began in the 1950s when engineering educators began to question the validity of what was being tested. They accepted that some forms of testing, notably standardized objective tests, could be reliable but began to appreciate that there were a whole lot of issues surrounding the question of validity (see Sections 1.4 and 1.9). Their thinking was anticipated by a group of American educators who questioned if we knew exactly what it was we were examining (testing) (Bloom et al., 1956). This group described the cognitive skills that they called behavioral objectives that educators should aspire to teach. In a *Taxonomy of Educational Objectives for the Cognitive Domain*, they gave examples of how these skills might be tested. As indicated in Chapter 2 (note 3), Dressel (1971), even after the second volume on the affective domain had been published, was vociferous in his criticism of *The Taxonomy* for its underestimation of the role of values in human behavior. Currently, it has been argued that ABET has equally misunderstood this point in its new proposals for engineering criteria (see Section 12.10). The Alverno curriculum shows that valuing cannot be separated from decision making (Mentkowski and Associates, 2000; see Section 12.7).

In 1962, in England a psychologist showed that in one of the élite mechanical engineering departments the examinations in the different engineering science subjects tested the "same thing" that he called the ability to pass examinations (Furneaux, 1962). Inspired by the *Taxonomy*, one of the Matriculation Boards allowed examiners to design an examination in engineering science that would obtain specified objectives especially one's related to what engineers do as well as to higher-order thinking (Appendix B). The idea of "objectives" was slowly embedded in educational systems but it began to be replaced by "outcomes" in the late 1980s although some thought there was no difference between the two concepts. The concept of competency introduced in medicine decades ago has also become part of the engineering vernacular and some authors now use it as an alternative to "outcomes."

But in the 1960s lists of skills that went beyond the cognitive domain of *The Taxonomy* and embraced what is loosely called the "affective," that is the motivation, attitudes, and values required of students and for work were also compiled in the United Kingdom (Exhibits 3.7 and 5.4) and in the United States (Exhibit 2.5). A second volume of *The Taxonomy* for the affective domain was published in 1964 (Krathwohl et al., 1964; Exhibit 4.5).

All of this is rehearsed in Chapters 1, 2, 3, and 5 in some detail. It is recalled here for two reasons. First, because subsequent lists that have been and continue to be described are but variants of the lists referred to above, a fact that raises the question as to whether anything has been achieved or changed. Clearly, very many teachers are engaged in trying to make their courses meet these requirements although their understanding of validity is not psychometrically robust. At the same time, considerable steps have been made in our understanding of the curriculum process, how students develop (Chapter 7), how they learn, and in the context of this book that assessment is a complex activity that has a major impact on learning and the design of the curriculum. But these understandings have yet to become part of an agreed corpus that every engineering educator should know.

The second reason is that these developments came in a period when higher education systems were expanding in response to political perceptions of the needs of the economy, and when the concept of market freedom was the dominant philosophy. The philosophy that governed education was primarily utilitarian and it was assumed by policy makers that there was no need for any substantial debate about the aims of education. Those who spoke up for the liberal tradition in the United Kingdom were ignored, whereas in the United States, engineering students continued to have to participate in programs that included general education and ethics.

In sum, while instruments can be designed that are reliable for them to be valid, they have to be based on a philosophy about the aims of education. In so far as the United States is concerned, the recent proposals to revise the ABET criteria make such a debate an imperative (Slaton and Riley, 2015). But the challenge is a world-wide one.

12.4 Screening the Aims of Engineering Education

In so far as engineering education is concerned, a major effect of the utilitarian philosophy of engineering education arises from the relentless search for profits by large organizations. It is that graduates should be prepared to work immediately in industry.

Complaints about the quality of graduates, often contradictory, have continued through-out the period covered by this book particularly in the United Kingdom and the United States (e.g., Chapters 2, 4, 5, and 9) and in spite of substantial attempts by some institu-tions to better prepare graduates for work these complaints continue (e.g., McCahan and Romkey, 2014). Within the framework of a utilitarian philosophy the matching of the outputs of education to the inputs (needs) of industry might be regarded as its moral pur-pose. A failure to meet this obligation would be a moral failure. In this case, the purpose of assessment is to ensure quality. These complaints have been made in the absence of any theory of integration of the practical (industrial) and the academic in student learn-ing. One reason for this is that the aims of education, such as they are, have not been exposed to what E. J. Furst (1958) calls screening (Heywood, 1981). By this he means the use of philosophy and other social sciences to "select a small number of important and consistent goals that can be attained in the time available." For example, if only one in four graduates from STEM courses takes up STEM jobs, what is the most appropriate engineering education for such graduates? Cheville (personal communication) reframes the question to ask what proportion of a graduating class needs to go to a specific job in order to have preparation for that job as a component in the curriculum? Answers to that question in the United States or wherever it holds might require substantial studies of the jobs engineers do, or it might be reframed more generally in terms of the aims of higher education. In any case, if the education they receive is to help them with their future careers, assessment should surely ensure that they are able to think critically and transfer knowledge and skill.

The whole of this text may be regarded as an exercise in screening. It has attempted to convey a picture of what is happening in classrooms and schools of engineering and also to document what are perceived to be the strengths and weaknesses of these activities. Impressionistic though it may be, it finds that the accrediting agencies are having a profound impact on what individual faculty and departments do. That is, much time is spent some would say an inordinate amount of time, by faculty showing how the program outcomes dictated by the agencies are being met which involves faculty in making substantial lists of outcomes to fit within domain categories. Others concentrate on drawing up lists of learning outcomes. The striking thing about many of the papers that describe such activities is that, as indicated above, little or no attention is paid to validity either of learning outcomes or assessor's assessments. It seems to be accepted that if a learning outcome is stated that it is necessarily valid. Yet one or two papers show that face validity judgments are questionable. That is, neither students and, in some cases, teachers perceive these (or some of these) statements of outcomes to have the same meaning that their designers intended (see Section 6.3; Squires and Cloutier, 2011). In the absence of evaluation, the validity of what is presented is questionable. Morally teachers have to be able to justify the outcomes they use by some other method than the acceptance of face validity judgments. There remain issues about the validity of the techniques of assessment used to ensure that the outcomes are obtained. In spite of many criticisms that examinations and tests do not assess the higher-order thinking skills, and in particular, the ability to solve "wicked problems" relatively little attention is paid to the design and evaluation of the items (questions) that make up such tests.

12.5 The Role of Educational Institutions in the Preparation for Industry (the Development of Professional Skills)

A great deal of activity is focused on developing professional skills within courses. The question arises as to whether they can prepare students immediately for work in industry. Overall the papers examined in this text suggest that there are limits to what college education can achieve and one paper suggests that even in project-based team activities students can acquire or reinforce attitudes already acquired that are inimical to industry's requirements (see Section 9.5; Leonardi, Jackson, and Diwan, 2011). For this reason, its authors argued that selectors should recruit students who already had experience of industry, which draws attention to the value of internships (see Section 2.2). Macmurray (1958) offers an epistemology as to why this might be as does Kolb's experiential model of learning (see Section 7.2; e.g., Chan, 2012). Both theories start with concrete experience. Macmurray argues that in contrast to Western philosophy, which places theory before practice, all our theoretical activities derive from our need to solve practical problems. Given that this is the case then there is a basis for a theory of integration. An internship prior to academic study should provide experience that leaves questions unanswered or explanations not fully understood that would be illuminated in college. It goes without saying that an arrangement in which the experience begins in college can serve the same purpose. In either event, this places an obligation on industrial organizations to work with colleges to ensure that the desired integration happens. Given that the self is agent, an obligation is also placed on the company to work with the student to achieve mutually agreed goals.

If students receive a prior induction course that indicates what they should look out for in an internship, then much more can be made of the academic course, for example management studies that address the person as well as the organization during the academic program. More can be made of the internship if the students are also introduced to the notion of reflection (see Chapter 7). It may be argued that if the curriculum follows the Kolb cycle of experiential learning that only a period in industry is likely to provide the experience that is necessary for reflection provided the academic course is designed to draw out that capacity (*educare*). This principle may be extended to cooperative courses. If an industrial period is arranged to follow an academic period, then it becomes possible to provide the students with insights that will help them understand more easily how the organization functions and how engineering theories play out in practice. Such insights cannot come by osmosis from working in teams in college.

An investigation by Korte (2009) of the experience of the workplace of young graduates shows how this might work. A student told him that he "wish(ed) someone had told him how to play the political game here." He needed to know about the informal organization of the company. "Informal Organization" is one of the earliest ideas of organizational theory that is not much considered today, but from the perspective of a person beginning a new job, it is a very useful concept. Understanding the informal networks and how individuals within them function (behave) leads to an understanding of how power is distributed and influences the movement of the organization toward or away from its goals, as well as how personality influences behavior (Heywood, 1984; Jackson, 1968). It has its equivalent in the hidden curriculum in school. Korte (personal communication) would argue that it is possible to use the students' experiences of the hidden curriculum to sensitize them to what they are likely to experience in industry and how to interpret it.

Educational activities may help prepare the students for industry and industry can contribute to them as well as to the provision of industrial experiences. For example, Barry et al. (2008) completely redesigned a traditional lecture course in water and waste water treatment so that the students would learn about professional competences in addition to content. They called the new approach "challenged-based," which is a project-based approach to problem-based learning in that "instruction begins with the presentation of a term project as a culminating event for the students' learning." Formative assessment involved the students in helping with the design of the rubric that, it was argued, would help them think about their own criteria for success. The final assessment was an oral defense of their work by the students individually with the instructors. "Each student acted as a representative of their consulting company and was asked to discuss the alternatives presented in the RFP (Request for Proposal) and the recommendations made by their consulting company. Students were asked to defend their company's recommendation or discuss why they personally feel their company may have proposed the wrong design recommendation." Two instructors graded the oral (viva) against predefined criteria. Apart from an evaluation of pre- and postcourse assessments, the authors reported that an exit survey revealed that the students thought the most significant aspects of the experience were related to professional competencies. In this case, assessment enhances learning particularly as students were involved in some of its design.

Elsewhere another consortium has developed an Integrated Design Engineering Assessment and Learning System (IDEALS), which targets professional development. The professional skills assessment instruments have been piloted in six different universities across the United States. Students perceived that the instruments added value to their program (McCormack et al., 2011). As indicated previously, McCahan and Romkey (2014) have proposed a Taxonomy for Teaching Engineering Practice, and Thomas and Izatt (2003) have proposed a Taxonomy of Engineering Design Tasks.

An unusual suggestion comes from Trevelyan (2010). His studies of engineers at work reveal another key work skill (competence), that of "technical coordination" that involves the skill of collaboration (see Section 4.6). He likens this activity to that of a teacher. Trevelyan says, *"first the engineer describes what needs to be done and when* (the process of lesson planning), *and negotiates a mutually agreeable arrangement with other people* (students), *who will be contributing their skills and expertise. Next, while the work is being performed, the engineer* (teacher) *keeps in contact with the people* (students) *doing the work to review the work and spot misunderstandings or differences of interpretation. The engineer* (teacher) *will also join in discussions of unexpected issues that arise and may need to compromise on the original requirements* (task objectives). *Third, when the work has been completed, the engineer* (teacher) *will carefully review the results* (assess) *and check that no further work or rectification is necessary."* One might add that an effective engineer (teacher) will continually evaluate his or her own performance (self-accountability). (The brackets are mine: for detailed descriptions of lesson planning; see Heywood, 2008.) I have suggested that there is much to be gained if engineering students are able to teach in high school or college a subject like technological and engineering literacy (Heywood, 2015). Elsewhere Davies and Rutherford (2012) have described how engineering students in the United Kingdom can learn from fellow students who already have experience of industry.

12.6 The Role of Industry in Professional Development

Fundamental to the outcomes approach is a particular Western view of learning. That is, outcomes are abilities/skills that can be developed in students. They are innate individual properties of the person (the "inside" model of competence). Criticisms of this position seem to have emanated at about the same time as the publication of the ABET Criteria. Sternberg's (2005) model of the development of abilities into competencies and competencies into expertise was published later in 2005. Nevertheless at the same time there was a substantial literature on competency-based assessment particularly in respect of medicine that should have been noted (Griffin, 2012). Had the term competency been used instead of outcomes, as it is now beginning to be used, a better understanding of what is achievable in college and what industry has to contribute might have emerged. As it is the "inside" model within a utilitarian philosophy leads to the treatment of individuals as commodities. Industry seeks from governments policies that allow it to treat people as commodities. It does not care whether engineers call themselves professionals or not. It wants them to do particular tasks and education is training them to perform these tasks which may be at a lower level than their qualifications merit.

However, a phenomenological study of engineers at work by Sandberg (2000) showed quite clearly that a particular competence required by a firm in the automobile industry was context dependent (See Chapter 11). It is representative of the "outside" model of competence development (Griffin, 2012). This study also found that this particular competence was possessed at different levels of skill among the engineers performing the task with which it was associated. The task had to be done before training needs could be identified and decisions made about the added-value of training, if any. There is always learning on the job and some of this is continuous, which is why firms are learning organizations and they can organize themselves so that they enhance or impede learning (Youngman et al., 1978; Senge, 1990). Those that impede learning will inevitably die. College programs can prepare students for organizational learning, but they cannot provide the situations in which engineers will find themselves because the competences required will be particular to the organization concerned (e.g., Mistree, Ifenthaler, and Siddique, 2013; Section 9.6).

Blandin (2011; see Chapter 11) gives a different meaning to "levels" to Sandberg. He writes, "[…] it can be said that 'competency' is an ingredient of 'competence'. But competence is more than the sum of its ingredients. In fact when compiling the literature, it appears that 'competence' develops and is recognized at three levels: at the level of the individual, at the level of the group in which the individual works, and at the level of the organization in which he/she works. At the level of the individual, competence has a cognitive dimension […] the result is a demonstrated level of proficiency and a feeling of self-efficacy." The importance of self-efficacy is to be found in a number of studies in the engineering education literature (e.g., Raelin et al., 2007).

The model of "competency" presented takes into account these three dimensions. Hence the knowledge, skills, and procedures relate to and are affected by the context and indicators of competence. Blandin and his colleagues found that the cognitive dimension of competence had five main competency indicators. One "acting as engineer in an organization" was found to be more important than the others, and was thus the core competency. It "develops only within the company and cannot exist without long experience within a company." Within the company, the students who were technicians seeking to graduate as engineers developed competencies that were specific to their job. The writer takes from this study support for the view that the interaction between periods of academic study and industrial work helps students acquire professional competence that is not available to courses of the traditional kind that have no industrial contact.

It is concluded that the development of professional competence is as much the responsibility of industry as it is educational establishments, and this requires a joint approach to assessment in which a portfolio is likely to have a major role.

12.7 Assessment and the Curriculum

A major issue is the extent to which systems of assessment similar to those developed by ABET cause creative and innovative approaches to assessment, and therefore, to the student's experience of learning. At the time of writing, ABET has expressed disappointment at the failure of EC 2000 to achieve that goal.[1] But such systems can lead to the generation of long lists of outcomes that too readily become checklists with the accompanying rigidities that checklists cause.

A Spanish investigation conducted surveys of teacher coordinators and teachers to determine the core competencies in computer engineering. Edwards, Tovar, and Soto (2008) found 37 core competencies. They ranked and rated the competencies for the degree to which they were taught in classes (high, medium, basic, or null). As a result, they were able to reduce the list to five core competencies. In their structure, analysis is considered to be of the same dimension as synthesis, and "soft" skills (now known as professional skills) are confined to forms of communication (see Section 10.6).

However, a reduction in the number of competencies is unlikely to solve the problem unless it takes into account how students learn and develop. Indeed, one response to ABET's lament is that it developed its framework in the absence of any well-supported framework of learning and development. The presentation of ABET's new proposals at the annual conference of the American Society for Engineering Education (June 2015)[1] suggests that this lesson has not been learnt. It revealed that the tension between educational research and practice has yet to be resolved so that it becomes normal practice to base policy on evidence-based analyses of practice. To be fair, ABET commissioned an evaluation study in 2002.[2] There may be a justifiable criticism of this research in that it was conducted before programs had had time to develop. Nevertheless, before EC 2000 there were published demonstrations of practice supported by theory as for example work at the Colorado School of Mines on the application of Perry's theory of development to the curriculum (see Chapter 7). Similarly, a great deal was known about the Alverno College eight-domain ability-led curriculum before *Learning That Lasts* was published in 2000 (Mentkowski and Associates, 2000).

As explained in Section 5.5, every course in the Alverno curriculum is structured so as to enable students to practice the broadly based subabilities of each domain so that they can master the subject matter. Furthermore, the curriculum and its instruction are designed so that the students can develop in each of these ability areas. This form is their general education that links with their professional studies. Within their professional studies, they expect to "hear their major described in terms of these abilities in language more specific to the field" [...] "Within each major, students continue to develop the required broad abilities redefined by the faculty in the disciplines and professions" (Mentkowski and Associates, 2000).

Alverno's reason for establishing this curriculum was the belief that women got a raw deal in the world of work and that they needed to be helped take on that world. The abilities were chosen because they believed this would help women do just that. The domains were (1) Communication; (2) Analyze; (3) Problem solving; (4) Valuing in decision making; (5) Social interaction; (6) Global perspectives; (7) Effective citizenship; and (8) Aesthetic responsiveness (see Exhibit 2.4).

A major feature of the Alverno curriculum is the use of volunteer assessors from outside the college and particularly the world of work. This arrangement helps links liberal education with professional study. There are three aspects of the curriculum that are of interest here. First is that in addition to a normal 4-year program, they also conduct week-end schools for those who are unable to undertake full-time study. Second, they regard anyone who is beyond secondary schooling as an adult, which is a reminder that in the engineering literature there are very few references to adult education (e.g., Harris, 2013; Pembridge, 2013). Yet, it should be evident that there is much that is of value in that literature that is pertinent to engineering education (see Section 11.6) not least changing patterns of employment, and the increasing costs of higher education that if incomes remain relatively static will become beyond the means of many people. Third, and perhaps the most important from the perspective of this book, is that a personal internalization and positive response to this assessment-driven curriculum is essential from each member of the community for the commitment required for the successful operation of such systems depends for their success on the intrinsic motivation of faculty. That is unlikely to be obtained from external systems that necessarily rely on extrinsic motivation and hope that faculty will become intrinsically motivated.

12.8 Changing Patterns in the Workforce, the Structure of Higher Education

In spite of a substantial flow of data to the contrary, many policymakers find it difficult to believe the model of technological employment that says for every job that technology makes redundant it will replace it by another is obsolete. Yet there is a continuing flow of information that supports that contention. Wheeler (2015), for example, writes "the digital economy is vaporizing the good jobs and replacing them with two kinds of jobs: minimum wage jobs (think Amazon warehouse employees) and so-called 'sharing-economy jobs' (think Uber drivers)." In the British Isles, think zero-hours contracts. If this is correct, the middle class will disappear and meaningful occupations will become a thing of the past and what to do about it is the major problem of our time (Lanier, 2013). One critical response to Wheeler's article was that globalization was the culprit and had taken too many highly paid low-skilled jobs from the United States, but the same can be said of Europe. However, the idea that technology has nothing to do with this is fanciful. The purpose of engineering is to make peoples' lives more comfortable and that implies that machines will do work, much of it routine and on occasion unpleasant, that was previously done by people: that is the direction that technology is traveling. Support for this contention is to be found in a report from The (Organization for Economic and Cultural Development 2012), which stated that automation and computerization accounted for as many as 80% of the decline in "labor share," that is the measurement of wages as a proportion of the total income generated in an economy. This was found to be true of 22 of 26 countries evaluated. It is then safe to assume that what is happening in the United States may be followed elsewhere among other industrialized nations.

More generally, in the United States recent evidence on the demand for personnel seems to suggest that the pattern of demand is changing. So, is there a changing pattern in the demand for personnel with technological skills?

If it is possible to extrapolate from the experience of Silicon Valley, then the demand for technological manpower is declining, irrespective of specific shortages. The US Bureau of Labor Statistics recorded for the decade ending 2010 that technoscientific employment fell by 19%, and that average wages in Silicon Valley fell by 14% (Zachary, 2011).

Zachary also wrote that often emerging technologies require far fewer workers (Zachary, 2011). The new titans of Silicon Valley employ far fewer workers than the older titans, and this is likely to apply equally to their offshore establishments. At the same time, some emerging technologies destroy jobs. He also draws attention to the phenomenon of "jobless" innovation. This occurs when an innovation is off-shored to countries where qualified manpower is much cheaper to employ. Zachary goes on to ask, "How can Americans capture more of the employment associated with job expanding innovations. They can start by examining their faith in the traditional equating of technological innovation with healthy markets?"

Related to employment in the software industry is the "E Mail" column of the November 2011 issue of *ASEE Prism*. It contains an exchange of letters between Professor Allen Plotkin and columnist Vive Wadwha about an article that Wadwha had written in the September issue of the magazine (Wadwha, 2011). He had asked, why should a company pay a 40-year-old engineer a considerable salary if it can get the same job done much more cheaply by an entry-level employee? He said that it was happening in the software industry. "After all the graduate is likely to have more up-to-date skills and work harder." An Irish Academic told this writer that firms had said to him that they wanted young graduates who could do the job immediately, but they would keep them only for 7–9 years! Relate that to Wadwha, who said that "if you listen to the heart-wrenching stories of older engineers" (who have become unemployed) "you learn they have a great many skills, but no one wants to hire them." Professor Plotkin questions whether or not anyone would want to work in an industry that treats its workers in the way described in this article. Nevertheless, it seems that there is a serious unemployment problem among middle-aged and older engineers in some sectors of the United States. Wadwha's response is to cite the metaphor of a roller coaster and suggest that the universities need to prepare students for that ride so that when the need arises they are able and interested to change jobs. Hence the need for the formative assessment of transferable skill and reflective judgment, which by its very nature embraces critical thinking. The concept of lifelong learning is being taken much more seriously as the development of MOOCS indicates (Carey, 2015). The possibilities engendered by a digital age are enormous, but courses that support continuing professional development are likely to be as much about personal development as they are about specific topics in engineering and that cannot be achieved simply by an online exchange.

The implications of these findings for educational policy makers are (1) that policy making should be undertaken from a systems perspective that embraces elementary education at one end of the spectrum and lifelong (permanent) education at the other; (2) that engineering educators, together with industrialists, should pay much more attention to lifelong education, and therefore in continuing professional development for engineers in both technical and personal dimensions; and (3) that engineering educators should better prepare students with the skills of flexibility and adaptability required to cope with ever-changing knowledge, that is "personal transferable skills." All of these have implications for the design and structure of assessment and the credentials that accompany it (see below).

12.9 Lifelong Education and Credentialing

The need for divergent visioning of the future of engineering education is based on the premise that the structure of the workforce is changing rapidly and that the model of manpower that argues that each innovation brings with it an increase in the workforce has broken down. This raises the question: What happens to those engineers whose companies wish to replace them with younger engineers after they have spent a decade in the organization? The subquestions are: Are they employable, and if not why not? How does the education they have received contribute, if at all, to their employability? Do they require a 4-year degree to do the work that firms want them to do in their first decade? And, what responsibility, if any, does an employer have for their future employability? What are the long-term effects on companies that adopt these employment models?

There is an assumption in the National Governors Association Report (Sparks and Waits, 2011) that employers know best, but do they? There is little doubt that in the Western world there is considerable dissatisfaction, some may say it is better called distress, with the system of free-market capitalism as it is practiced (Reich, 2012). While it is generally agreed that the Marxist experiment has failed and most of us would not want to be ruled by a committee, many people consider that the concept of a firm as an agent of profit in a free market is irresponsible. They understand that people (individuals) and organizations (firms) exist in a social system and are mutually obligated to each other. Given that this is the case then organizations (firms) have an obligation to accept that they have a social responsibility for an individual's development while he or she is in their care and a broader responsibility to help foster skills in employees that will guarantee them a basis for transferability.

In the twenty-first century, skills and knowledge will become rapidly outdated. Individuals are likely to have to change jobs more quickly than they have in the past. Unfortunately, a recent American study shows that firms require entrants to have specific skills suitable for the particular jobs they have (Rothwell, 2014), and firms complain when universities do not produce such persons. They have to learn that that is not the university's function to meet their immediate needs. Rather, the task of the university is to produce graduates who will be both adaptable and flexible and contribute as much to society as they do to the firm. That means taking a holistic view of education and this view changes not only the way ethics is taught but its content and focus as Kallenberg (2013) shows. All else follows from a focus on the person since it is the person who is agent. Educations first concern is with the moral agency that helps persons realize themselves within the competing systems (networks) they occupy one of which, and a very important one, is work. Clearly, self-assessment will play a major role in such a system.

There may be a considerable *occupational transfer gap* between the current job and jobs sought. This places an obligation on current employers to reduce this "occupational transfer gap" and enables employees to obtain new jobs in other spheres of knowledge. In the past in Britain and Ireland, employers have been unwilling to accept entrants whose qualifications are not directly related to the jobs they have (Thomas and Madigan, 1974). Firms have been unwilling to accept that individuals have transferable skills that can be of value in any job. Coupled with the need to acquire a new work identity, this can lead to the self-fulfilling hypothesis that employees come to believe that individuals are only suitable for work in the areas for which they have been specifically trained. The case for *labor arenas* was argued in Section 4.4.

Is this "occupational transfer gap" real or imagined? This is an issue that Wadwha did not discuss but it is at the heart of Professor Plotkin's complaint. Wadwha had asked, "why are there so many middle-aged engineers in the ranks of the unemployed?" And, he had asserted that "most engineering professors don't understand the dynamics of the real world and they don't prepare their students adequately" (p. 10). But should he have asked, what is it that they, together with the institutions of education, have to do to prepare engineers for work at 50? It is a combined social responsibility that will prevent this wasteful use of resources. It is clear that this has implications for assessment and credentialing.

Just as the *clichés* of "adaptability" and "change" are rapidly becoming a reality, so the platitudes about lifelong education are becoming equally real and individuals will have to take responsibility for a continuing engagement with learning throughout their lives, and for this they will need employer support. The implications for the institutional structures of higher education and their relations with the world of work are profound. Change will require divergent visioning and there is a major question for assessment—namely, how does it show (predict) that an individual is adaptable?

Alan Cheville (personal communication) has taken this thinking somewhat further. He suggests that students should take out an insurance policy for a lifelong engagement with their university so that they can either return to their university at intervals or use e-learning to obtain immediately required knowledge, or knowledge for further personal and professional development. He envisages that there will be many pathways. The implications for credentialing are profound. First, credentials should no longer signify the end of education but should simply be indicators of personal and professional progress. Second, this implies that assessment is a record of progress that indicates a *labor arena* or *arenas* covered by the skills a person possesses. Clearly, the most appropriate way to provide these indicators is through a validated portfolio, that validation being a responsibility of both employer and educational institution. Self-assessment necessarily has a major role to play in such development, how else can a person judge which future path to take?

Denham (2014) has suggested in respect of the UK system of higher education that at least 30% of students need only 2 years of university study amounting to 78 weeks. It is clear that such an education must be liberal if students are to develop the personal transferable skills that most concede they need. How can engineering, as part of that program, prepare students for their first jobs? This has been and is being discussed elsewhere. It is an uncomfortable problem belonging to the future with a view to changing the present. In this context, the role of competency-based assessment is to help people to learn, and that will cost. It will be apparent from the foregoing that self-assessment is a major competency to be developed.

12.10 Conclusion

Overall this study challenges the utilitarian model of education that has governed think- ing since the end of the Second World War. The tension between the liberal and the utilitarian curricula (Carl Mitcham calls it the two cultures problem in engineering) was also present when ABET presented its proposals at ASEE's 2015 annual conference.[1] Some of those present felt that the proposals for Criterion 3 played down the liberal element of education and some went further and suggested a conspiracy between ABET and industry so that industry thinks its needs will be met. Whether or not they would be assuaged by the inclusion of a statement of attitudes, motives, and interests that the curriculum (such as that described in Exhibit 3.7) should foster is a matter for debate. Trevelyan (2014) would likely argue that this list does not cause students to answer the question, "Why engineering provides value and what this means for you?" Some may argue that the overarching term "values" needs to be used to give the conversation more depth and breadth, and this to be about personal values as well as professional values. Such discussion seems to be predicated by the category of professionalism and the meaning ascribed to the term (Heywood and Cheville, 2015). The evidence presented in this text suggests that industry has much to learn about the effects that its organization has on attitudes, individual learning, and development.

The ABET representatives at the ASEE conference were not asked to give examples of what they meant by creativity and innovation. My reaction was to wonder why, if there is a drive toward integrated study, use was not made of comprehensive examinations (Resnick and Goulden, 1987; Heywood, 1989). Given the emphasis on teamwork, it is surprising that faculty do not work together in teams to produce comprehensive systems of assessment that integrate the knowledge obtained during their various courses.

At a more fundamental level, ideas as to what might be achieved are to be found in the study of the Alverno College curriculum that, as we have seen, focuses on student learning outcomes. Alverno College takes the view that such a curriculum has to be thought through continually so that educators are continually questioning what ought to happen and how to make it happen in practice (Mentkowski and Associates, 2000, Chapter 9). Patricia Cross (1986) argued in the United States that teachers should learn to use their classrooms as they do their laboratories, and in Ireland, Heywood (2008) argued that effective teaching is research. Cross and Angelo (1993) showed how this can be accomplished at a basic level through assessment, and with Steadman (1996) presented an alternative approach to classroom research that teachers could do. In this context, it is worrying that one of ABET's initial findings was that some outcomes have proven difficult to assess in a useful and repeatable manner. This suggests a tension not only between research and practice but between innovation research in higher education more generally and practice and the trade-offs to be made in the design of assessment (Heywood, 2013).

In the Alverno study, the terms "inquiry" and "scholarship" were used rather than research. The latter was defined in terms of Boyer's fourfold vision of scholarship as discovery, integration and synthesis, application, and teaching (Boyer, 1990). In this way, teaching becomes both exciting and scholarly. However, progress will not be made without a fundamental and philosophical debate about the purposes of education, be it for professionals or for all. Uncompromising questions will have to be faced and answered such as what is the purpose of education? And, why do we educate engineers?

If it is correct that in the future many individuals will have to make radical changes in their career paths it follows that they will have to be very adaptable and flexible. Realistically this means that the skills they develop in formal education will have to be transferable. The ability to make such transfers can come only from a curriculum that is in the first place general in order for them to see the possibilities of transfer. But in today's understanding "transfer" will not take place if these subjects are taught independently of each other. Since transfer will only occur to the extent we expect to occur, the curriculum has to show how it can occur in what might best be described as interdisciplinary or transdisciplinary situations. The failure to approach study in this way is the reason why a general education that comprises the study of a number of independently organized subjects is not liberal. It is the reason why in subject specialisms like engineering so many students are unable to combine knowledge from the subdisciplines to solve complex problems. Taught in a way that overcomes this problem, that is, in a spirit of universality, engineering is as much a liberal study as any other. Any detailed analysis of the activity (process) of engineering will demonstrate that this is so. One model that satisfies the constraints of the future is for full-time university (higher education) to be no longer than a couple of years, its objectives being an introduction to the world of knowledge, the development of "being," and skill in the transfer of knowledge. Some component of that education would necessarily have a component of independent (student designed) study (see Section 5.4). Assessment would focus on how well a person is able to transfer knowledge in a variety of new situations. In such a curriculum, engineering and technological literacy would have a major role to play, intrinsically for their own merits, but also as a building block for those who wish to go forward to study engineering. In this model, higher education is a new beginning and assessment reflects the circumstance of life. That is, it is always formative.

Change on the scale demanded, as Wankat (2013) has concluded, will only be possible if engineering educators receive substantial training in pedagogy and all that that entails.

Notes

1. From the slides presented at the ABET talk at the ASEE Annual Conference June 2015. Available from http://www.abet.org/wp-content/uploads/2015/04/EAC-proposed-Revisions-to-Criteria 3-and-5.pdf.

   Slide 1. ABET. Engineering Accreditation Commission—proposed Revisions to Criteria 3 and 5.

   Slide 4. Categories of outcomes *Technical: the specialized skills that are required by a practitioner in the discipline. *Business: the skills required to function within a larger enterprise. *Communication: the skills to convey information effectively using a variety of methods and media. *Professionalism: the personal and professional conduct and qualities expected for a practicing engineer. *Individual—Skills such as creativity, leadership, innovation, and practical ingenuity are desirable qualities that can be emphasized to the degree that meets a program's mission.

   Slide 7 Draft Outcomes. Criterion 3 Student Outcomes.

   The program must have documented student outcomes that prepare graduates to enter the engineering profession. Student outcomes are outcomes (1) through (6) plus any additional outcomes that may be articulated by the program.

   1. An ability to use principles of science and mathematics to identify, formulate, and solve engineering problems.
   2. An ability to apply both analysis and synthesis in the engineering design process, resulting in designs that meet constraints and specifications. Constraints and specification include societal, economic, environmental, and other factors as appropriate to the design.
   3. An ability to develop and conduct appropriate experimentation and testing procedures, and to analyze and draw conclusions from data.
   4. An ability to communicate effectively with a range of audiences through various media.
   5. An ability to demonstrate ethical principles in an engineering context.
   6. An ability to establish goals, plan tasks, meet deadlines, manage risk and uncertainty, and function effectively on teams.

2. In 2002, ABET commissioned the Center for the Study of Higher Education at Pennsylvania State University to undertake a 3.5-year study to assess whether the implementation of the new Ec 2000 criteria was having the intended effects. Their report was published in 2006.

   Lattuca, L. R., Terenzini, P. T., and Volkwein, J. F. (2006). *Engineering Change: A Study of the Impact of Ec 2000*. Baltimore, MD: ABET Inc. The findings based on a national study were generally positive. "The findings from this study strongly suggest improvements in student learning have indeed resulted from changes in engineering program curricula, teaching methods, faculty practices, and student experiences inside and outside the classroom. Although many dimensions of engineering programs shape learning, the findings of this study indicate that students' classroom experiences are the most powerful and consistent influences." However, a later academic study acknowledges that there is a need for more complex research designs. Lambert, A. D., Terenzini, P. T., and Lattuca, L. R. (2007). More than meets the eye: curricular and programmatic effects on student learning. *Research in Higher Education*, 48(2), 141–168.

References

Barry, B. E., Brophy, S. P., Oakes, W. C., Banks, K. M., and Sharvelle, S. E. (2008). Developing professional competencies through challenge to project experiences. *International Journal of Engineering Education*, 24(6), 1148–1162.

Blandin, B. (2011). The competence of an engineer and how it is built through an apprenticeship program: a tentative model. *International Journal of Engineering Education*, 28(1), 57–71.

Bloom, B. S., Engelhart, M. D., Furst, E. J., Hill, W. H., and Krathwohl, D. R. (1956). Taxonomy of Educational Objectives. Handbook 1. Cognitive Domain. New York. David Mackay.

Boyer, E. L. (1990). *Scholarship Reconsidered: Priorities for the Professoriate*. Special report of the Carnegie Foundation for the Advancement of Teaching. Princeton, NJ: Princeton University Press.

Carey, K. (2015). *The End of College: Creating the Future of Learning and the University of Everywhere*. New York, NY: Riverhead Books.

Chan, C. K. Y. (2012). Exploring an experiential learning project through Kolb's learning theory using a qualitative research method. *European Journal of Engineering Education*, 37(4), 404–415.

Cheville, A., and Heywood, J. (2015). Work in progress. Drafting a code of ethics for engineering education. In: ASEE/IEEE Proceedings frontiers in Education Conference.

Cross, K. P. (1986). A proposal to improve teaching or "what taking teaching seriously" should mean. *American Association of Higher Education Bulletin (AAHE)*, September 9–14.

Cross, K. P., and Angelo, T. A. (1993). *Classroom Assessment Techniques*. San Francisco, CA: Jossey-Bass.

Cross, K. P., and Steadman, M. H. (1996). *Classroom Research. Implementing the Scholarship of Teaching*. San Francisco, CA: Jossey-Bass.

Davies, J. W., and Rutherford, U. (2012). Learning from fellow engineering students who have current professional experience. *European Journal of Engineering Education*, 37(4), 354–365.

Denham, J. (2014). Labour plan to fix student funding by solving loan gap. http://theguardian .com/education/2014/jan/14/universities-debt-cancellation-teaching-john-denham. Accessed January 16, 2014.

Dressel, P. L. (1971). Values, cognitive and affective. *Journal of Higher Education*, 42(5), 500–502.

Edwards, M., Tovar, E., and Soto, O. (2008). Embedding core competence curriculum in computer in engineering. In: ASEE/IEEE Proceedings Frontiers in Education Conference, S2E-15 to 20.

Furneaux, W. D. (1962). The psychologist and the university. *Universities Quarterly*, 17, 33–47.

Furst, E. J. (1958). *The Construction of Evaluation Instruments*. New York: David MacKay.

Griffin, C. (2012). A longitudinal study of portfolio assessment to assess competence of undergraduate student nurses. Doctoral dissertation, University of Dublin, Dublin.

Harris, G. L. (2013). Incorporating adult learning methods in project based learning in laboratory metrology courses. In: Proceedings of the Annual Conference of the American Society for Engineering Education. Paper 6902.

Heywood, J. (1981). The academic versus the practical debate. A case study in screening. *Institution of Electrical Engineers Proceedings* Part A, 128(7), 511–519.

Heywood, J. (1984). *Considering the Curriculum during Student Teaching*. London: Kogan Page.

Heywood, J. (1989). *Assessment in Higher Education*, 2nd edition. Chichester, UK: John Wiley & Sons.

Heywood, J. (2008). *Instructional and Curriculum Leadership. Towards Inquiry Oriented Schools*. Dublin: National Association of Principals and Deputies/Original Writing.

Heywood, J. (2013). Trade-offs in multiple objective (strategy) assessment and learning. A retrospective and comparative examination of examinations. *European Studies in Educational Management*, 2(1), 56–75.

Heywood, J. (2015). Teaching, education, engineering and technological literacy. In: Proceedings Annual Conference of the American Association for Engineering Education. Paper 12900.

Heywood, J., and Cheville, A. (2015). Is engineering education a professional activity? In: Proceedings of Annual Conference of the American Society for Engineering Education, June 2015. Paper 12907.

Jackson, P. W. (1968). *Life in the Classroom*. New York: Holt Reinhart and Winston.

Kallenberg, B. J. (2013). *By Design. Ethics, Technology and the Practice of Engineering*. Eugene, Oregon: Cascade Books.

Korte, R. F. (2009). How newcomers learn the social norms of an organization: a case study of the socialization of newly hired engineers. *Human Resource Development Quarterly*, 20(3), 285–306.

Krathwohl, D. R., Bloom, B. S., and Masia, B. B. (1964). Taxonomy of Educational Objectives. The Classification of educational Goals; Handbook II. The Affective Domain. New York, David McKay.

Lanier, J. (2013). *Who Owns the Future?* London, UK: Allen Lane Press (Penguin).

Leonardi, S. M., Jackson, M. H., and Diwan, A. (2009). The enactment–externalization dialectic rationalization and the persistence of counterproductive technology design practices in student engineering. *Academy of Management Journal*, 52(2), 400–420.

McCahan, S., and Romkey, L. (2014). Beyond Bloom's. A taxonomy for teaching engineering practice. *International Journal of Engineering Education*, 30(5), 1176–1189.

McCormack, J., Beyerlein, S., Brackin, P., Davis, D., Trevisan, M., Davis, H., Lebeau, J., Gerlick, R., Thompson, P., Khan, M. J., Leiffer, P., and Howe, S. (2011). Assessing professional skill development in capstone design courses. *International Journal of Engineering Education*, 27(6), 1308–1323.

Macmurray, J. (1958). *The Self as Agent*. London, UK: Faber and Faber.

Mentkowski, M., and Associates (2000). *Learning That Lasts. Integrating Learning, Development, and Performance in College and Beyond*. San Francisco, CA: Jossey-Bass.

Mistree, F., Ifenthaler, D., and Siddique, Z. (2013). Empowering engineering students to learn how to learn: a competency based approach. In: Proceedings of Annual Conference of the American Society for Education. Paper 7234.

Organization for Economic and Cultural Development. (2012). *Employment Outlook 2012*. Paris: Organization for Economic and Cultural Development (Summarized in *The Times* 11, 07:2012, p. 32).

Pembridge, J. J. (2013). A comparison of adult learning characteristics between first-year and senior capstone students. In: Proceedings of the Annual Conference of the American Society for Engineering Education. Paper 9819.

Raelin, J., Reisberg, R., Whitman, D., and Harmann, J. (2007). Cooperative education as a means to self-efficacy among sophomores (with particular attention to women) in undergraduate engineering. In: ASEE/IEEE Proceedings Frontiers in Education Conference, F1G-20 to 24.

Reich, R. B. (2012). *Beyond Outrage. What Has Gone Wrong with Our Economy and Our Democracy and How to Fix It.* New York: Vinatage Books.

Resnick, D. P., and Goulden, M. (1987). Paper presented at the American Association for Higher Education 2nd National Conference on Assessment in Higher Education, Denver, CO. An earlier version is found in D. Halpern (ed.), *Student Outcomes Assessment. A Tool for Improving Teaching and Learning.* San Francisco, CA: Jossey-Bass.

Rothwell, J. (2014). *Still Searching. Job Vacancies and STEM Skills.* Washington, DC: Brookings Institute.

Sandberg, J. (2000). Understanding human competence at work. An interpretive approach. *Academy of Management Journal,* 43(3), 9–25.

Senge, P. M. (1990). The Fifth Discipline. The Art and Practice of Learning. New York. Doubleday.

Slaton, A. E., and Riley, D. M. (2015). The wrong solution for STEM education. *Inside Higher Ed.* https://www.insidehighered.com/views/2015/07/08/essay-criticzes-proposed-changes-engineering-accrediatation-standards. Accessed July 10, 2015.

Sparks, E., and Waits, M. J. (2011). *Degrees for What Jobs? Raising Expectations for Universities and Colleges in a Global Economy.* National Governors Association.

Squires, A. F., and Cloutier, A. J. (2011). Comparing perceptions of competency knowledge development in systems engineering curriculum: a case study. In: Proceedings of Annual Conference of American Society for Engineering Education. Paper 1162.

Sternberg, R. J. (2005). Intelligence, competence and expertise. In: A. J. Elliot and C. S. Dweck (eds.), *Handbook of Competence and Motivation.* New York, NY: Guilford Press.

Thomas, R., and Izatt, J. (2003). A taxonomy of engineering design tasks and its applicability to university engineering education. *European Journal of Engineering Education,* 28(4), 535–547.

Thomas, B., and Madigan, C. (1974). Strategy and job choice after redundancy: a case study in the aircraft industry. *Sociological Review,* 22, 83–102.

Trevelyan, J. (2010). Restructuring engineering from practice. *Engineering Studies,* 2(3), 175–195.

Trevelyan, J. (2014). *The Making of an Expert Engineer.* London: CRC Press/Taylor and Francis Group.

Wadwha, V. (2011). Leading edge: Over the hill at 40. *ASEE Prism,* p. 32.

Wankat, P. C. (2013). progress in reforming chemical engineering education. *Annual Review of Chemical and Biomolecular Engineering,* 2013, 423–443.

Wheeler, D. R. (2015). Silicon valley to millenials: drop dead. CNN.com. http://edition.cnn.com/2015/03/18/opinions/wheeler-silicon-valley-jobs/index.html.

Youngman, M. B., Oxtoby, R., Monk, J. D., and Heywood, J. (1978). *Analysing Jobs.* Aldershot: Gower Press.

Zachary, G. P. (2011). Jobless innovations? *IEEE Spectrum,* April, 8.

A Quick Guide to the Changing Terminology in the Area of "Assessment"

A.1 Objectives and Outcomes

1. Although the idea of stating educational objectives can be traced back to the end of the nineteenth century, a useful start is with R. W. Tyler's (1949) description of the four basic tasks an educator has to undertake. These were as follows:

    (i) the determination of the objectives which the course (class lecture) should seek to obtain,

    (ii) the selection of the learning experiences that will help to bring about the attainment of those objectives,

    (iii) the organization of those learning experiences so as to provide continuity and sequence for the student and to help him integrate what might otherwise appear as an isolated experience, and

    (iv) the determination of the extent to which the objectives are being achieved.

    Tyler defined an objective as a change in behavior, ways of acting learning, and feeling. The fourth task is the activity of assessment.

2. The group led by Benjamin Bloom that met to develop the *Taxonomy of Educational Objectives* based its work on Tyler's definition of an objective. In general, the *"group believed that some common framework be used by all college and university examiners could do much to promote the exchange of test materials*

The Assessment of Learning in Engineering Education: Practice and Policy, First Edition. John Heywood.
© 2016 The Institute of Electrical and Electronics Engineers, Inc. Published 2016 by John Wiley & Sons, Inc.

and ideas for testing. They also believed that such a framework could be useful in stimulating research on examinations and the relation between examinations and education. After considerable discussion, there was agreement that the framework might best be obtained through a system of classifying the goals of the educational process using the educational objectives" (Bloom, 1994). The group distinguished between three domains—cognitive, affective, and psychomotor. The first volume was published in 1956 for the cognitive domain (Bloom et al., 1956). The Complete title was *Taxonomy of Educational Objectives: The Classification of Educational Goals, Handbook I. Cognitive Domain*. The *goals* were six classes of educational behaviors arranged in order of complexity. Following a substantial critique (Anderson and Sosniak, 1994) it was substantially revised in 2001 (Anderson and Krathwohl, 2001). The knowledge dimension comprises factual, conceptual, procedural and meta-cognitive knowledge. The cognitive process dimension comprises Remember, Understand, Apply, Analyze, Evaluate and Create

To confuse matters, Heywood (2000) calls such groupings "domains." More recently, Larkin and Uscinski (2013) wrote "*Goals* express intended learning outcomes in general terms and *objectives* express them in specific terms." Within the literature, goals are preceded by such terms as "course," "instructional," "learning," "long-term," "overarching," and "project."

In the statement on "vision," the committee wrote, "[..] this taxonomy is designed to be a classification of the student behaviors which represent the intended outcomes of the educational process." There is no escape from the fact that in the taxonomy educational objectives and intended outcomes are congruent although some authorities have attempted to differentiate between them.

The committee "recognized that the **actual behavior** of the students after they completed the unit of instruction may differ in degree as well as kind from the **intended behaviors** specified by the **objectives**."

For some years after the publication of the *Taxonomy*, the term "behavioral objective" was used. Some authorities continue to use it (see Felder and Brent below). In the United Kingdom, Cohen and Mannion (1977) in their bestselling book on teaching practice, suggested that students should at the beginning of planning a lesson declare its *Aim, nonbehavioral objective, and behavioral objectives. Aim* in this sense is something that is much more limited than the aims of education that are discussed by philosophers (e.g., Whitehead, 1932; Wringe, 1988). *Nonbehavioral objective corresponds* with Larkin and Uscinski's (2013) *goal*. In some texts, it is difficult to distinguish between the aim and nonbehavioral objective (Heywood, 2008). Some authors use *learning objective* in a nonbehavioral sense (e.g., "the student will develop an understanding of how to…"; Choudhury, 2013). Some authors used the term *terminal objective*, the intention of which should be self-evident to the reader.

Aim (Cohen and Mannion, 1977): To introduce the principles of objective testing (two lectures).

Nonbehavioral objective (Cohen and Mannion, 1977): **Goals** (Larkin and Uscinski, 2013).

At the end of three classes, the students will be introduced to the principal types of objective item, how they are written, and how objective test results are analyzed.

Behavioral objectives (Cohen and Mannion, 1977): **Learning Outcomes** (Larkin and Uscinski, 2013). Competencies.

At the end of the exercise, the students will be able to:

(a) Construct objective items into a test in the subject taught by the student.

(b) Conduct a short classroom test.

(c) Recognize the limitations of such tests.

(d) Conduct an item analysis of the test.

(e) Evaluate the analysis and suggest changes in the items where this is thought necessary.

The committee said that it had created a taxonomy and not a classification "a taxonomy must be validated by demonstrating its consistency with theoretical views in research findings in the field it attempts to order."

The first known attempt to apply the taxonomy in an engineering curriculum was in the advanced level of the General Certificate of Education in Engineering Science in the United Kingdom (from 1968 to 1988; Carter, Heywood, and Kelly, 1986; see Chapter 3 and Appendix B). It was not found to be consistent with the needs of the curriculum and new goals (e.g., communication and creativity) were required. The idea of specific objectives was followed for coursework assessment and its associated rubrics, and their validation was discussed by Heywood and Kelly (1973) at the Frontiers in Education Conference of that year.

The higher-order categories of *The Taxonomy* have been associated generically with "critical thinking" and what came to be known as "Higher Order Thinking Skills" (HOTS). This notion of "critical thinking" is acknowledged in the section on "Turning Plan into Action. The Big Picture." This activity is labeled "intellectual abilities, and skills." "The most general operational definition of these abilities and skills is that the individual can find appropriate information and techniques in his previous experience to bring to bear on new problems and situations." It is interesting to note that authors of *The Taxonomy* felt they had to justify the development of intellectual abilities and skills! They were also adamant that *The Taxonomy* covered problem solving.

3. The first direct attempt to describe the value of *The Taxonomy* for engineering educators seems to be due to Stice (1976) in what must be regarded as a seminal paper in *Engineering Education*. But he arrives at *The Taxonomy* via a discussion of *Instructional Objectives*, which had become important for the designers of programmed instruction. He followed the advice of Mager (1962) who had published a much-read book on the topic. He cites Mager's definition of an objective as an "intent communicated by a statement describing a proposed change in the learner." But the literature often drops the intended and simply uses learning outcome.

Mager evidently thinks that the student demonstrates a *competence* because his instruction reads "describe the important conditions under which the learner will demonstrate competence." Stice follows Gronlund (1970) who writes about objectives in each of three domains of the cognitive, affective, and psychomotor.

4. In 1984, R. G. Carter (1984) published a paper on engineering curriculum design and in the following year (1985) published a more general paper that described a taxonomy for professional education. Others have also developed taxonomies, as for example, by L. D. Fink, which has been used in at least one engineering study (see Section 6.2). Engineering educators at the University of Brighton have developed a taxonomy of engineering design tasks in which each of the eight task levels is characterized by a greater degree of design freedom (Thomas and Izatt, 2003).

5. In 1989, at the Frontiers in Education Conference, Heywood (1989) compared "comprehensive examinations" in the United States with Engineering Science at the Advanced Level in the United Kingdom. He pointed out that it was only possible for faculty to consider a limited number of ability domains for which reason he sometimes called them "focussing objectives." The domains selected have to have a high level of significance for the ongoing purposes of a student's education. *Focussing objectives* are derived from the key learning skills that a person will need in the job for which he or she is trained. Necessarily, they incorporate values. They are the process skills that lead to the product. In later work, he refers to them as *ability domains* following the practice of Alverno College.

6. *The Taxonomy* was significantly revised in 2001 but retained its general characteristics (Anderson et al., 2001). The domain of "*synthesis*" was omitted and a sixth domain of "*create*" added. The domain of knowledge was radically changed. See Section 6.2 for a discussion of the differences between the two.

7. *Outcomes* seems to have come to be preferred to *objectives* in the United Kingdom in the late 1980s and to have some association with the United Kingdom's Employment Department who sponsored the Enterprise in Higher Education Initiative in that they sponsored work in this area undertaken by Otter (1991, 1992; see Section 7.3). She preferred to differentiate between objectives and outcomes and describe objectives as intentions. The 1991 publication contained a statement of learning outcomes for engineering education.

8. By the middle of the 1990s, *learning outcomes* seem to have come to be preferred by the authorities (e.g., ABET) although many engineering educators continue to refer to *objectives*. ABET distinguishes between program objectives and program learning outcomes.

9. Felder and Brent (2003, and in their latest book, which is in the press) tried to make sense of this problem. They understand that "ABET means by '*Program Educational Objectives*' the collection of knowledge, skills, and attitudes program graduates should have several years after they graduate, and '*Program Learning Outcomes*' are the knowledge skills and attitudes they should have at

the time they graduate which will equip them to attain the educational objectives. We (Felder and Brent) also said that *'Course learning Objectives'* were equivalent to Mager and Gronlund's *'instructional objectives'* or *'behavioral objectives'*—observable actions a student could take to demonstrate his or her mastery of the knowledge, skills, and attitudes being taught in the course. The key is 'observable' as in "The student will be able to define, explain, calculate, derive, model, critique, design...]" (which as I understand it was the intention of the authors of *The Taxonomy*). Felder continues, "That necessary condition rules out such words as 'know', 'learn', 'understand,' and 'appreciate' which are worthwhile statements of goals but since they're not directly observable you have no way of knowing whether the students have attained them" (Focused statements including such words are what Cohen and Mannion call *non-behavioral objectives;* I pointed out that they were important because they were part of the emotional vocabulary of teachers; Heywood, 1977). Felder continues, " *You have no way of knowing whether students have attained them—you have to ask them to do something observable to demonstrate attainment, which gets you back to learning objectives*" (Rich Felder, personal communication, December 2014).

10. Elliot Eisner (1979, see Chapter 10) distinguished between three kinds of objectives—*behavioral, problem solving,* and *expressive*. He attached the term *objectives* to the first two and *expressive* to the latter. This was because the term *objective* implied a preformulated goal whereas outcome was the result of what happened. He argued that while there was a case for preformulated goals there were many activities for which we did not preformulate specific goals. We undertook such activities in the anticipation that something would happen. For example, we do not think much beyond the data, even though we could predict from the ample criteria at our disposal. What we do is to evaluate retrospectively what happened against these criteria. From this, he deduces that teachers should be able to plan activities that do not have any specific objectives. Associated with this view is the point that very important *unintended outcomes* often accompany the attainment of stated learning outcomes.

11. Most recently, some engineering educators in the United States have begun to talk about competencies which they align with learning outcomes.

12. The idea of *competency-based testing* has a long history. Attempts to design measures of the *competency* of teachers date back to the 1960s in the United States. The first domains that Alverno College described were called competencies (see Section 5.5). Subsequently, they were called "abilities." A comprehensive reason for their using the term *abilities* is given in Mentowski et al. (2000). "*Abilities are complex combinations of motivations, dispositions, attitudes, values, strategies, behaviors, self-perceptions, and knowledge of concepts and procedures*" (p. 10). There was probably some political motivation in the change in that the competency-based testing of teachers had rather brought competency-based testing into disrepute. In this respect Youngman et al. (1978) in their analysis of the

work done by engineers, decomposed each task category into a number of sub-tasks. In their first report, they called them *abilities*. Later because they thought psychologists would object they changed them to *operations* (see Chapter 5). They would have favored Anastasi's (1980) approach from which the Alverno definition is derived.

13. There are at least two definitions of *competency* in use. The first is quite narrow and is clear statement of *mastery*, for example. This person is able to switch any computer on. That can be easily tested and the answer is either "yes" or "no." But then competence may be applied to categories such as the domains of the *Taxonomy*. Both were used to assess the coursework component of engineering science at "A" level in the United Kingdom (see Chapter 3). They were called *assessment criteria* and the scheme reflected a concern for *behavioral objectives*. No mention is made of the term *competence*. The term *criteria* is used through-out their final study. That is not surprising since at the time the term *criterion referenced testing* was in common use. Clearly, some items on the assessment schedule are *criterion referenced* since they seek *mastery and a person* can only be right or wrong, correct or incorrect. But it is easy to recognize the need for some domains to recognize different levels of performance, for example, eval-uation. In subsequent papers about this subject's development, these have been called *semi-criterion referenced*. Choudhury (2013) citing Friedlan (1995), who defines *competency* "as an understanding of information, skills, and approaches needed to perform a specific task effectively and efficiently at a defined level of performance," which seems to be as good a description of the second level as any.

14. With respect to the second usage in recent engineering literature from Europe, *competency* seems to have become the favored term. Recently some American authors have been using the term in discussions about outcomes for which it is also sometimes a substitute.

15. However, Griffin (2012) points out that the way in which competency is used is dependent on the way the user understands learning. She distinguishes between two positions that she calls "inside" and "outside," which depend on the user's view of learning and a distinction between *ways of knowing* and *ways of being* (see Chapter 11).

16. There has been a substantial debate about the relative meaning of *competence* and *competency* and the two are often confused. There has been an equally substantial debate about differences between *competency* and *performance*.

17. In the late 1980s, the concept of "personal transferable skills" was adopted by the Employment Department in the United Kingdom. This seems to have been a substitute for generic competencies, for example, creativity, communication, problem-solving, and the subdimensions that make up these dimensions. It was argued that students could be provided with learning experiences that would not only help students develop these skills but be able to transfer them to a variety of different situations. Hence, the idea of *transferable skill*.

A.2 Assessment and Evaluation

1. In the 1960s, the term *evaluation* was applied to a program or a course. It sought to answer questions such as: Does the program or course obtain its objectives: if not, why not? Examination (test) results were not the sole criteria for answering such questions. The American Educational Research Association produced a series of books on the topic. The term was also used in the sense of the category in *The Taxonomy*. Students had to show they could *evaluate* what they had done, as for example, in a project. Now, however, *assessment* is often used instead of *evaluation* and like it has several meanings.

2. Chapter 1.8 outlines how it came to be used in England as a descriptor for coursework in the 1960s.

3. In the mid-1980s, the US Federal Department of Education began to take an interest in assessment in higher education and Cliff Adelman (1986) organized a conference and publication on the topic (1986). A paper by T. W. Hartle (1986) in that publication suggested that in the United States six uses of assessment were commonly deployed. These were:

    (i) State-mandated requirements to evaluate quality. At the time there was a lot of interest in indicators of quality.

    (ii) The use of testing for counselling (which is one of its original uses in the psychology of testing).

    (iii) The use of tests for placement.

    (iv) The use of tests for admission into higher education.

    (v) The use of tests in licensing examinations.

    (vi) Tests of knowledge and skills.

    In the United States, faculty sometimes use *standardized tests* to test knowledge and skills. Such tests are not used by faculty in the United Kingdom.

4. Assessment is also used for the measurement of student attitudes and values, and the evaluation of institutional goals. Internationally it seems more generally to have replaced *evaluation*. Thus *program assessment* means program evaluation in the earlier language. But some faculty still use the term *evaluation*.

5. 1986 seems to have been an important year for thinking about examinations. Carter, Heywood, and Kelly called the approach adopted for the advanced level examination in engineering science *Multiple Strategy*. They had derived this from Heywood's *multiple objective* approach to assessment described in his 1977 book on *Assessment in Higher Education* in which he had argued that no one measure will satisfactorily assess a student's performance because there are many outcomes that have to be assessed. Therefore, for each significant objective, there would likely be preferred methods of assessment and instruction. It seemed that in the United States, the National Association of Governors held a similar view, which was discussed in their 1986 report on education.

6. For a number of years, the American Association for Higher Education held a series of conferences on assessment commencing in 1986. To the foreign observer, the distinction that was made between *program assessment* and the *assessment of learning* was difficult to grasp (see Section A 9). More attention seemed to be paid to the former than the latter. A minor difficulty was the realization that *rubric* in the United States referred to a technique of assessment (e.g., the profiles in engineering science) whereas *rubric* as used in examinations in the United Kingdom referred to the instructions on the paper (e.g., answer four out of six questions).

   The work of some of the contributors to the AAHE conferences is being recognized by engineering educators (e.g., Walvoord, 2004).

7. The terms *formative* and *summative evaluation* are now often called *formative and summative assessment* (see George and Cowan, 1999).

References

Adelman, C. (ed.) (1986). *Assessment in Higher Education*. Washington, DC: US Department of Education.

Anastasi, A. (1980). Abilities and the measurement of achievement. In: Schrader, W. B. (ed.), Measuring Achievement. Progress over a Decade. San Fransisco. Jossey Bass.

Anderson, L. W., Krathwohl, D. R., Airasian, P. W., Cruikshank, K. A., Mayer, R. E., Pintrich, P. R., Raths, J., and Wittrock, M. C. (eds.) (2001). *A Taxonomy for Learning Teaching and Assessing. A Revision of Bloom's Taxonomy of Educational Objectives*. Addison Wesley/Longman.

Anderson, L. W., and Sosniak, L. A. (eds.) (1994). *Bloom's Taxonomy a Forty-Year Retrospective*. Chicago: National Society for the Study of Education, University Press of Chicago.

Bloom, B. S., Englehart, M. D., Furst, E. J., Hill, W. H., and Krathwohl, D. R. (eds.) (1956). *The Taxonomy of Educational Objectives. Handbook 1. Cognitive Domain*. New York David Mackay. (1964). London: Longmans Green.

Bloom, B. (1994). Ch's 1 and 2 of Anderson and Krathwohl et al. (2001).

Carter, G., Heywood, J., and Kelly, D. T. (1986). *A Case Study in Curriculum Assessment. GCE Engineering Science (Advanced)*. Manchester: Roundthorn Publishing.

Carter, R. G. (1984). Engineering curriculum design. *Institution of Electrical Engineers Proceedings*, 131, Part A, 678.

Carter, R. G. (1985). Taxonomy of objectives for professional education. *Studies in Higher Education*, 10(2).

Choudhury, I. (2013). Appraisal of learning objectives of a course in construction science. In: Proceedings of Annual Conference of American Society for Engineering Education, June 2013. Paper 6068.

Cohen, L., and Mannion, L. (1977). *A Guide to Teaching Practice*. London: Methuen.

Eisner, E. W. (1979). *The Educational Imagination. On the Design and Evaluation of School Programs*. New York: Macmillan.

Felder, R. M., and Brent, R. (2003). Designing and teaching course to satisfy the ABET engineering criteria. *Journal of Engineering Education*, 92(1), 7–25.

Friedlan, J. M. (1995). The effects of different teaching approaches on student perceptions of the skills needed for success in accounting courses and by practicing accountants. *Issues in Accounting Education*, 10(1), 47–63.

George, J., and Cowan, J. (1999). *A Handbook of Techniques for Formative Evaluation: Mapping the Students Learning Experience*. London: Kogan Page.

Griffin, C. (2012). A longitudinal study of portfolio assessment to assess competence of undergraduate student nurses. Doctoral dissertation, University of Dublin, Dublin.

Gronlund, N. F. (1970). *Stating Behavioral Objectives for the Classroom*. New York: Macmillan.

Hartle, T. W. (1986). The growing interest in measuring the education achievement of college students. In: C. Adelman (ed.), *Assessment in Higher Education*. Washington, DC: US Department of Education.

Heywood, J. (1977). *Assessment in Higher Education*. Chichester, UK: John Wiley & Sons.

Heywood, J. (1989). Problems in the evaluation of focusing objectives and their implications for the design of systems models of the curriculum with special reference to comprehensive examinations. In: ASEE/IEEE Proceedings Frontiers in Education Conference, pp. 235–241.

Heywood, J. (2000). *Assessment in Higher Education. Student Learning, Teaching, Programmes and Institutions*. London: Jessica Kingsley.

Heywood, J. (2008). *Instructional and Curriculum leadership. Towards Inquiry Oriented Schools*. Dublin: National Association of Principals and Deputies/Original Writing.

Heywood, J., and Kelly, D. T. (1973). The evaluation of course work—a study of engineering science among schools in England and Wales. In: ASEE/IEEE Proceedings Frontiers in Education Conference, pp. 269–276.

Larkin, T. L., and Uscinski, J. (2013). The evolution of curriculum assessment within the physics program at American University. In: Proceedings of Annual Conference of American Society for Engineering Education, June 2013. Paper 6739.

Mager, R. F. (1962). *Preparing Instructional Objectives*. New York: Fearon Publishers.

Mentkowski, M., and Associates. (2000). *Learning That Lasts. Integrating Learning, Development, and Performance in College and Beyond*. San Francisco, CA: Jossey-Bass.

Otter, S. (1991). *What Can Graduates Do? A Consultative Document*. Sheffield: Employment Department. Unit for Continuing Adults Education.

Otter, S. (1992). *Learning Outcomes in Higher Education*. London: HMSO for the Employment Department.

Stice, J. E. (1976). A first step toward teaching. *Engineering Education*, 67, 394–398.

Thomas, R., and Izatt, J. (2003). A taxonomy of engineering design tasks and its applicability to university engineering education. *European Journal of Engineering Education*, 28(4), 535–547.

Tyler, R. W. (1949). Achievement testing and curriculum construction. In: E. G. Williamson (ed.), *Trends in Student Personnel Work*. Minneapolis: University of Minnesota.

Walvoord, B. E. (2004). *Assessment Clear and Simple. A Practical Guide for Institutions, Departments and General Education*. San Francisco, CA: Jossey-Bass.

Whitehead, A. N. (1932). *The Aims of Education*. London: Benn.

Wringe, C. (1988). *Understanding Educational Aims*. London: Unwin Hyman.

Youngman, M. B., Oxtoby, R., Monk, J. D., and Heywood, J. (1978). *Analysing Jobs*. Aldershot: Gower Press.

B

Extracts from the *Syllabus and Notes for the Guidance of Schools for GCE Engineering Science (Advanced) 1972* Joint Matriculation Board, Manchester

B.1 Extract 1 (pp. 2–6)

Introduction

The reviewers of chapter 3 suggested that the reader would be helped if they could see the complete syllabus. Unfortunately the documents referenced are all out of print and the Joint Matriculation Board (JMB) no longer exists following reorganizations of the system of public examining in England and Wales. Traditionally the Matriculation Boards only published the syllabus and example questions. In this case the JMB published substantial notes for guidance on the assessment and teaching of Engineering Science. The extracts which follow from the 1972 notes expand on what is written in pages 60 to 73. Readers will find the published view of the differences between the approach of physics and that of engineering science of interest. (B3). It adds to the philosophy set out in the aims (&pp 60 & 61). The syllabus concludes the extract in B1. A detailed evaluation of the syllabus is given in Carter, G., Heywood, J., and Kelly, D. T. (1986). *A Case Study in Curriculum Assessment. GCE Engineering Science (Advanced).* Roundthorn. Manchester.

The Assessment of Learning in Engineering Education: Practice and Policy, First Edition. John Heywood.
© 2016 The Institute of Electrical and Electronics Engineers, Inc. Published 2016 by John Wiley & Sons, Inc.

The aims of the syllabus

In order to solve problems, the engineering scientist must be able to consider how different solutions will affect the outcome of the task. His solution will be derived from an optimisation of the resources available which takes into account the restraints imposed on the problem by technological, economic and social factors and the opportunities for innovation. The solution will generally involve the engineering scientist in the specification of a plan of action, a design, an investigation or a process, and its execution by way of administration and/or management and/or construction of an artefact or scheme. In pursuit of these activities the engineering scientist may be called on to make an original and creative contribution both in respect of the task at hand and communication with those directly concerned with the problem who may or may not be engineering scientists. The syllabus is designed to enable schools to provide courses which give an opportunity for pupils to begin to develop the abilities and attitudes required of engineering scientists. The principles and some of the factual content essential to the understanding of engineering science are outlined in the syllabus. The examination questions and coursework assessment procedures illustrate how the abilities required for the definition and solution of engineering problems may be developed in the teaching/learning context through study exercises, independent reading, experimental investigations, and a major project.

The objectives of the written examination and coursework

A. GENERAL

Assessment of the achievement and development of the abilities is obtained through the written examination and moderated coursework which is internally assessed. The Board's *Notes for the Guidance of Schools* (ES/N2, July 1972) on Engineering Science should be read in conjunction with this statement of the objectives of the examination and the list of subject matter.

B. THE WRITTEN EXAMINATION

The recall of factual knowledge, though essential to any examination, is only one of several major abilities which candidates will be required to demonstrate. In order to make these general objectives clear, there follows a detailed breakdown of the abilities to be tested. The overall aim of the examination is to test knowledge and understanding of the subject matter in general terms and the ability to apply this knowledge and understanding to particular systems situations and problems. It is not suggested that it is always possible or even desirable to design an examination question to test one particular facet or ability as detailed below or that there is no overlap between the abilities tested. Questions will be selected whenever possible to test both knowledge and one or more of the abilities of comprehension, communication, application of principles, analysis, synthesis and design, evaluation and judgment.

C. COURSEWORK

Coursework is designed to assist in the development of many of the abilities noted under B above and in particular those in paragraphs 3, 4 and 5 of section

D (ii). The more important general aims associated with coursework are marked with an asterisk. It is not the intention that the preparation for the written papers and coursework activities should be entirely complementary since it is possible to measure some of the abilities developed by coursework through the written examination as well as by the use of moderated assessment of coursework.

A developed statement of these abilities together with the procedure for the assessment of coursework is detailed in *Notes for the Guidance of Schools* (ES/N2, July 1972) on Engineering Science. The *Engineering Science Course Work Assessment* booklet (ES/CWA) also gives details of the procedure for assessment. Both these publications are obtainable from the Secretary on request.

D. KNOWLEDGE, UNDERSTANDING AND ABILITIES TO BE TESTED

(i) *Knowledge and understanding of*

(a) terms, conventions and units commonly used in engineering science;

(b) particular principles (or laws) and generalisations of engineering science and their effects and interrelationships;

(c) specialist apparatus and techniques used for the demonstration of the principles referred to in (b); the limitations of such apparatus and techniques;

(d) the use of different types of apparatus and techniques in the solution of engineering problems.

(ii) *Abilities*

1. *Comprehension*

The ability to

(a) understand and interpret scientific and other information presented verbally, mathematically, graphically or by drawing;

(b) appreciate the amount of information required to solve a particular problem or the fact that sufficient information may not exist;

(c) understand how the main facts, generalisations and theories of engineering science can provide explanations of familiar phenomena;

(d) recognise the scope, specification and requirements of a problem;

(e) understand the operation and use of scientific apparatus and equipment;

(f) recognise the analogue of a problem in other related fields of engineering science and practice.

2. *Communication*

The ability to

(a) explain principles, phenomena, problems and applications adequately in simple English;

(b) formulate relationships in verbal, mathematical, graphical or diagrammatic terms;

(c) translate information from one form to another;

(d) present the results of practical work in the form of reports which are complete, readily understandable and objective.

3. *Analysis*

The ability to

(a) break down a problem into its separate parts;

(b) recognise unstated assumptions;

(c) acquire, select and apply known information, laws and principles to routine problems and problems that are unfamiliar or presented in a novel manner.

4. *Synthesis and design*

The ability to

(a) design the manner in which an optimum solution may be obtained efficiently and to propose alternative solutions taking into account the restraints imposed by material, economic and social considerations;

(b) make a formal specification, having decided on the design or scheme;

(c) make a plan for the execution or manufacture of the design or scheme;

(d) use observations to make generalisations or formulate hypotheses;

(e) suggest the new questions and predictions which arise from the hypotheses formulated;

(f) suggest methods of testing these questions and predictions;

(g) find the optimum solution to an engineering design or other problem and give valid reasons for the rejection of alternatives.

5. Evaluation and judgment

The ability to

(a) check that hypotheses are consistent with given information, to recognise the significance of unstated assumptions and to discriminate between hypotheses;

(b) assess the validity and accuracy of data, observations, statements and conclusions;

(c) assess the design of apparatus or equipment in terms of the results obtained and the effects upon the environment and suggest means of improvement;

(d) judge the relative importance of all the factors that comprise an engineering situation;

(e) appreciate the significance of social, economic or design considerations in an engineering situation.

The form of the examination

Paper I. Objective test: to determine the candidate's knowledge of the subject matter of the syllabus and to test his depth of understanding.

Paper II. Section A. Comprehension and communication test: to test the ability of the candidate to comprehend a technical paper on subject matter not necessarily included within the syllabus and to test his ability to interpret and to communicate in his own words.

Section B. Project design: to test the development achieved by the candidate in the field of coursework with special reference to the planning, design and specification of projects within the time imposed by the examination.

Paper III. This paper is composed of two sections. Section 1 consists of short-answer questions and Section 2 long-answer questions. In Section 2 an attempt is made to pose issues of choice relating to practical feasibility, design and cost. Questions are set, wherever possible, which attempt to determine the candidate's understanding of the syllabus through the application of physical principles to devices and practical situations.

Coursework: Candidates are required to satisfy the examiners in the reports submitted on their coursework. Coursework is intended to enable the candidate to demonstrate his ability in design, problem finding and solving and experimental technique.

Allocation of marks between papers and sections

The allocation of marks between papers and sections will be approximately as follows.

Paper I		13 per cent
Paper II	Section A	13 per cent
Section B		13 per cent
Paper III	Section 1	20 per cent
Section 2		20 per cent

20 per cent of the total marks will be allocated to coursework.

Subject matter to be tested in the examination

The syllabus which follows gives details of the subject matter on which the abilities listed above will be tested. A descriptive understanding only is required of the parts of the syllabus printed in italics.

1. MECHANICS

   Definition resolution and composition of co-planar vectors. Concepts of mass, force and weight. Moment of force, couple. Friction. Equilibrium of co-planar forces; application to simple structures including shear force and bending moment diagrams.

   Motion on linear and circular paths, linear and angular acceleration. Newton's laws of motion. Rotation about a fixed axis, moment of inertia. Relation between moment of a couple, moment of inertia, and angular acceleration. Applications to flywheels and machines.

Concepts of potential energy and kinetic energy, momentum and angular momentum, impulse, mechanical work, conservation of energy and momentum, effects of friction.

2. MATERIALS SCIENCE

2.1 *Evidence for electrical charge. Properties of materials which suggest atomic structure. Nuclear and electronic structure of atoms. Qualitative description of energy levels and ionisation potential. Periodic arrangement of atoms. Forces between atoms; elementary description of gases, liquids and crystalline solids*

2.2 Mechanical properties of materials.

The relationship between mechanical properties and structure. Hooke's law. Young's modulus. Stress/strain relationships. An elementary treatment of shear forces.

2.3 Electrical properties of materials.

Electron energy levels in solids. Nature of conduction in conductors and semi-conductors. Resistance and thermal effects. Hall effect. Work function, thermionic and photo emission. Qualitative understanding of dielectric and magnetic materials. Physics of p-n junctions and junction transistors.

2.4 Thermal properties of materials.

Conduction. Specific heat capacity. Thermal expansion.

2.5 Properties of liquids.

Viscosity, surface tension.

3. TRANSPORT PROCESSES
Conduction processes in mass, energy and charge transport. Concepts of pressure, temperature and potential and their gradients. Flow rate, resistance, conductance and capacitance.

Flow of an inviscid fluid. Bernoulli's equation. The pitot-static tube. Simple applications such as flow through nozzles.

Heat transfer by conduction. Fourier's law. Applications to uni-axial heat flow.

Descriptive treatment of heat transfer by convection.

E.M.F., internal resistance, Ohm's and Kirchhoff's laws.

4. FIELD PHENOMENA AND APPLICATIONS
Uniform fields. Concepts of field strength, flux, flux density and potential.

Capacitance of parallel plates. Capacitors in series and parallel. Polarisation and dielectrics. The permittivity of free space. Field due to point charge.

Magnetic flux and flux density, the value of these quantities in a toroid. The simple magnetic circuit, magnetomotive force and its gradient, reluctance.

Detection and measurement of flux by electromagnetic induction. The laws of Faraday and Lenz. Flux distribution due to current in a long solenoid, a plane coil and a straight conductor. The permeability of free space.

Force between two parallel current-carrying conductors, the standard ampere. Self and mutual inductance, transformer action.

Deflection of electrons in electric and magnetic fields. The cathode ray oscilloscope.

Ferromagnetic materials, hysteresis and saturation.

Permanently magnetised materials in a magnetic circuit.

5. PERIODIC AND WAVE PHENOMENA

Elementary treatment of simple harmonic motion. The mass-spring system.

The effect of damping. The concepts of free and forced vibration, resonance. Applications to engineering structures.

Alternating current, r.m.s. value. Current-voltage relations for R, L and C and phase difference.

The concept of wave generation, propagation and reflection. Velocity, frequency, wave length, and their relationship. Amplitude, intensity and phase of waves. Longitudinal and transverse waves. The electromagnetic spectrum.

6. ANALYSIS OF SYSTEMS AND ELECTRONICS

Basic systems concepts; function, control, systems elements; their characteristics and inter action. Sources of power, efficiency.

Analysis of d.c. electrical circuits.

Analysis of an a.c. single loop circuit containing R, L and C or a combination of these elements.

Transfer characteristics of the semi-conductor diode and transistor I/V curves; resistive loads, load lines, application to amplification and switching. Equivalent circuits illustrated by the operational amplifier.

Systems analysis—characteristics of a total system; open and closed loop feedback the use of analogue in systems analysis.

7. THERMODYNAMICS

The thermodynamic system; zero'th law of thermodynamics. Simple kinetic theory of gases. Expression for pressure in terms of molecular mass, velocity and density. Boyle's law. Relation between molecular energy and temperature. Specific heat at constant pressure and constant volume. Heat engines. Theoretical and practical pressure-volume cycles in a heat engine and their use in predicting and measuring power and efficiency.

B.2   Extract 2 (p. 9)

Coursework

The examiners will assume that candidates have carried out, during their course of study, the following three types of laboratory work.

(a) *Controlled assignments,* short undertakings which are closely controlled by the teacher.

(b) *Experimental investigations,* longer undertakings in which the candidate is given the opportunity to exercise considerable discretion within the general terms of the investigation.

(c) *Projects,* major undertakings in which candidates are encouraged to exercise the greatest possible freedom in planning and carrying out the investigation.

Each candidate will be required to submit laboratory reports on *two* experimental investigations on topics chosen from separate sections of the syllabus and on a project undertaken during the course of study. These reports will be marked initially by the staff of the centre in accordance with detailed instructions available from the Secretary; the marks awarded will be subject to moderation by the examiners, who may wish to visit centres for this purpose. For the purposes of internal assessment and subsequent moderation by the Board's examiners, each candidate will be required to maintain, in the form of a journal, a continuing record of all his written work connected with the practical course. All candidates will be required to prepare an outline of the project to be presented for examination purposes, whether or not it is to be submitted for comment by the Board's moderators. The outlines are to be prepared in special booklets which are available on application to the Secretary. Centres entering candidates for the first time will be required to submit to the Board, for comment by the moderators, the outline of each project to be presented by their candidates for examination purposes. These outlines are to be submitted by not later than 1 May in the year preceding the examination. Subsequently centres are not required to submit outlines for comments but may do so if they wish; in this case the last date for submission is 1 November in the year preceding the examination; this date also applies to centres entering candidates for the first time who are completing the course in one year. When the final project report is submitted by the centre to the examiners for moderation it is to be accompanied by the project outline whether or not the outline was previously submitted to the Board for preliminary approval.

B.3   Extract 3 (pp. 13–16)

1. The purpose of the course

1.1 THE COURSE COMPARED WITH PHYSICS (ADVANCED)

Engineering Science is offered as an alternative to Physics (Advanced) and is rooted in material that will be known to most science teachers. The difference in style and content of the syllabus compared with Physics (Advanced) is threefold.

(i) The syllabus attempts to place the study of physical material in an Engineering Science context.

(ii) The syllabus material is grouped around focal points which enable the material to be integrated and reinforced from an engineering point of view.

(iii) The syllabus attempts to develop attitudes and a range of abilities which are used in engineering practice and to recognise the educational advantages of linking the engineering and the scientific approaches in the same syllabus.

1.2 THE INTENDED EDUCATIONAL ADVANTAGES OF THE ENGINEERING SCIENCE APPROACH

The intended educational advantages of the Engineering Science approach are many but the following deserve special mention.

(a) It employs methods and techniques related to the solution of engineering and other problems which are grounded in experience. These both contrast with and complement the methods and approaches of pure science.

(b) It helps to increase the student's range of thought processes by including elements drawn from both pure and applied science.

(c) It shows that design is primarily a logical rather than an accidental or intuitive process.

(d) It provides motive and interest for the student by giving a socially useful context to the teaching of science.

1.3 THE WIDER ASPECTS OF ENGINEERING SCIENCE

An important aspect of the study of Engineering Science is that it inevitably leads to the consideration of social, economic and aesthetic factors. The application of scientific knowledge to the requirements of man implies that there exists a real relationship between science and man. The process of design also requires the consideration of social, economic, aesthetic and time-dependent considerations, and the professional engineer is involved in the ethical problems created as a result of the impact of science on society. Moreover he needs to be aware of the environmental effects of the products he designs and of the management process required for their production.

Hence the study of Engineering Science leads the student to a consideration of many aspects in addition to the purely material; the subject can justifiably be regarded as contributing to a liberal education. Although no explicit mention of these topics is given in the syllabus it is hoped that pupils will be encouraged to include them whenever possible in their project. Reference to past examination papers will enable teachers and students to observe how these topics influence the form and content of some examination questions.

1.4 THE RATIONALE USED IN PRESENTING THE FACTUAL MATERIAL

In working out the content of the syllabus, the procedure has been to move from basic core material towards more specific concepts and relationships. It is in dealing with these specific concepts that the links with everyday articles become more apparent and the treatment becomes increasingly more practical and less theoretical. As it will not always be possible to discover the magnitude of all the factors involved in a given situation, a measure of judgment will be demanded of the student when studying these concepts. It must be emphasised that the order of the syllabus, although it helps the teacher and student to appreciate the way in which these ideas have been developed, does not necessarily represent a suitable or desirable teaching or learning order. It may be found that by beginning with real situations and developing the abilities of judgment at the same time as the abilities of analysis, the teaching and learning will be more effective in achieving some of the objects of the course.

1.4.1 *The grouping of the syllabus content*

The content of the syllabus has been grouped under the headings of basic types of phenomena and their applications.

  (i) The interaction of forces and masses in static and dynamic situations. (Mechanics)

 (ii) The properties of matter and their use in predicting the behaviour of materials. (Materials science)

(iii) The displacement of fluid mass and various forms of energy under the action of an electrical field or gradients of pressure or temperature. (Transport processes)

(iv) The nature and properties of fields and their variation in space. (Field phenomena and applications)

 (v) The regular oscillation of a quantity between upper and lower limits. (Periodic and wave phenomena).

(vi) The concept of a system and the ways in which systems can be represented in terms of electrical and electronic circuits and components. (Analysis of systems and electronics)

(vii) The release, translation, degradation and application of energy. (Thermodynamics)

All these sections follow the same development pattern and are derived from the same basic concepts. Real or analogous links can connect a part of one section with a part of another. For example, the idea of resistance can be used in connection with the transfer of heat, electrical charge, and fluid mass.

2. The aims and objectives of the course and the examination

2.1 THE AREAS OF ABILITY TO BE DEVELOPED

The course seeks to bring about the development of the student in three broad areas: mental skills, physical skills, and attitudes. The balance between these three areas is basic to Engineering Science, and is achieved by the development of skills required for various forms of problem solving. The course is concerned in particular with the form of problem solving usually associated with engineering design. This activity is characterised by:

  (i) identifying the problem in terms of the function to be performed, the resources available and the limits set by material and other considerations;

(ii) acquiring the necessary information;

(iii) providing several possible solutions and comparing them with the situation in which the problem occurs;

(iv) planning the approach so as to make the best use of resources and time.

The design process involves the general abilities to:

(v) communicate with other disciplines;

(vi) judge the amount of information required;

(vii) recognise problem situations;

(viii) engage in self-criticism;

(ix) discriminate between alternative courses of action.

2.2 COGNITIVE ABILITIES

The course is also concerned in part with the acquisition, retention and absorption of knowledge and the associated mental abilities. It will be recognised that the abilities involved vary in difficulty. For example, it is easier for a student merely to remember facts than it is for him to organise facts in order to postulate an original hypothesis. Thus objectives can be organised into ascending levels of difficulty depending on the degree of mental complexity involved. These mental processes are tested in the examination and a statement of the objectives of the written examination and the abilities to be tested is given in the syllabus which appears at the front of these Notes.

The statement of the abilities to be tested is preceded by a statement of the knowledge to be tested, which may be regarded as basic to any Engineering Science course. The list of abilities to be tested is not only given in ascending order of difficulty but represents the order in which an engineering problem would be tackled. The initial *comprehension* of the problem must be followed by the *communication* of its essential requirements to other people and must be translated, if necessary, into terms by which the problem can be solved. This is followed by a detailed *analysis* of the problem and the exploration of all the implications and possibilities which the problem presents. At the *synthesis and design* stage the route to the solution of the problem should be apparent and the shape and requirements of the artefact or system should become clear. The final stage of *evaluation and judgment* is necessary to "prove" the design or solution and calls for decisions based upon experience and careful consideration of all the factors involved.

2.3 ATTITUDES, MOTIVES AND INTERESTS

In addition to the objectives referred to in 2.2 there is a group of aims associated with attitudes, motives and interests: the outlook and personality of the individual. Once again, these can be classified more precisely; at the lowest level the individual may be aware of the existence of an attitude without necessarily accepting it as being of value or importance to him, while at the highest level he might subscribe to it completely. The development of the engineering attitude to its highest level is not easy, and a course in Engineering Science cannot expect a high achievement in this respect. Nevertheless, if the list is borne in mind it may directly help the teacher and student as well as indicate a balance between different aspects of the course. The following attitudes are relevant:

   (i) The recognition of the need for a method which is organised, careful, and intellectually honest-particularly in respect of experimental observation.

  (ii) The acceptance of the need to consider the parallel social and economic bases of engineering.

 (iii) An awareness of the advantages of deriving the more particular relationships from the basic concepts.

 (iv) An awareness of the advantage of seeking parallels in other fields to relate one kind of phenomenon to another.

  (v) An awareness of the advantage of attempting to reduce a social, economic or scientific situation to a simple system.

 (vi) The recognition of the fact that it may be necessary to exercise judgment as well as reason when dealing with a problem.

(vii) The recognition of the fact that a perfect answer to a problem may not exist, and that the best available answer must be sought.

(viii) The recognition of the fact that not all the information necessary to tackle a problem may be available, and that some which is available may not be relevant.

 (ix) The acceptance of the fact that more than one way of thinking exists, and that different ways may be more appropriate to different problems or different stages of the same problem.

  (x) The recognition of the fact that the required exactness of calculation may vary from case to case (for example, from a preliminary, quick, "order of magnitude" estimate to a precise forecast of performance).

Unlike the examination objectives related to the knowledge and the abilities to be tested, not all of these attitudes can be directly measured. They can, however, be detected in the way students tackle problems based both on syllabus content and on coursework.

2.4 PHYSICAL SKILLS

A course in Engineering Science should be concerned with physical skills: the co-ordination of hand and eye. These are traditionally associated with basic craft work, and their relationship to this syllabus in terms of constructional project work is obvious enough, although no emphasis on workshop practice is given in the syllabus. However, it is easy to forget that physical skills are also associated with experimental work in the laboratory. Skills too can be organised into levels of difficulty characterised by the degree of control required by the student.

The relevant skills might be specified as follows.

(i) The ability to make observations as accurately as the means allow, and to estimate the margin of error.

(ii) The "feel" for handling a range of apparatus and equipment arising from familiarity with such devices and knowledge of their possible uses.

(iii) The art of maintaining concise and accurate records under laboratory and workshop conditions.

(iv) The art of sketching quickly and meaningfully.

Author Index

Aanstoos, T. A., 146, 152
Abdulwahed, M., 162, 183
Abercrombie, M. L. J., 13, 27
Abraham, D. M., 280
Abraham, J., 41, 54
Abro, S., 78, 85
Abu-Idayil, B., 19, 27
Adair, J., 196, 211
Adams, R. D., 199, 214
Adams, R. S., 223, 236
Adelman, C., 307–308
Agano, A., 238
Ager, M., 16, 27
Agogino, A., 237
Aguiar, E. M. D., 213
Ahern, H., 171, 183
Ahmed, K., 163, 186
Ahn, B-K., 218, 239, 251, 263
Airasian, P. W., 54, 308
Akasheh, F., 86
Al-Asaf, Y., 27
Al-Attar, H., 19, 27
Alexander, P. A., 120, 130
Allen, T., 14, 27
Alley, M., 59
Almarshoud, A. F., 137, 152
Al-Nashash, H., 19, 27
Alpay, A., 200, 211
Amabile, T. M., 47, 54
Ambrose, G. A., 213
Amengual, J-C., 145, 155, 205, 215
Anakwa, W. K. N., 213
Anastasi, A., 92, 120, 130, 306, 308
Anderson, J., 135–136, 153

Anderson, L. W., 204, 302, 308
Andres, C., 39, 58
Angelo, T. A., 76, 85, 296, 298
Ann-Abel, R., 13, 30
Anson, C. M., 22, 28
Arasian, P. W., 153
Archer, L. B., 43, 54
Argyris, C., 159, 183
Armarego, J., 247–248, 260
Ashour, O. M., 149–150, 155
Ashworth, P., 113, 130, 211, 246, 260
Assalah, K., 27
Astin, A., 12, 28
Atman, C. J., 153, 174, 183, 272, 280
Ayres, R., 211

Baer, J., 44–45, 47, 54
Bailey, B. D., 125, 130
Baillie, C., 46, 54
Baker, D., 253, 260
Baker, Lord, 115–116
Balancha, L., 238
Balascio, C. C., 247, 260
Baldock, T. E., 201, 214
Banks, K. M., 298
Bannister, D., 92, 107
Barkataki, A., 220, 238
Barnard, C., 94, 107
Barnes, L. B., 36, 46, 49, 91, 107
Baron, F., 43, 59
Bar-On, R., 191, 211
Baron-Maldonado, M., 31
Barr, R. E., 146, 152–153
Barreiros, B. V., 213

The Assessment of Learning in Engineering Education: Practice and Policy, First Edition. John Heywood.
© 2016 The Institute of Electrical and Electronics Engineers, Inc. Published 2016 by John Wiley & Sons, Inc.

Subject Index

ABET, 7, 10, 16–17, 27, 63, 65, 74, 84, 96, 134, 137, 141, 144–146, 207, 220, 244–245, 284, 289–290, 295–297

abilities, 10, 18–20, 32, 119–120, 134, 291, 305. *See also* competencies

ability, 7, 9, 32, 92, 119, 244

academic/vocational divide, 41, 115

accountability, 8, 25–26

accreditation/accreditation boards, 133, 252

achievement, 3–5, 8, 16, 23, 29, 78

ACT test, 171

adaptability/adaptive expertise/flexibility, 180–181, 243, 254, 294

adult development, 267, 291

adult learning, 172, 275–291

adult literacy, 254–255

affective domain, 26, 82, 96, 98, 152, 224, 248–249, 255, 260

aims (goals) of (higher) education, 26, 90, 118, 182, 286

and assessment, 22, 26, 283, 285

screening, 285, 286

"A" Level (of General Certificate of Education), 3, 21

alignment, 244

alumni, 24, 244

Alverno College, 26–28, 75–76, 112, 119–120, 122, 126, 128–130, 277–278, 291, 296

American Association for Higher Education, 308

American College Testing Program (ACT), 16, 74, 112, 130

American Educational Research Association (AERA), 307

American Society for Engineering Education (ASEE), 290

American Society of Civil Engineers, 47

apprentice (ship)

Commission on Apprenticeship, 273

Approaches to Study Inventory, 24

aptitude, 19, 21

aptitude tests, 16

argument maps, 168

ASSESS, 77

assessment, *see also* evaluation, examinations, testing

and anxiety (stress), 16

balanced system of, 61, 74–75, 85

continuous, 17, 19, 27–28. *See also* coursework

criterion referenced, 7, 61, 74

critical incidents, 80

definitions of, 307

of development, 172–174

direct, 17, 77, 143

of entrepreneurship, 145–147

formative, 308. *See also* formative assessment

functions, 1, 8

of general practice, 79–83

indirect, 77

of information gathering, 81

innovation in, 84

issues and terms, 301–308

laboratory practicals, 69–70

The Assessment of Learning in Engineering Education: Practice and Policy, First Edition. John Heywood.
© 2016 The Institute of Electrical and Electronics Engineers, Inc. Published 2016 by John Wiley & Sons, Inc.